Lecture Notes in Biomathematics

Vol. 1: P. Waltman, Deterministic Threshold Models in the Theory of Epidemics. V, 101 pages. 1974.

Vol. 2: Mathematical Problems in Biology, Victoria Conference 1973. Edited by P. van den Driessche. VI, 280 pages. 1974.

Vol. 3: D. Ludwig, Stochastic Population Theories. VI, 108 pages. 1974.

Vol. 4: Physics and Mathematics of the Nervous System. Edited by M. Conrad, W. Güttinger, and M. Dal Cin. XI, 584 pages. 1974.

Vol. 5: Mathematical Analysis of Decision Problems in Ecology. Proceedings 1973. Edited by A. Charnes and W. R. Lynn. VIII, 421 pages. 1975.

Vol. 6: H. T. Banks, Modeling and Control in the Biomedical Sciences. V, 114 pages. 1975.

Vol. 7: M. C. Mackey, Ion Transport through Biological Membranes, An Integrated Theoretical Approach. IX, 240 pages. 1975.

Vol. 8: C. DeLisi, Antigen Antibody Interactions. IV, 142 pages. 1976.

Vol. 9: N. Dubin, A Stochastic Model for Immunological Feedback in Carcinogenesis: Analysis and Approximations. XIII, 163 pages. 1976.

Vol. 10: J. J. Tyson, The Belousov-Zhabotinskii Reaktion. IX, 128 pages. 1976.

Vol. 11: Mathematical Models in Medicine. Workshop 1976. Edited by J. Berger, W. Bühler, R. Repges, and P. Tautu. XII, 281 pages. 1976.

Vol. 12: A. V. Holden, Models of the Stochastic Activity of Neurones. VII, 368 pages. 1976.

Vol. 13: Mathematical Models in Biological Discovery. Edited by D. L. Solomon and C. Walter. VI, 240 pages. 1977.

Vol. 14: L. M. Ricciardi, Diffusion Processes and Related Topics in Biology. VI, 200 pages. 1977.

Vol. 15: Th. Nagylaki, Selection in One- and Two-Locus Systems. VIII, 208 pages. 1977.

Vol. 16: G. Sampath, S. K. Srinivasan, Stochastic Models for Spike Trains of Single Neurons. VIII, 188 pages. 1977.

Vol. 17: T. Maruyama, Stochastic Problems in Population Genetics. VIII, 245 pages. 1977.

Vol. 18: Mathematics and the Life Sciences. Proceedings 1975. Edited by D. E. Matthews. VII, 385 pages. 1977.

Vol. 19: Measuring Selection in Natural Populations. Edited by F. B. Christiansen and T. M. Fenchel. XXXI, 564 pages. 1977.

Vol. 20: J. M. Cushing, Integrodifferential Equations and Delay Models in Population Dynamics. VI, 196 pages. 1977.

Vol. 21: Theoretical Approaches to Complex Systems. Proceedings 1977. Edited by R. Heim and G. Palm. VI, 244 pages. 1978.

Vol. 22: F. M. Scudo and J. R. Ziegler, The Golden Age of Theoretical Ecology: 1923–1940. XII, 490 pages. 1978.

Vol. 23: Geometrical Probability and Biological Structures: Buffon's 200th Anniversary. Proceedings 1977. Edited by R. E. Miles and J. Serra. XII, 338 pages. 1978.

Vol. 24: F. L. Bookstein, The Measurement of Biological Shape and Shape Change. VIII, 191 pages. 1978.

Vol. 25: P. Yodzis, Competition for Space and the Structure of Ecological Communities. VI, 191 pages. 1978.

Lecture Notes in
Biomathematics

Managing Editor: S. Levin

36

Angelika Wörz-Busekros

Algebras in Genetics

Springer-Verlag
Berlin Heidelberg New York 1980

Author

Angelika Wörz-Busekros
Universität Tübingen
Lehrstuhl für Biomathematik
Auf der Morgenstelle 28
D-7400 Tübingen 1
Federal Republic of Germany

AMS Subject Classifications (1980): 17DXX, 92A10

ISBN 978-3-540-09978-9 ISBN 978-3-642-51038-0 (eBook)
DOI 10.1007/978-3-642-51038-0

Library of Congress Cataloging in Publication Data. Wörz-Busekros, Angelika, 1946- Algebras in genetics. (Lecture notes in biomathematics; 36) Bibliography: p. Includes index. 1. Genetics--Mathematics. 2. Algebra. I. Title. II. Series. [DNLM: 1. Genetics. 2. Mathematics. W1 LE334 v.36 / QH430 W843a] QH438.4M33W63 575.1'01'512 80-13681

2141/3140-543210

PREFACE

The purpose of these notes is to give a rather complete presentation of the mathematical theory of algebras in genetics and to discuss in detail many applications to concrete genetic situations. Historically, the subject has its origin in several papers of Etherington in 1939-1941. Fundamental contributions have been given by Schafer, Gonshor, Holgate, Reiersøl, Heuch, and Abraham. At the moment there exist about forty papers in this field, one survey article by Monique Bertrand from 1966 based on four papers of Etherington, a paper by Schafer and Gonshor's first paper. Furthermore Ballonoff in the third section of his book "Genetics and Social Structure" has included four papers by Etherington and Reiersøl's paper. Apparently a complete review, in particular one comprising more recent results was lacking, and it was difficult for students to enter this field of research. I started to write these notes in spring 1978. A first german version was finished at the end of that year. Further revision and translation required another year. I hope that the notes in their present state provide a reasonable review and that they will facilitate access to this field.

I am especially grateful to Professor K.-P. Hadeler and Professor P. Holgate for reading the manuscript and giving essential comments to all versions of the text. I am also very grateful to Dr. I. Heuch for many discussions during and after his stay in Tübingen. I wish to thank Dr. V. M. Abraham and Professor Holgate for sending material prior to publication. I am indebted to Miss Brigitte Pölter for preparing altogether three typescripts. Finally I wish to thank my husband Wolfgang and our son Markus for their patience during the preparation of this and preceding versions of this monograph.

Tübingen, February, 1980

CONTENTS

0. INTRODUCTION

Algebras arise in population genetic models in a quite natural way. If populations with one or several genetic traits are considered then each particular population is characterized by the frequencies of these traits, e.g. gene or genotype frequencies. These frequencies are collected into a vector. Convex combinations of such vectors describe mixtures of populations, thus the addition of vectors and the multiplication with constants have a meaning. Sexual reproduction leads to a multiplicative structure, as it is suggested by a formal representation of Mendel's laws, e.g.

$$Aa \times Aa = \frac{1}{4}AA + \frac{1}{2}Aa + \frac{1}{4}aa \quad .$$

A wide variety of genetic situations with random mating and absence of differential selection can be described by genetic algebras. The algebraic setting leads to a simple and lucid description of the reproduction laws and the development of the population in successive generations. In particular in complex situations as in polyploidy the algebraic approach is very useful, when compared with difference equations, say.

As other major achievements of theoretical population genetics, e.g. the Hardy-Weinberg law or the Fisher - Wright - Haldane selection model, genetic algebras were introduced in the twenties and thirties of this century. Apparently Serebrowsky [103] was the first to give an algebraic interpretation of the sign "x" indicating sexual reproduction and to give a mathematical formulation of the Mendelian laws. Glivenko [44] pursued these efforts and introduced the so called Mendelian "algebras" for diploid populations with one locus or with two uncoupled loci. Independently Kostitzin [82] introduced a "symbolic multiplication" to express the Mendelian laws.

In 1939 Etherington [27] succeeded in giving a precise mathematical formulation of Mendel's laws in terms of algebras. He introduced baric algebras and train algebras (see below) and he showed that many algebras modeling concrete biological situations are special train algebras. Furthermore he introduced the concept of commutative duplication by which with the gametic algebra of a randomly mating population a zygotic algebra is associated.

Already in these early papers it appeared that genetic algebras would develop into a field also of independent mathematical interest. These algebras are in general commutative but not associative, furthermore they do not belong to any of the well-known classes of non associative algebras such as Lie - , Jordan - or alternative algebras. Their additional properties "baric algebra" (non-trivial representation over the underlying field), "train algebra" (the coefficients of the rank equation are only functions of the images under this representation), "Schafer genetic" (the characteristic equation of the elements of the transformation algebra has such property) lead to many interesting mathematical results.

Since the beginnings of the theory in Etherington's paper [27] the mathematical investigation of these algebras has been further pursued and these algebras have been applied to more and more genetic situations (Etherington [28] - [31] , Schafer [101] , Raffin [96] , Gonshor [45] , Reiersøl [98] and Holgate [61]).

Particulary important from a mathematical point of view are the papers by Schafer and Gonshor. The first introduced the concept of "genetic algebra" (a definite mathematical terminus with a somewhat narrower meaning than "algebra related to genetics"), the latter proved two fundamental theorems on the existence and uniqueness of idempotents and on the convergence behavior of sequences of plenary powers in special train (essentially genetic) algebras.
In 1966 Monique Bertrand [16] presented some of the results of [27] ,[28], [29] ,[31] ,[45] under new aspects.

More recent results are due to Abraham [1] - [4] , Gonshor [46] - [48] , Heuch [52] - [60] and Holgate [62] - [72] and others mentioned in the references, where all papers or books with direct relationship to algebras in genetics are indicated by a "G" in front of the reference number.

The aim of the present monograph is to give a rather complete presentation of algebras in genetics and to apply the theory and the results to various concrete biological situations.

In section 1 we explain how algebras arise in genetics, we provide several simple examples which we take up again in the sections on applications 7,8. In sections 2 to 6 we develop the theory of these algebras, thereby frequently indicating the significance of given statements with respect to applications.

In section 1 we further discuss the most elementary though fundamental properties of algebras in genetics.

In section 2 we outline the general theory of non associative algebras, in particular we deduce the rank equation.

In section 3 we discuss commutative, non associative algebras which are train algebras, Schafer genetic algebras, Gonshor genetic algebras or special train algebras. Under new aspects we clarify implications and equivalences between these properties.
The results are collected in the implication diagram 3.31.

In section 4 we explain Gonshor's theorem on the existence of idempotents, then we prove two generalizations of this theorem. Furthermore we show that all algebras with genetic realization possess an idempotent. If in addition the algebra is genetic then all train roots except $\lambda_o = 1$ do not exceed 1/2 in modulus. Both these theorems can be expressed in terms of matrices. The result on the train roots implies that in every genetic algebra with genetic realization for every element of unit weight the sequence of principal powers converges. Furthermore we mention Gonshor's theorem on the convergence of sequences of plenary powers of elements of unit weight in genetic algebras and we give a generatization. This theorem will be applied in many concrete situations in section 7.

In section 5 we discuss how the properties of commutative algebras mentioned in section 3 can be carried over to the non commutative case. In table 5.15 we collect those properties which remain valid in the non commutative case. We give sufficient conditions for the existence of idempotents and we discuss the convergence behavior of sequences of plenary powers in non commutative Gonshor genetic algebras.

The theory of algebras in genetics enables one to reduce the investigation of a "complex" genetic situation to the study of simpler ones. Suppose each single feature can be described by an algebra. Then, in many cases, it is possible to construct a new algebra from these which comprises the features. With this aim we consider several constructions of algebras from given ones in section 6, e.g. linear combinations and tensor products, and we discuss which properties are preserved. It turns out that the tensor product of commutative special train algebras again has this property and analogous statements are true for Gonshor genetic algebras as well as weakly left - and left Gonshor genetic algebras (theorem 6.11). Beyond this we discuss in section 6 the non commutative and the commutative duplication of an algebra, where in the latter case the algebra itself has to be commutative.

In sections 7 and 8 we apply the theory to genetics. After a general discussion about those genetic situations which already have been or which possibly could be investigated by means of algebras we consider genetic features which can be described by commutative algebras. We discuss in detail the cases of additive segregation rates, partially coupled loci with recombination, autopolyploidy with an arbitrary amount of double reduction. Furthermore we mention some algebraic models for the inheritance of populations with overlapping and continuously overlapping generations. Frequently we can apply the theorems on the bounds for the train roots of a genetic algebra with genetic realization and on the convergence behavior of sequences of plenary powers in genetic algebras.

In section 8 we consider genetic features which lead to not necessarily commutative algebras. First we investigate the case of asymmetric additive segregation rates.
Then we consider the inheritance of a finite number of sex linked characters. With this aim we construct the sex linked duplicate of a baric algebra with respect to a certain linear mapping of this algebra into itself. It turns out that the sex linked duplicate of a genetic algebra contains a subalgebra which again is genetic provided that the said linear mapping is "compatible" with the multiplication.
Finally we can also cover some cases of assortative mating: Diploid or polyploid individuals in which in every generation fixed portions of the population reproduce by several types of inbreeding such as selfing, brother-sister-mating, etc., while the remaining portion undergoes random mating. Again this situation can be described by an algebra which in many cases is Gonshor genetic.

In section 9, the last section of this monograph, we mention some open problems and finally we discuss the work of Ljubič [84] - [88] and Holgate [68] on Bernstein algebras. These algebras describe the inheritance of populations which are in equilibrium after one generation.

The first (german) version of this monograph has been accepted as Habilitationsschrift by the "Fakultät für Biologie der Universität Tübingen".

1. ALGEBRAS IN GENETICS

In this section we provide a short introduction to "algebras in genetics". We explain how algebras arise in population genetics and we construct some examples of gametic algebras. These examples will be reconsidered and discussed in a wider frame-work in section 7. We introduce zygotic and copular algebras and we discuss the construction of these algebras from gametic algebras. The exposition will reveal various elementary though fundamental properties of these algebras.

A. The occurence of algebras in genetics

Consider an infinitely large random mating population of diploid (or more generally 2r-ploid) individuals which differ genetically at one or several autosomal loci. Let $a_1,\dots,a_n$ be the genetically distinct gametes produced by this population. The state of the population can be described by the vector $(\alpha_1,\dots,\alpha_n)^T$ of gamete frequencies

$$0 \le \alpha_i \le 1,\ i=1,\dots,n,\quad \sum_{i=1}^{n} \alpha_i = 1 \quad .$$

By random union of gametes a_i and a_j zygotes of type a_ia_j are formed, $i,j=1,\dots,n$.
Assume that a zygote a_ia_j produces a number $\tilde{\gamma}_{ijk}$ of gametes of type a_k, which survive the next generation $k=1,\dots,n$, $i,j=1,\dots,n$. Let μ be the maximal number of surviving offspring produced by a zygote, i.e.
$\mu := \max_{i,j} \sum_{k=1}^{n} \tilde{\gamma}_{ijk}$. Define $\gamma_{ijk} := \mu^{-1}\tilde{\gamma}_{ijk}$, then

$$0 \le \gamma_{ijk} \le 1, \quad i,j,k=1,\dots,n. \tag{1.1}$$

In the absence of selection, i.e. all zygotes have the same fertility, every zygote produces the same number of surviving gametes, i.e.
$\sum_{k=1}^{n} \tilde{\gamma}_{ijk} = \mu$ for all $i,j=1,\dots,n$, hence

$$\sum_{k=1}^{n} \gamma_{ijk} = 1 \qquad \text{for all } i,j=1,\dots,n. \tag{1.2}$$

In this case γ_{ijk} is the probability that a zygote a_ia_j produces a gamete a_k, $k=1,\dots,n$, $i,j=1,\dots,n$. Thus the frequency of gametes a_k produced by the total population is

$$\sum_{i,j=1}^{n} \alpha_i \gamma_{ijk} \alpha_j \; , \qquad k=1,\dots,n \; .$$

If (1.2) is satisfied then the real numbers γ_{ijk}, $i,j,k=1,\dots,n$, are called segregation rates or segregation coefficients (Bernstein [15], Kostitzin [82] and Ljubič [84] term them inheritance coefficients).

In the zygote $a_i a_j$ the gametes a_i or a_j may be of maternal or paternal origin. If this origin does not play a role in the production of gametes by the zygote considered, i.e. in the case of symmetric inheritance, the formally distinct zygotes $a_i a_j$ and $a_j a_i$, $i \neq j$, will produce gametes a_k with the same probability, i.e.

$$\gamma_{ijk} = \gamma_{jik} \; , \; i,j,k=1,\dots, \qquad (1.3)$$

then the segregation rates are called symmetric.

So far we have explained how in one population the transition from gene frequencies in one generation to the frequencies in the next generation is described by the segregation rates. There is no algebraic structure yet.

Now let the vectors of gamete frequencies $(\alpha_1,\dots,\alpha_n)^T$ and $(\beta_1,\dots,\beta_n)^T$ represent two populations A and B, respectively. Then the filial population resulting from random mating of populations A and B is represented by the vector

$$\begin{pmatrix} \sum_{i,j=1}^{n} \alpha_i \gamma_{ij1} \alpha_j \\ \vdots \\ \sum_{i,j=1}^{n} \alpha_i \gamma_{ijn} \alpha_j \end{pmatrix} \; .$$

Each component of this vector is a bilinear form over $\mathbb{R}$. It can be assumed that this formula (1.4) was the starting point for the following formal representation (cf. Etherington [27]):

The gametes $a_1,\dots,a_n$ are considered as abstract elements which are free over the field $\mathbb{R}$ of real numbers. They span an n-dimensional vector space

$$\mathcal{W} := \left\{ \sum_{i=1}^{n} \alpha_i a_i \;\middle|\; \alpha_i \in \mathbb{R} \; , \; i=1,\dots,n \right\} \; .$$

With the assistance of the segregation constants γ_{ijk} which satisfy relations (1.1)-(1.3) a multiplication is defined in $\mathcal{V}$ by

$$a_i a_j := \sum_{k=1}^{n} \gamma_{ijk} a_k, \quad i,j=1,\ldots,n, \tag{1.4}$$

and the bilinear extension onto $\mathcal{V}\times\mathcal{V}$. The resulting algebra $\mathcal{G} := \mathcal{V}(+, \mathbb{R}, \cdot)$ is called a <u>gametic algebra</u>.

Each element $a = \sum_{i=1}^{n} \alpha_i a_i$ in $\mathcal{G}$ with $0 \leq \alpha_i \leq 1$, $i=1,\ldots,n$, and $\sum_{i=1}^{n} \alpha_i = 1$ corresponds to an actual population A, in which α_i is the frequency of gamete a_i. If $b = \sum_{j=1}^{n} \beta_j a_j$ corresponds to a second population B, then the filial population resulting from random mating of A and B is represented as the product

$$ab = \left(\sum_{i=1}^{n} \alpha_i a_i\right)\left(\sum_{j=1}^{n} \beta_j a_j\right) = \sum_{k=1}^{n} \left(\sum_{i,j=1}^{n} \alpha_i \gamma_{ijk} \beta_j\right) a_k .$$

From the biological interpretation it is already clear, that in general the product of populations A and B mated with a population C gives a result different from the mating of population A with the product of populations B and C. This fact is depicted in the following graph:

Thus the algebra $\mathcal{G}$ is not in general associative, i.e. the polynomial identity

$$(xy)\,z = x(yz)$$

is, in general, not true.
On the other hand the algebra $\mathcal{G}$ is commutative in view of (1.3), i.e. the polynomial identity

$$xy = yx$$

is satisfied.

The study of the inheritance of non sex linked traits leads to certain commutative, in general non associative, algebras. This class of algebras is rather wide, and few relevant statements can be made for the behavior of populations without further insight into the peculiar structure of these algebras. Before discussing such additional properties we shall consider further examples and reveal more connections between algebras and genetics.

Example 1.1. Gametic algebra for simple Mendelian inheritance

Consider a random mating population of diploid individuals which differ in an autosomal locus with alleles $a_1,\dots,a_n$. Assume that all zygotes have equal fertility and that homozygotes a_ia_i produce gametes a_i only and heterozygotes a_ia_j, $i \neq j$, produce gametes a_i and a_j in equal portions $i,j=1,\dots,n$. Using Kronecker's symbol

$$\delta_{ik} := \begin{cases} 1 & \text{if } i = k \\ 0 & \text{if } i \neq k \end{cases} \qquad , \; i,k=1,\dots,n.$$

we can write the segregation rates as

$$\gamma_{ijk} = \frac{1}{2}(\delta_{ik} + \delta_{jk}), \quad i,j,k=1,\dots,n. \tag{1.5}$$

Assume the zygotes are formed by random union of gametes. Then the genetics can be described by the gametic algebra $\mathcal{G}$ spanned by $a_1,\dots,a_n$ whose multiplication is given by (1.4) with the segregation rates γ_{ijk}, $i,j,k=1,\dots,n$, given in (1.5). The product of two basis elements simplifies to

$$a_ia_j = \frac{1}{2}(a_i+a_j), \quad i,j=1,\dots,n .$$

$\mathcal{G}$ is called the gametic algebra for simple Mendelian inheritance and the segregation rates γ_{ijk} are called simple Mendelian segregation rates. (We shall consider this example in more detail in section 7A.) □

Example 1.2. Gametic algebras for simple Mendelian inheritance with mutation

Assume in the above example that in addition gene mutation may occur. Let σ_{ik} be the probability that the allele a_i mutates into the allele a_k,

$$0 \leq \sigma_{ik} \leq 1, \quad i,k=1,\dots,n, \; \sum_{k=1}^{n} \sigma_{ik} = 1, \; i=1,\dots,n.$$

Then the gamete a_k is obtained from a zygote a_ia_j with probability

$$\sum_{s=1}^{n} \frac{1}{2}(\delta_{is} + \delta_{js})\sigma_{sk} = \frac{1}{2}(\sigma_{ik} + \sigma_{jk}), \quad i,j,k=1,\dots,n.$$

Define

$$\gamma_{ijk} := \frac{1}{2}(\sigma_{ik} + \sigma_{jk}), \qquad i,j,k=1,\dots,n. \tag{1.6}$$

Then the genetics of this population is described by the gametic algebra $\mathfrak{G}$ spanned by $a_1,\dots,a_n$ with multiplication table given by (1.6). $\mathfrak{G}$ is called the <u>gametic algebra</u> for <u>simple Mendelian inheritance</u> with <u>mutation</u>. □

<u>Example 1.3. Gametic algebra with recombination</u>

Assume two coupled autosomal loci with alleles A,a and B,b. The population produces four gametes $a_1 := AB$, $a_2 := Ab$, $a_3 := aB$, $a_4 := ab$. Let Θ, $0 \leq \Theta \leq \frac{1}{2}$, be the recombination rate, i.e. the probability that a zygote a_1a_4 undergoes a transition into a_2a_3 or conversely. With this assumption and classical Mendelian segregation the multiplication constants γ_{ijk} of the gametic algebra $\mathfrak{G}$ spanned by a_1,a_2,a_3,a_4 are obtained from the multiplication table of the basis elements.

	a_1	a_2	a_3	a_4
a_1	a_1	$\frac{1}{2}(a_1+a_2)$	$\frac{1}{2}(a_1+a_3)$	$\frac{1}{2}(a_1+a_4) - \frac{1}{2}\Theta d$
a_2		a_2	$\frac{1}{2}(a_2+a_3) + \frac{1}{2}\Theta d$	$\frac{1}{2}(a_2+a_4)$
a_3			a_3	$\frac{1}{2}(a_3+a_4)$
a_4				a_4

where $d := a_1 - a_2 - a_3 + a_4$. □

<u>Example 1.4. Gametic algebra for autopolyploidy with chromosome segregation.</u>

We consider a population of autopolyploid individuals each with a set of chromosomes consisting of groups of 2r homologous chromosomes, in short: 2r-ploid individuals. We assume that the chromosomes of one group differ in one locus with alleles A and a. Then the r+1 genetically different gametes produced by the population can be represented by

$$a_o := A^r, \quad a_1 := A^{r-1}a, \ \dots \ , \ a_{r-1} := Aa^{r-1} \ , \quad a_r := a^r .$$

If only chromosome segregation occurs, i.e. only chromosomes can be distributed independently among the offspring, then a zygote a_ia_j produces gametes a_k with probability

$$\frac{\binom{i+j}{k}\binom{2r-i-j}{r-k}}{\binom{2r}{r}}, \quad k=0,1,\dots,r \ .$$

This is, of course, the hypergeometric distribution. Again we construct the algebra $\mathcal{G}$ as the span of the gametes $\mathcal{G}:=\left\{\sum_{i=0}^{r}\alpha_i a_i \,\middle|\, \alpha_i\in\mathbb{R}\right\}$, and the multiplication is defined as

$$a_i a_j=\binom{2r}{r}^{-1}\sum_{k=0}^{r}\binom{i+j}{k}\binom{2r-i+j}{r-k}a_k, \quad i,j=0,1,\dots,r \ .\square$$

Paralleling the above argument one can derive the zygotic algebra:
Let $a_{ij}:=a_i a_j$, $i\leq j$, $i,j=1,\dots,n$, the genetically different zygotes of a population. Random mating of zygotes a_{ij} and a_{pq} yields zygotes a_{ks} with probability $\gamma_{ij,pq,ks}$,

$$0\leq\gamma_{ij,pq,ks}\leq 1,$$

$$i\leq j,\ p\leq q, k\leq s,\ i,j,p,q,k,s,=1,\dots,n,$$

$$\gamma_{ij,pq,ks}=\gamma_{pq,ij,ks},$$

$$\sum_{\substack{k,s=1\\ k\leq s}}^{n}\gamma_{ij,pq,ks}=1\ ,\ i\leq j,\ p\leq q,\ \ i,j,p,q=1,\dots,n.$$

We introduce the vector space $\mathcal{Z}$ as the span of the zygotes,

$$\mathcal{Z}:=\left\{\sum_{\substack{i,j=1\\ i\leq j}}^{n}\alpha_{ij}a_{ij}\ \middle|\ \alpha_{ij}\in\mathbb{R},\ \ i\leq j,\quad i,j=1,\dots,n\right\},$$

and define a multiplication by

$$a_{ij}a_{pq}:=\sum_{k\leq s}\gamma_{ij,pq,ks}a_{ks}, \quad i\leq j,\ p\leq q,\ i,j,p,q=1,\dots,n, \qquad (1.7)$$

and bilinear extension onto $\mathcal{Z}\times\mathcal{Z}$. Then the vector space $\mathcal{Z}$ becomes a commutative algebra which in general is non associative. $\mathcal{Z}$ is called the <u>zygotic algebra</u>.

If zygotes are formed by random mating of gametes, then the zygotic algebra $\mathcal{Z}$ can be constructed from the gametic algebra in a canonical way.

In this case the probability that the mating of zygotes a_{ij} and a_{pq} yields a zygote a_{ks} is the product of the probabilities that the zygote a_{ij} formed from gametes a_i and a_j produces the gamete a_k and that the zygote a_{pq} formed from gametes a_p and a_q produces a gamete a_s. Since we have identified zygotes a_{sk} and a_{ks} we have to add these probabilities appropriately. Then the multiplication constants of $\mathfrak{Z}$ are obtained as

$$\gamma_{ij,pq,ks} = \begin{cases} \gamma_{ijk}\gamma_{pqs} + \gamma_{ijs}\gamma_{pqk}, & k < s \\ \gamma_{ijk}\gamma_{pqs}, & k = s \end{cases} \tag{1.8}$$

This process which attributes to a gametic algebra a zygotic algebra in a canonical way according to (1.8) is called commutative duplication (cf. Etherington [27] , [30]).

Gonshor [47] has given a basis-free definition of commutative duplication:

Let $\mathfrak{G} \otimes \mathfrak{G}$ be the tensor product of the vector space $\mathfrak{G}$ with itself and $\mathfrak{J}$ the following subspace

$$\mathfrak{J} := \left\{ \sum_{i\in I} (x_i \otimes y_i - y_i \otimes x_i) \;\middle|\; x_i, y_i \in \mathfrak{G},\ i \in I,\ |I| < \infty \right\}.$$

The equivalence class of $x \otimes y$ in $\mathfrak{G} \otimes \mathfrak{G}/\mathfrak{J}$ is denoted by $x \circledast y$. Then in $\mathfrak{G} \otimes \mathfrak{G}/\mathfrak{J}$ a multiplication is defined by

$$\left(\sum_{i\in I} x_i \circledast y_i\right)\left(\sum_{j\in J} x_j \circledast y_j\right) := \sum_{(i,j)\in I\times J} x_i y_i \circledast x_j y_j \ ,$$

here I,J denote arbitrary, finite index sets. With this multiplication $\mathfrak{G} \otimes \mathfrak{G}/\mathfrak{J}$ becomes a commutative algebra which is isomorphic to the commutative duplicate of $\mathfrak{G}$, cf. section 6C.

Another type of algebra pertaining to genetics is the copular algebra. In the biological interpretation the relevant elements of the copular algebra are mating types, i.e. unordered pairs of zygotes. With random mating it is obvious how the frequencies of mating types in the filial generation are obtained from the parental populations.

One can show that the resulting copular algebra is obtained from the zygotic algebra by duplication (cf. Etherington [27]).

Again consider the zygotic algebra $\mathfrak{Z}$ of dimension $p=n(n+1)/2$ over $\mathbb{R}$ whose multiplication table is given by (1.7). We enumerate the basis elements in lexicographic order

$$b_1 := a_{11}, \quad b_2 := a_{12}, \ \dots\ , \quad b_n := a_{1n}$$
$$b_{n+1} := a_{22}, \ \dots\ , b_{2n-1} := a_{2n}$$
$$\dots\dots\dots\dots\dots\dots$$
$$b_p := a_{nn}$$

and redefine the multiplication constants appropriately. Then the algebra $\mathfrak{Z}$ assumes the form of a gametic algebra, i.e. the multiplication constants with respect to the basis $b_1, \dots, b_p$ satisfy (1.1)-(1.3). Of course these basis elements do not correspond to the gametes of the population considered. In a similar manner also the copular algebra can be transformed into a formal gametic algebra. For this reason we restrict our considerations to such algebras which assume the form of a gametic algebra with an appropriately chosen basis.

B. Algebras with genetic realization and baric algebras

In the following let $\mathfrak{A}$ be an n-dimensional algebra over a field $\mathbb{K}$, which is not necessarily associative or commutative.

Definition 1.5. Let $\mathfrak{A}$ be an algebra over $\mathbb{K} = \mathbb{R}$ or $\mathbb{C}$. Assume $\mathfrak{A}$ admits a basis $a_1, \dots, a_n$ such that the multiplication constants γ_{ijk} with respect to this basis,

$$a_i a_j = \sum_{n=1}^{n} \gamma_{ijk} a_k, \qquad i,j=1,\dots,n,$$

are real and satisfy relations (1.1) and (1.2), i.e.

$$0 \leq \gamma_{ijk} \leq 1, \quad i,j,k=1,\dots,n, \quad \sum_{k=1}^{n} \gamma_{ijk} = 1, \quad i,j=1,\dots,n.$$

Then $\mathfrak{A}$ is called an algebra with genetic realization and this basis is called a natural basis.

(In a recently published paper [88] Ljubič denotes these algebras stochastic algebras.)

All algebras which have been constructed earlier from concrete genetic situations have a genetic realization. On the other hand if an algebra $\mathfrak{A}$ has a genetic realization then the elements $a_1, \dots, a_n$ of a natural basis can be interpreted as genotypes of a population and the real number γ_{ijk} can be considered as the probability that random mating of genotypes a_i and a_j produces a genotype a_k, $k=1,\dots,n$, $i,j=1,\dots,n$.

The following trivial example shows that the natural basis of an algebra with genetic realization may not be unique, there may be continuously many natural bases.

Example 1.6. Let $\mathfrak{G} = \left\{ \sum_{i=1}^{n} \alpha_i a_i \,\middle|\, \alpha_i \in \mathbb{R},\ i=1,\dots,n \right\}$ be the gametic algebra for simple Mendelian segregation from example 1.1. Then any basis

$$b_i = \sum_{k=1}^{n} \beta_{ki} a_k, \quad i=1,\dots,n, \quad \text{with} \sum_{k=1}^{n} \beta_{ki} = 1, \quad i=1,\dots,n,$$

is a natural one.

Proof. Let $b_1,\dots,b_n$ be a basis of the assumed form. The multiplication constants with respect to this basis follow from

$$\begin{aligned} b_i b_j &= \Big(\sum_{k=1}^{n} \beta_{ki} a_k\Big)\Big(\sum_{l=1}^{n} \beta_{lj} a_l\Big) \\ &= \sum_{k,l=1}^{n} \beta_{ki}\beta_{lj} a_k a_l \\ &= \sum_{k,l=1}^{n} \beta_{ki}\beta_{lj} \tfrac{1}{2}(a_k + a_l) \\ &= \tfrac{1}{2} \sum_{k=1}^{n} \beta_{ki} a_k \Big(\sum_{l=1}^{n} \beta_{lj}\Big) + \tfrac{1}{2} \sum_{l=1}^{n} \beta_{lj} a_l \Big(\sum_{k=1}^{n} \beta_{ki}\Big) \\ &= \tfrac{1}{2} b_i + \tfrac{1}{2} b_j, \qquad i,j=1,\dots,n. \ \square \end{aligned}$$

Many authors require that the algebras considered have a genetic realization, cf. Heuch [59] , but in most cases, except in [112] , [113] , not many essential consequences are drawn from this hypothesis. We collect some simple facts about algebras with a genetic realization; in section 4 we shall discuss this property in more detail.

Etherington [27] gave the following definition.

Definition 1.7. An algebra $\mathfrak{A}$ over $\mathbb{K}$ is baric if it admits a non trivial algebra homomorphism $\omega : \mathfrak{A} \longrightarrow \mathbb{K}$. The homomorphism ω is called the weight function or weight homomorphism.

From the homomorphism theorem for algebras follows

Theorem 1.8. Let $\mathfrak{A}$ be an n-dimensional baric algebra over $\mathbb{K}$ with weight homomorphism ω . Then $\mathfrak{N} := \ker\omega$ is an (n-1)-dimensional ideal of $\mathfrak{A}$ and the factor algebra $\mathfrak{A}/\mathfrak{N}$ is isomorphic to the field $\mathbb{K}$.

Remark. There is a similarity with the theory of representations of finite groups. Let $\mathbb{K}[G] := \left\{ \sum_{i=1}^{|G|} \alpha_i g_i \,\middle|\, \alpha_i \in \mathbb{K} \right\}$ be the group algebra of a finite group over $\mathbb{K}$. Then $\sum \alpha_i g_i \longmapsto \sum \alpha_i$ defines a homomorphism of $\mathbb{K}[G]$ onto $\mathbb{K}$.

The kernel $\left\{ \sum \alpha_i g_i \,\middle|\, \sum \alpha_i = 0 \right\}$ of this homomorphism is the augmentation ideal.

If the characteristic of the field $\mathbb{K}$ is a divisor of the group order $|G|$, then the augmentation ideal is the radical (in the sense of Jacobson) of $\mathbb{K}[G]$, cf. e.g. Herstein [51]. In section 3 we shall see that also in our situation we have in many cases that $\mathfrak{N} := \ker\omega$ is the radical of $\mathfrak{A}$.

Theorem 1.9. Let $\mathfrak{A}$ be an algebra over $\mathbb{K} = \mathbb{R}$ or $\mathbb{C}$ which has a genetic realization with respect to a natural basis $a_1,\ldots,a_n$. Then the following is true

(1) $\mathfrak{A}$ is baric

(2) $\mathfrak{A}$ is a (non associative) Banach algebra with respect to the norm

$$\|x\| := \sum_{i=1}^{n} |\xi_i| \qquad \text{for } x = \sum_{i=1}^{n} \xi_i a_i \in \mathfrak{A} \ . \tag{1.9}$$

Proof. The multiplication constants with respect to the natural basis $a_1,\ldots,a_n$ satisfy relations (1.1) and (1.2).

I. Define $\omega : \mathfrak{A} \longrightarrow \mathbb{K}$ by $\omega(a_i)=1$, $i=1,\ldots,n$, and the linear extension onto $\mathfrak{A}$. In view of (1.2) the mapping ω is compatible with the multiplication in $\mathfrak{A}$, i.e.

$$\omega(xy) = \omega(x)\omega(y) \qquad \text{for all } x,y \in \mathfrak{A} \ .$$

II. From (1.1) and (1.2) follows that this particular norm is submultiplicative.

Let $x = \sum_{i=1}^{n} \xi_i a_i$, $y = \sum_{j=1}^{n} \eta_j a_j \in \mathfrak{A}$, then

$$\|xy\| = \left\|\sum_{k=1}^{n}\left(\sum_{i,j=1}^{n}\xi_i\gamma_{ijk}\xi_j\right)a_k\right\| = \sum_{k=1}^{n}\left|\sum_{i,j=1}^{n}\xi_i\gamma_{ijk}\xi_j\right|$$

$$\leq \sum_{k=1}^{n}\sum_{i,j=1}^{n}|\xi_i\gamma_{ijk}\xi_j| = \sum_{k=1}^{n}\sum_{i,j=1}^{n}|\xi_i|\gamma_{ijk}|\xi_j|$$

$$= \sum_{i=1}^{n}|\xi_i|\left(\sum_{k=1}^{n}\gamma_{ijk}\right)\sum_{j=1}^{n}|\xi_j| = \|x\|\,\|y\|. \square$$

Finally we collect some characterizations of baric algebras, cf. Schafer [101] , and of algebras with genetic realization.

Lemma 1.11. Let $\mathfrak{A}$ be an n-dimensional algebra over $\mathbb{R}$. The following conditions are equivalent.

(1) $\mathfrak{A}$ is baric

(2) $\mathfrak{A}$ has a basis $a_1,\dots,a_n$, such that the multiplication constants defined by $a_ia_j = \sum_{k=1}^{n}\gamma_{ijk}a_k$ satisfy $\sum_{k=1}^{n}\gamma_{ijk} = 1$, $i,j=1,\dots,n$.

(3) $\mathfrak{A}$ has an (n-1)-dimensional ideal $\mathfrak{N}$ and $\mathfrak{A}^2 \nsubseteq \mathfrak{N}$.

Lemma 1.12. Let $\mathfrak{A}$ be an n-dimensional algebra over $\mathbb{R}$. Then the following conditions are equivalent.

(1) $\mathfrak{A}$ has a genetic realization

(2) $\mathfrak{A}$ is baric and the linear manifold of all elements of weight 1 contains an (n-1)-dimensional simplex $\mathfrak{r}$ with $\mathfrak{r}^2 \subseteq \mathfrak{r}$.

Proof of lemma 1.10. I. (1) $\Longrightarrow$ (2): Let $\omega : \mathfrak{A} \longrightarrow \mathbb{K}$ be a non trivial algebra homomorphism. Then $\ker\omega =: \mathfrak{N}$ is an (n-1)-dimensional ideal of $\mathfrak{A}$. Let $d_2,\dots,d_n$ be a basis of $\mathfrak{N}$. Since ω is non trivial there is an $a \in \mathfrak{A}$ with $\omega(a) \neq 0$. The elements

$a_1 := \frac{1}{\omega(a)}\, a$, $a_i := a - d_i$, $i=2,\dots,n$, form a basis of $\mathfrak{A}$ with $\omega(a_i)=1$, $i=1,\dots,n$.

In view of $\omega(a_ia_j) = \omega(a_i)\omega(a_j)$, the multiplication constants γ_{ijk} of $\mathfrak{A}$ with respect to this basis satisfy the relation $\sum_{k=1}^{n}\gamma_{ijk} = 1$, $i,j=1,\dots,n$.

II. (2) $\Longrightarrow$(3): Define $d_i := a_1 - a_i$, $i=2,\dots,n$. One can easily check that $\mathfrak{N} := \langle d_2,\dots,d_n\rangle$ is an (n-1)-dimensional ideal of $\mathfrak{A}$.

In addition $a_1^2 = a_1 - \sum_{k=2}^{n}\gamma_{11k}d_k \notin \mathfrak{N}$.

III. (3) $\Longrightarrow$ (1): By $\varphi : x \longmapsto x + \mathfrak{N}$ we define a linear mapping of $\mathfrak{A}$ onto $\mathfrak{A}/\mathfrak{N}$. The vector space $\mathfrak{A}/\mathfrak{N}$ is isomorphic with $\mathbb{K}$. Since $\mathfrak{N}$ is an ideal of $\mathfrak{A}$ the mapping φ is an algebra homomorphism. Thus $\mathfrak{A}/\mathfrak{N}$ is either isomorphic with $\mathbb{K}$ or with the zero algebra over $\mathbb{K}$. The latter possibility can be excluded in view of $\mathfrak{A}^2 \nsubseteq \mathfrak{N}$, therefore $(\mathfrak{A}/\mathfrak{N})^2 \neq \langle 0 \rangle$. □

Proof of lemma 1.11. (1) $\Longrightarrow$ (2): Let $a_1, \ldots, a_n$ be a basis of $\mathfrak{A}$ with (1.1) and (1.2). Define $\omega : \mathfrak{A} \longrightarrow \mathbb{R}$ by $\omega(a_i)=1$, $i=1,\ldots,n$. Then $\mathfrak{A}$ is baric with the algebra homomorphism ω. Let $\mathfrak{T}$ be the convex hull of $a_1, \ldots, a_n$. In view of (1.1) and (1.2) we have $a_i a_j \in \mathfrak{T}$, $i,j=1,\ldots,n$, hence $\mathfrak{T}^2 \subseteq \mathfrak{T}$.

II. (2) $\Longrightarrow$ (1): Let $\omega : \mathfrak{A} \longrightarrow \mathbb{R}$ be any weight function. By assumption this simplex $\mathfrak{T}$ is the convex hull of n linearly independent elements $a_1, \ldots, a_n$ with $\omega(a_i)=1$, $i=1,\ldots,n$.
These elements form a basis of $\mathfrak{A}$. In view of $\mathfrak{T}^2 \subseteq \mathfrak{T}$ the multiplication constants γ_{ijk} of $\mathfrak{A}$ with respect to this basis satisfy the relations $0 \leq \gamma_{ijk} \leq 1$, $i,j,k=1,\ldots,n$.
Moreover $\sum_{k=1}^{n} \gamma_{ijk} = 1$ holds for all $i,j=1,\ldots,n$, as has been shown in Lemma 1.10. □

2. ALGEBRAIC PRELIMINARIES

Before we collect definitions specifically concerning non associative algebras we list the following

Notations.

$\mathbb{N}$ - the set of natural numbers

$\mathbb{N}_0 = \mathbb{N} \cup \{0\}$

$\mathbb{K}$, $\mathbb{L}$, $\mathbb{F}$ - commutative fields

$\mathbb{R}$ - field of real numbers

$\mathbb{C}$ - field of complex numbers

Elements of fields are denoted by lower case greek letters, except 0 and 1.

$\mathbb{K}[X]$ - polynomial algebra in one associative indeterminate X over $\mathbb{K}$.

$\mathbb{K}(X_1,\dots,X_n)$ - field of rational functions in n associative commuting indeterminates over $\mathbb{K}$.

$\mathbb{K}\{X_1,\dots,X_n\}$ - algebra of all "polynomials" in n associative, non commuting indeterminates over $\mathbb{K}$, in which two monomials are multiplied by writing them side by side.

$\mathbb{K}\langle X\rangle$ - vector space of all "polynomials" in right principal powers of a non associative, non commutative indeterminate over $\mathbb{K}$, cf. section 2B.

Matrices with coefficients in a field $\mathbb{K}$ are denoted by capital latin or greek letters. The transpose of a matrix is denoted by a superscript T.

Vector spaces are denoted by capital gothic letters.

$\mathfrak{M}_n(\mathbb{K})$ is the algebra of all n-square matrices with elements in $\mathbb{K}$.

The elements of the n-dimensional vector space $\mathbb{K}^n$ are written as columns.

If $\mathfrak{V}$ is a vector space over $\mathbb{K}$ and $a_1,\dots,a_r \in \mathfrak{V}$, then $\langle a_1,\dots,a_r\rangle_{\mathbb{K}}$ is the linear span of these elements over $\mathbb{K}$.
If misinterpretations can be excluded we write simply $\langle a_1,\dots,a_r\rangle$.

For vector spaces $\mathfrak{V}_1,\dots,\mathfrak{V}_r$ over $\mathbb{K}$ let $\mathfrak{V}_1 \otimes_{\mathbb{K}} \mathfrak{V}_2 \otimes_{\mathbb{K}} \dots \otimes_{\mathbb{K}} \mathfrak{V}_r$, or simply $\mathfrak{V}_1 \otimes \mathfrak{V}_2 \otimes \dots \otimes \mathfrak{V}_r$, denote the tensor product of these vector spaces, cf. Greub [49].
If $\mathfrak{V}$ is a vector space over $\mathbb{K}$ and $\mathbb{L}$ is a field extension of $\mathbb{K}$ then $\mathfrak{V}_{\mathbb{L}} := \mathfrak{V} \otimes_{\mathbb{K}} \mathbb{L}$ is the vector space over the extension field.

A. Non associative algebras - definitions and notations

In this section we collect some standard definitions and notations for not associative algebras, cf. Braun and Koecher [19], Jacobson [76], Schafer [101].

In the following let $\mathfrak{A}$ be an n-dimensional non commutative, non associative algebra over a field $\mathbb{K}$ (i.e. the algebra need not be commutative or associative).

Let $\mathfrak{Z}(\mathfrak{A})$ be the set of elements of $\mathfrak{A}$ which commute with all elements of $\mathfrak{A}$, i.e.

$$\mathfrak{Z}(\mathfrak{A}) := \{y \in \mathfrak{A} \,|\, yx = xy \text{ for all } x \in \mathfrak{A}\}.$$

$\mathfrak{Z}(\mathfrak{A})$ is a subspace of $\mathfrak{A}$. However, in general it is not a subalgebra of $\mathfrak{A}$ since $\mathfrak{A}$ is not associative.

$\mathfrak{M}(\mathfrak{A})$ or simply $\mathfrak{M}$ denotes the algebra of all linear mappings of $\mathfrak{A}$ into itself.
$\mathfrak{M}$ is associative and isomorphic with the n^2-dimensional matrix algebra $\mathfrak{M}_n(\mathbb{K})$ over $\mathbb{K}$.

To each $x \in \mathfrak{A}$ correspond the linear mappings

$$\mathcal{L}_x : \mathfrak{A} \longrightarrow \mathfrak{A} \quad , \; a \longmapsto xa \quad , \qquad a \in \mathfrak{A} \, ,$$
$$\mathcal{R}_x : \mathfrak{A} \longrightarrow \mathfrak{A} \quad , \; a \longmapsto ax \quad , \qquad a \in \mathfrak{A} \, ,$$

these are the <u>left</u>- and <u>right</u> <u>transformation</u> induced by $x \in \mathfrak{A}$. The equality $\mathcal{L}_x = \mathcal{R}_x$ holds if and only if $x \in \mathfrak{Z}(\mathfrak{A})$.
Linear mappings of $\mathfrak{A}$ into $\mathfrak{M}$ are defined by $\Phi : x \longmapsto \mathcal{L}_x$ and $\Psi : x \longmapsto \mathcal{R}_x$, $x \in \mathfrak{A}$. These mappings are algebra homomorphisms if and only if $\mathfrak{A}$ is associative.

Let $\mathfrak{T}(\mathfrak{A})$ be the subalgebra of $\mathfrak{M}$ which is generated by all $\mathcal{L}_x, \mathcal{R}_x$, $x \in \mathfrak{A}$, and the identity $\mathfrak{I}$ of $\mathfrak{A}$. The subalgebra $\mathfrak{T}(\mathfrak{A})$ of $\mathfrak{M}$ is called the <u>transformation</u> <u>algebra</u> of $\mathfrak{A}$.
The transformation algebra $\mathfrak{T}(\mathfrak{A})$ is associative, since it is a subalgebra of $\mathfrak{M}$. The associativity is the essential advantage of working in $\mathfrak{T}(\mathfrak{A})$ instead of in $\mathfrak{A}$.
Every transformation $\mathcal{T} \in \mathfrak{T}(\mathfrak{A})$ can be represented as a linear combination of the identity and products of certain left and right transformations of $\mathfrak{A}$, i.e.

$$\mathcal{T} = f(\mathcal{L}_{x_1}, \mathcal{R}_{x_2}, \ldots, \mathcal{L}_{x_{s-1}}, \mathcal{R}_{x_s}) \quad \text{for an } f \in \mathbb{K}\{X_1, \ldots, X_s\}.$$

In general this representation is not unique.
$\mathcal{T}_n(\mathbb{K})$ denotes any subalgebra of the full matrix algebra $\mathfrak{M}_n(\mathbb{K})$ which is isomorphic with $\mathcal{T}(\mathfrak{A})$.
Let $\mathfrak{B}$ be any subalgebra of $\mathfrak{A}$. Then $\mathfrak{B}^*$ is the subalgebra of $\mathcal{T}(\mathfrak{A})$ generated by all left- and right transformations $\mathcal{L}_b$ and $\mathcal{R}_b$, $b \in \mathfrak{A}$.
We emphasize that $\mathfrak{A}^*$ may be a proper subalgebra of $\mathcal{T}(\mathfrak{A})$ if $\mathfrak{A}$ does not contain a unit element.

Let $a_1,\dots,a_n$ be a $\mathbb{K}$-basis of $\mathfrak{A}$ with multiplication table

$$a_i a_j = \sum_{k=1}^{n} \gamma_{ijk} a_k, \qquad i,j=1,\dots,n. \tag{2.1}$$

With respect to this basis the left transformation $\mathcal{L}_x : a \longmapsto xa$, $x = \sum_{i=1}^{n} \xi_i a_i$ $\mathfrak{A}$ is represented by the matrix

$$L_x := \begin{pmatrix} \sum_{i=1}^{n} \xi_i \gamma_{i11} & \cdots & \sum_{i=1}^{n} \xi_i \gamma_{in1} \\ \cdot & & \cdot \\ \cdot & & \cdot \\ \cdot & & \cdot \\ \sum_{i=1}^{n} \xi_i \gamma_{i1n} & \cdots & \sum_{i=1}^{n} \xi_i \gamma_{inn} \end{pmatrix} = \left(\sum_{i=1}^{n} \xi_i \gamma_{ijk} \right)^T_{\substack{j=1,\dots,n \\ k=1,\dots,n}} \tag{2.2}$$

The matrices which correspond to the left transformations $\mathcal{L}_{a_1},\dots,\mathcal{L}_{a_n}$ are denoted by

$$\Gamma_1 := (\gamma_{1jk})^T, \dots, \Gamma_n := (\gamma_{njk})^T . \tag{2.3}$$

Then L_x can be written as

$$L_x = \sum_{i=1}^{n} \xi_i \Gamma_i . \tag{2.4}$$

Similarly the matrix

$$R_x := \begin{pmatrix} \sum_{i=1}^{n} \xi_i \gamma_{1i1} & \cdots & \sum_{i=1}^{n} \xi_i \gamma_{ni1} \\ \cdot & & \cdot \\ \cdot & & \cdot \\ \cdot & & \cdot \\ \sum_{i=1}^{n} \xi_i \gamma_{1in} & \cdots & \sum_{i=1}^{n} \xi_i \gamma_{nin} \end{pmatrix} = \left(\sum_{i=1}^{n} \xi_i \gamma_{jik} \right)^T_{\substack{j=1,\dots,n \\ k=1,\dots,n}} \tag{2.5}$$

corresponds to the right transformation $\mathcal{R}_x$: $a \longmapsto ax$. The algebra $\mathfrak{A}$ is commutative if and only if the matrices $\Gamma_1,\dots,\Gamma_n$ have the property that the i-th column of Γ_j equals the j-th column of Γ_i, $i,j=1,\dots,n$.

Let the algebra $\mathfrak{A}$ have a genetic realization in the sense of definition 1.4 and let $a_1,\dots,a_n$ be a natural basis of $\mathfrak{A}$. Then the matrices $\Gamma_1,\dots,\Gamma_n$ are column stochastic.

We consider again the general case. Let $c_1,\dots,c_n$ be a second basis of $\mathfrak{A}$ with multiplication table

$$c_i c_j = \sum_{k=1}^{n} \lambda_{ijk} c_k, \qquad i,j=1,\dots,n. \tag{2.6}$$

Let $S = (\sigma_{ij})$ be the transformation carrying the basis $\{c_j\}$ into the basis $\{a_j\}$, then

$$a_i = \sum_{j=1}^{n} \sigma_{ji} c_j, \qquad i=1,\dots,n \quad .$$

Let $S^{-1} = T = (\tau_{ij})$, thus

$$c_i = \sum_{j=1}^{n} \tau_{ji} c_j \quad , \qquad i=1,\dots,n \quad .$$

The multiplication constants of $\mathfrak{A}$ with respect to these bases satisfy the relations

$$\gamma_{ijk} = \sum_{p,q,r=1}^{n} \sigma_{pi}\sigma_{qj}\lambda_{pqr}\tau_{kr}, \qquad i,j,k=1,\dots,n, \tag{2.7}$$

$$\lambda_{ijk} = \sum_{p,q,r=1}^{n} \tau_{pi}\tau_{qj}\gamma_{pqr}\sigma_{kr}, \qquad i,j,k=1,\dots,n. \tag{2.8}$$

In matrix notation these equations read

$$\Gamma_i = T\Big(\sum_{p=1}^{n} \sigma_{pi}\Lambda_p\Big)S \quad , \qquad i=1,\dots,n, \tag{2.9}$$

$$\Lambda_i = S\Big(\sum_{p=1}^{n} \tau_{pi}\Gamma_p\Big)T \quad , \qquad i=1,\dots,n, \tag{2.10}$$

where

$$\Lambda_i := (\lambda_{ijk})^T_{\substack{j=1,\dots,n\\ k=1,\dots,n}} \quad , \qquad i=1,\dots,n. \tag{2.11}$$

The "classical" non associative algebras do not play a central role in the theory of algebras in genetics. For this reason we restrict ourselves to definitions and well known constructions for Lie and Jordan algebras from associative algebras which will be used later. In the following we drop the condition that $\mathfrak{A}$ should be finite-dimensional.

The algebra $\mathfrak{A}$ is called a Lie algebra if the following two conditions hold

$$x^2 = 0 \qquad \text{for all } x \in \mathfrak{A}$$

and

$$(xy)z + (yz)x + (zx)y = 0 \qquad \text{for all } x,y,z \in \mathfrak{A}, \tag{2.12}$$

the second relation is called the Jacobi identity.

A commutative algebra $\mathfrak{A}$ over a field of characteristic unequal 2 is called a Jordan algebra if the Jordan identity

$$(xy)x^2 = x(yx^2) \tag{2.13}$$

is satisfied for all $x,y \in \mathfrak{A}$.

Let $\mathfrak{A}$ be an arbitrary algebra over $\mathbb{K}$, then the commutator product

$$(x,y) \longmapsto [x,y] := xy - yx \tag{2.14}$$

and the symmetric product

$$(x,y) \longmapsto x \cdot y := \frac{1}{2}(xy + yx) \tag{2.15}$$

define two new multiplications, in the second case we require char $\mathbb{K} \neq 2$. The algebras which are obtained with these multiplications are denoted by $\mathfrak{A}^-$ and $\mathfrak{A}^+$.

If the algebra $\mathfrak{A}$ is associative then $\mathfrak{A}^-$ is a Lie algebra and $\mathfrak{A}^+$ is a Jordan algebra.

A Jordan algebra $\mathfrak{B}$ is called a special Jordan algebra if it is isomorphic with a subalgebra of a Jordan algebra $\mathfrak{A}^+$ corresponding to an associative algebra $\mathfrak{A}$.

An algebra $\mathfrak{A}$ is called an alternative algebra if the left and right alternative laws

$$x^2y = x(xy) \quad , \quad yx^2 = (yx)x$$

are satisfied for all $x,y \in \mathfrak{A}$.

In a non associative, non commutative algebra $\mathfrak{A}$ the product of more than two factors is not well defined by the factors and their ordering alone. The association of the factors must be specified. Etherington [26] has given a classification of products in a non associative algebra. We shall not discuss this classification here since we shall use only a few special types of products and powers.

Definition 2.1. Let $x \in \mathfrak{A}$ and $\mathfrak{U} \leq \mathfrak{A}$ be a subalgebra of $\mathfrak{A}$.

(1) Then

$$ {}^{1}x := x, \qquad {}^{r+1}x := x({}^{r}x), \qquad r \in \mathbb{N}, $$
$$ x^{1} := x, \qquad x^{r+1} := (x^{r})x, \qquad r \in \mathbb{N}, $$

are called the left and right principal powers of x.
Similarly left and right principal powers of a subalgebra $\mathfrak{U}$ of $\mathfrak{A}$ are defined by

$$ {}^{1}\mathfrak{U} := \mathfrak{U}, \qquad {}^{r+1}\mathfrak{U} := \langle \mathfrak{U}({}^{r}\mathfrak{U}) \rangle, \qquad r \in \mathbb{N}, $$
$$ \mathfrak{U}^{1} := \mathfrak{U}, \qquad \mathfrak{U}^{r+1} := \langle (\mathfrak{U}^{r})\mathfrak{U} \rangle, \qquad r \in \mathbb{N}. $$

(2) The principal powers of a subalgebra $\mathfrak{U}$ of $\mathfrak{A}$ are defined by

$$ \mathfrak{U}^{(1)} := \mathfrak{U}, \quad \mathfrak{U}^{(r+1)} := \langle \mathfrak{U}\mathfrak{U}^{(r)}, \mathfrak{U}^{(r)}\mathfrak{U} \rangle, \qquad r \in \mathbb{N}. $$

(3) The plenary powers of an element $x \in \mathfrak{A}$ or a subalgebra $\mathfrak{U}$ of $\mathfrak{A}$, respectively, are defined as

$$ x^{[1]} := x, \qquad x^{[r+1]} := x^{[r]}x^{[r]}, \qquad r \in \mathbb{N}, $$
$$ \mathfrak{U}^{[1]} := \mathfrak{U}, \qquad \mathfrak{U}^{[r+1]} := \langle \mathfrak{U}^{[r]}\mathfrak{U}^{[r]} \rangle, \qquad r \in \mathbb{N}. $$

(4) An element $x \in \mathfrak{A}$ is called power associative if the relation holds

$$ x^{i}x^{j} = x^{i+j} \qquad \text{for all } i,j \in \mathbb{N}. $$

(Remark. These notations for powers of subalgebras $\mathfrak{U}$ of $\mathfrak{A}$ deviate from those of other authors, e.g. Etherington [31] denotes by $\mathfrak{U}^{(r)}$ the subalgebra of all elements of altitude r-1 or the (r-1)th generation of $\mathfrak{U}$ and $\mathfrak{U}^{[r]}$ the subalgebra of all elements of degree r. On the other hand in the theory of Lie algebras $\mathfrak{U}^{[r]}$ denotes the r-th derivative of $\mathfrak{U}$, cf. Jacobson [76].

We prefer the above notations because they are consistent with the notations for powers of an element.)

The left and right principal powers of a subalgebra $\mathfrak{U}$ of $\mathfrak{A}$ are left or right ideals, respectively, of $\mathfrak{U}$, and the principal powers are ideals of $\mathfrak{U}$. We have

$$^{r}\mathfrak{U}\,,\ \mathfrak{U}^{r} \subseteq \mathfrak{U}^{(r)} \qquad \text{for all } r \in \mathbb{N}, \tag{2.17}$$

in particular

$$^{r}\mathfrak{U} = \mathfrak{U}^{r} = \mathfrak{U}^{(r)} \qquad \text{for } r = 1,2.$$

If the subalgebra $\mathfrak{U}$ of $\mathfrak{A}$ is commutative, then the r-th left and r-th right principal power of $\mathfrak{U}$ coincide with the r-th principal power of $\mathfrak{U}$.

Since every element of the r-th plenary power of $\mathfrak{U}$ is also an element of $\mathfrak{U}$, the plenary power $\mathfrak{U}^{[r]}$ is contained in the r-th left and r-th right principle power of $\mathfrak{U}$, i.e.

$$\mathfrak{U}^{[r]} \subseteq {}^{r}\mathfrak{U} \cap \mathfrak{U}^{r} \qquad \text{for all } r \in \mathbb{N}. \tag{2.18}$$

Furthermore for each of the four types of powers of a subalgebra $\mathfrak{U}$ of $\mathfrak{A}$ defined above it is true that the (t+1)th power is contained in the r-th power of the same type, i.e.

$$^{r+1}\mathfrak{U} \subseteq {}^{r}\mathfrak{U}\,,\ \mathfrak{U}^{r+1} \subseteq \mathfrak{U}^{r},\ \mathfrak{U}^{(r+1)} \subseteq \mathfrak{U}^{(r)},\ \mathfrak{U}^{[r+1]} \subseteq \mathfrak{U}^{[r]}\,, \quad r \in \mathbb{N}\,. \tag{2.19}$$

<u>Definition 2.2.</u>

(1) An element $x \in \mathfrak{A}$ is called <u>left nilpotent</u> (<u>right nilpotent</u>) of index k if $^{k}x = 0$ but $^{k-1}x \neq 0$ ($x^{k} = 0$ but $x^{k-1} \neq 0$).

(2) A subalgebra $\mathfrak{U}$ of $\mathfrak{A}$ is called <u>left nil</u> (<u>right nil</u>) if all elements of $\mathfrak{U}$ are <u>left nilpotent</u> (<u>right nilpotent</u>).

(3) A subalgebra $\mathfrak{U}$ of $\mathfrak{A}$ is called <u>left nilpotent</u> (<u>right nilpotent</u>, <u>nilpotent</u>) of index k if $^{k}\mathfrak{U} = \langle 0\rangle$ but $^{k-1}\mathfrak{U} \neq \langle 0 \rangle$ ($\mathfrak{U}^{k} = \langle 0\rangle$ but $\mathfrak{U}^{k-1} \neq \langle 0\rangle$, $\mathfrak{U}^{(k)} = \langle 0\rangle$ but $\mathfrak{U}^{(k-1)} \neq \langle 0\rangle$, respectively). Algebras which are nilpotent of index 2 are called <u>zero algebras</u>.

(4) A subalgebra $\mathfrak{U}$ of $\mathfrak{A}$ is called <u>solvable</u> of index k if $\mathfrak{U}^{[k]} = \langle 0\rangle$ but $\mathfrak{U}^{[k-1]} \neq \langle 0\rangle$.

(Remark. Other authors, cf. Schafer [102], call an algebra nilpotent of index r if all products with r factors, no matter how the factors are associated, are zero but at least one product with r-1 factors does not vanish. Etherington [31] calls these algebras nilpotent of degree r while he calls $\mathfrak{U}$ nilpotent of altitude t-1 if $\mathfrak{U}^{(t)} = \langle 0 \rangle$ but $\mathfrak{U}^{(t-1)} \neq \langle 0 \rangle$. The index of nilpotency defined above differs from the altitude by 1. Every algebra which is nilpotent of index k is nilpotent of degree at most 2^{k+1}. Since the degree and the index of nilpotency do not play a role in our considerations we prefer the above definition 2.2(2).)

Of course, in a commutative algebra $\mathfrak{A}$ the properties left nilpotent, right nilpotent and nilpotent coincide.

From (2.7) and (2.18) we obtain the following

Corollary 2.3. If a subalgebra $\mathfrak{U}$ of $\mathfrak{A}$ is nilpotent of index k then it is left and right nilpotent and solvable of indices not greater than k.

On the other hand Suttles [110] has given an example of a solvable algebra which is not nilpotent.
In the following we assume again that $\mathfrak{A}$ is finite-dimensional.
In this case the idea of the radical has been carried over from associative algebras to non associative algebras by Albert [6].

Definition 2.4.

(1) An algebra $\mathfrak{A}$ which is not a zero algebra is called simple if it has no ideals except $\mathfrak{A}$ and $\langle 0 \rangle$.
(2) An algebra $\mathfrak{A}$ is called semi-simple if it is a direct sum of simple algebras.

Albert [6] proved the following

Theorem 2.5. Let the algebra $\mathfrak{A}$ be homomorphic to a semi-simple algebra. Then the family of ideals $\mathfrak{N}$ of $\mathfrak{A}$ with the property that $\mathfrak{A}/\mathfrak{N}$ is semi-simple has a smallest element $\mathfrak{R}$. This ideal $\mathfrak{R}$ is called the radical of $\mathfrak{A}$.

Albert [6] shows furthermore that any algebra $\mathfrak{A}$ contains an ideal $\mathfrak{N}$ such that either $\mathfrak{N}$ is the radical of $\mathfrak{A}$ or $\mathfrak{A}/\mathfrak{N}$ is a zero algebra.
The following theorem has also been proved by Albert [8].

Theorem 2.6. Let $\mathfrak{A}$ be an algebra with radical $\mathfrak{R}$. Then $\mathfrak{R}$ contains all solvable ideals of $\mathfrak{A}$.

The radical of an algebra $\mathfrak{A}$ need not to be the maximal solvable ideal as has been shown by a counter-example of Albert[7].

Since every nilpotent ideal is solvable the radical of an algebra contains all nilpotent ideals.

B. Polynomials in non-associative powers

The concept of a rank equation is fundamental for the study of algebras in genetics. Apparently the notation of a rank equation for non-associative algebras goes back to a paper by Dickson [21]. In his paper [27] Etherington refers to Dickson's "Linear Algebras" [22], however in § 19 of that book only the associative case is covered. Dickson has shown that in a non-associative algebra with unit element there is also an equation of minimal degree in right (left) principal powers which is satisfied by any element of this algebra, i.e. that a right (left) rank equation exists.

In this and the following section we shall reformulate Dickson's ideas and generalize them to the case of algebras without unit element.

We start with the vector space

$$\mathbb{K}\langle X\rangle := \left\{ \sum_{i=0}^{\infty} \alpha_i X^i \;\middle|\; \alpha_i \in \mathbb{K} \text{ with only finitely many } \alpha_i \neq 0 \right\}. \qquad (2.20)$$

of all polynomials in right principal powers in one indeterminate X which need be neither associative nor commutative,

$$X^0 := 1 \in \mathbb{K}, \qquad X^1 := X, \qquad X^{r+1} := X^r X, \qquad r \in \mathbb{N} .$$

In this section we prefer to use right instead of left principal powers since formulas come closer to traditional notation. Of course similar considerations are valid for left principal powers and later on we shall use the results for right and left principal powers as well.

As in the associative case we define

Definition 2.7. Let $0 \neq p(X) = \sum_{i=0}^{\infty} \alpha_i X^i \in \mathbb{K}\langle X\rangle$. Then the greatest number $n \in \mathbb{N}_0$ with $\alpha_n \neq 0$ is called the degree of p. A polynomial of degree 1 is called a linear polynomial and a polynomial of degree 0 a constant polynomial.

A polynomial p of degree n can of course be written as

$$p(X) = \sum_{i=0}^{n} \alpha_i X^i \qquad \text{with } \alpha_n \neq 0 .$$

A polynomial p of degree $n \in \mathbb{N}_0$ is said to be normalized, if $\alpha_n = 1$.

Every polynomial of degree n is a scalar multiple of a normalized polynomial of degree n.

The following subspaces of $\mathbb{K}\langle X\rangle$ are of particular interest:
The 2-dimensional subspace of all linear and constant polynomials

$$\mathbb{K}_1\langle X\rangle := \left\{\alpha_1 X^1 + \alpha_0 \;\middle|\; \alpha_0, \alpha_1 \in \mathbb{K}\right\}, \tag{2.21}$$

and the infinite-dimensional subspace of all polynomials with vanishing constant term

$$\mathbb{K}_1\langle X\rangle := \left\{\sum_{i=1}^{\infty} \alpha_i X^i \;\middle|\; \alpha_i \in \mathbb{K}, \text{ only finitely many } \alpha_i \neq 0\right\}. \tag{2.22}$$

The formal product of two polynomials in $\mathbb{K}\langle X\rangle$ is in general not an element of $\mathbb{K}\langle X\rangle$, since the powers of X are non associative. However we can define the multiplication of a polynomial in $\mathbb{K}\langle X\rangle$ with a linear polynomial from the right. In particular for $X = X^1 \in \mathbb{K}_1\langle X\rangle$ the following lemma holds which can be easily verified.

Lemma 2.8. The multiplication from the right

$$\varphi : p(X) = \sum_{i=0}^{n} \alpha_i X^i \longmapsto p(X)X := \sum_{i=0}^{n} \alpha_i X^{i+1} \tag{2.23}$$

defines a linear mapping of $\mathbb{K}\langle X\rangle$ onto $\mathbb{K}_1\langle X\rangle$.

Let φ^r be the r-th iterate of φ, $r \in \mathbb{N}$. Because of the associativity of the addition in $\mathbb{N}_0$ we have

$$\varphi^r p(X) = \sum_{r=0}^{n} \alpha_i X^{i+r} \quad , \qquad r \in \mathbb{N} \quad . \tag{2.24}$$

Furthermore we define

$$\varphi^{o} := \mathrm{id}_{\mathbb{K}\langle X\rangle} . \qquad (2.25)$$

Following Dickson [21] we define a multiplication in $\mathbb{K}\langle X\rangle$.

Let $p(X) = \sum_{i=0}^{n} \alpha_i X^i$, $q(X) = \sum_{j=0}^{m} \beta_j X^j \in \mathbb{K}\langle X\rangle$, then

$$(p,q) \longmapsto pq$$

where

$$(pq)(X) := \sum_{j=0}^{m} \beta_j \varphi^j p(X) \qquad (2.26)$$

defines a mapping of $\mathbb{K}\langle X\rangle \times \mathbb{K}\langle X\rangle$ into $\mathbb{K}\langle X\rangle$. Since φ is linear this mapping is bilinear and hence defines a multiplication in $\mathbb{K}\langle X\rangle$. We have the following theorem.

Theorem 2.9. $\mathbb{K}\langle X\rangle$ with the multiplication defined by (2.26) is a commutative, associative algebra which is isomorphic with the ordinary polynomial algebra $\mathbb{K}[X]$ in one indeterminate over $\mathbb{K}$.

Proof. It is easily checked that the multiplication defined by (2.26) acts commutatively and associatively on the basis elements X^o, X^1, X^2,... of $\mathbb{K}\langle X\rangle$. Thus the multiplication has these properties in general. Then the isomorphism is trivial. □

After this insight has been obtained it is obvious that the Euclidian algorithm and the unique factorization of a polynomial into prime factors can be carried over to the algebra $\mathbb{K}\langle X\rangle$ of polynomials in right principal powers. In particular we have

Theorem 2.10. Let $f(X) = \sum_{i=0}^{n} \alpha_i X^i \in \mathbb{K}\langle X\rangle$, let $\bar{f} \in \mathbb{K}[X]$ be the polynomial with the same coefficients as f. Let $\mathbb{L}$ be a field extension of $\mathbb{K}$ which contains the splitting field of $\bar{f}$. Then f can be written as the product (in the sense of (2.26)) of normed linear factors

$$f(X) = \alpha_n (X-\lambda_1)(X-\lambda_2)\dots(X-\lambda_n) \quad ,$$

where $\lambda_1,\dots,\lambda_n$ are the roots of $\bar{f}$ in $\mathbb{L}$.

In the following we study the replacement of the indeterminate X by elements of a non associative algebra.

Definition 2.11. Let $\mathfrak{A}$ be an algebra with (or without) unit element. Let $a \in \mathfrak{A}$ and let a^i be the i-th right principal power of a, $i \in \mathbb{N}_o$ ($i \in \mathbb{N}$, respectively) where a^o is the unit element of $\mathfrak{A}$. The mapping

$$E_a : \mathbb{K}\langle X\rangle \longrightarrow \mathfrak{A} \qquad (\mathbb{K}_1\langle X\rangle \longrightarrow \mathfrak{A})$$

defined by $X^i \longmapsto a^i$, $i \in \mathbb{N}_o$ ($i \in \mathbb{N}$) and the linear extension onto $\mathbb{K}\langle X\rangle$ (or $\mathbb{K}_1\langle X\rangle$, respectively) is called the replacement process. The image of $p \in \mathbb{K}\langle X\rangle$ ($\mathbb{K}_1\langle X\rangle$) under E_a is denoted by $p(a)$.

The mapping E_a is by construction linear but in general it is not an algebra homomorphism. In fact E_a is a homomorphism if and only if the element $a \in \mathfrak{A}$ is power associative. In the following we discuss the properties of the kernel of E_a. This kernel is the set of all polynomials in $\mathbb{K}\langle X\rangle$ ($\mathbb{K}_1\langle X\rangle$) which annihilate the given element a. Although E_a in general is not an algebra homomorphism the kernel is an ideal.

Theorem 2.12. Let $\mathfrak{A}$ be an algebra with (or without) unit element. Let $a \in \mathfrak{A}$, then the kernel of the linear mapping E_a is an ideal in $\mathbb{K}\langle X\rangle$.

Proof. Since E_a is a linear mapping, $\ker E_a$ is a subspace of $\mathbb{K}\langle X\rangle$. Now let $p \in \ker E_a$ and $q \in \mathbb{K}\langle X\rangle$ be arbitrary polynomials. We have to show that $pq \in \ker E_a$. For each $f(X) = \sum_{i=0}^{n} \alpha_i X^i \in \mathbb{K}\langle X\rangle (\mathbb{K}_1\langle X\rangle)$ we have

$$\left(\sum_{i=0}^{n} \alpha_i a^i\right) a = \sum_{i=0}^{n} \alpha_i a^{i+1} \quad ,$$

i.e.

$$f(a)a = E_a \varphi f(X) \quad ,$$

or, if we use the right transformation $\mathcal{R}_a : \mathfrak{A} \longrightarrow \mathfrak{A}$ this equation can be written

$$\mathcal{R}_a f(a) = E_a \varphi f(X) \quad .$$

By induction it follows that

$$\mathcal{R}_a^j f(a) = E_a \varphi^j f(X) \quad , \qquad j \in \mathbb{N} \quad .$$

Hence for $p \in \ker E_a$

$$E_a \varphi^j p(X) = \mathcal{R}_a^j p(a) = \mathcal{R}_a^j 0 = 0, \qquad j \in \mathbb{N}.$$

Let

$$q(X) = \sum_{j=0}^{m} \beta_j X^j \in \mathbb{K}\langle X\rangle \ .$$

By the definition of the multiplication

$$(pq)(X) = \sum_{j=0}^{m} \beta_j \varphi^j p(X)$$

and the linearity of E_a it follows that

$$E_a(pq)(X) = \sum_{j=0}^{m} \beta_j E_a \varphi^j p(X) = 0 \ ,$$

hence $(pq)(a) = 0$. □

Definition 2.13. Let $\mathfrak{A}$ be an algebra over $\mathbb{K}$ and $a \in \mathfrak{A}$. A right minimal polynomial of a is a normalized polynomial $m_a \in \mathbb{K}\langle X\rangle$ in right principal powers which annihilates $a \in \mathfrak{A}$ and has minimal degree.

The existence of a right minimal polynomial is simply a consequence of the fact that in an n-dimensional algebra $\mathfrak{A}$ over $\mathbb{K}$ the right principal powers $a^1, a^2, \ldots, a^n, a^{n+1}$ of any element $a \in \mathfrak{A}$ are linearly dependent. The next lemma provides uniqueness.

Lemma 2.14. Let $\mathfrak{A}$ be an algebra over $\mathbb{K}$. Let $a \in \mathfrak{A}$ and $m_a \in \mathbb{K}\langle X\rangle$ be a right minimal polynomial of a. Then m_a is a divisior of any polynomial $f \in \ker E_a$.

Proof. Let $\deg m_a = s$ be the degree of m_a. In view of theorem 2.9 and $\deg f \geq \deg m_a$ there are uniquely determined polynomials $q, r \in K\langle X\rangle$ with $\deg r \leq s-1$ such that

$$f(X) = (qm_a)(X) + r(X) \qquad .$$

If we apply the operator E_a then theorem 2.12 yields

$$0 = f(a) = (qm_a)(a) + r(a) = 0 + r(a) \qquad ,$$

thus $r \in \ker E_a$. Since the degree of m_a is minimal and $\deg r < \deg m_a$ we have $r = 0$. □

Existence and uniqueness of the minimal polynomial lead to the following

Theorem 2.15. Let $\mathfrak{A}$ be an algebra over $\mathbb{K}$. For any element $a \in \mathfrak{A}$ $\ker E_a$ is the principal ideal of $\mathbb{K}\langle X\rangle$ generated by the right minimal polynomial of a.

The next theorem going back to Dickson [18] establishes the well-known relation between the minimal polynomial and the characteristic polynomial.

Theorem 2.16. Let $\mathfrak{A}$ be an n-dimensional algebra with (without) unit element over $\mathbb{K}$. Let $a \in \mathfrak{A}$ and let f_a be the characteristic polynomial of the right transformation $\mathcal{R}_a$. Let

$$m_a(X) = X^s + \beta_1 X^{s-1} + \ldots + \beta_{s-1} X + \beta_s X^o \quad (\beta_s = 0)$$

be the right minimal polynomial of a.
Then

(1) The coefficients β_i are homogeneous polynomials of degree i in the coordinates of a, $i=1,\ldots,s$ $(s-1)$.

(2) $m_a(X)$ is a divisor of $f_a(X)$ ($f_a(X)X$, respectively).

Proof. We choose a fixed basis in $\mathfrak{A}$. Then to the right transformation $\mathcal{R}_a$ corresponds a matrix R_a, cf. (2.5). We interpret the characteristic polynomial $f_a(\lambda) = \det(\lambda I - R_a)$ as a polynomial in right principal powers

$$p(X) =: X^n + \alpha_1 X^{n-1} + \ldots + \alpha_{n-1} X + \alpha_n X^o.$$

Since the coefficients of R_a are homogeneous of degree 1 in the coordinates of a, the coefficients α_i of the characteristic polynomial f_a are homogeneous of degree i in the coordinates of a, $i=1,\ldots,n$. By the theorem of Cayley and Hamilton f_a annihilates R_a. Since the algebra $\mathfrak{M}$ of all linear mappings of $\mathfrak{A}$ into itself and the matrix algebra $\mathfrak{M}_n(\mathbb{K})$ are isomorphic we have

$$f_a(\mathcal{R}_a) = \mathcal{R}_a^n + \alpha_1 \mathcal{R}^{n+1} + \ldots + \alpha_{n-1} \mathcal{R}^1 + \alpha_n \mathcal{I} = 0.$$

Now we distingiush two cases:
If $\mathfrak{A}$ has a unit element 1 we apply the zero transformation $f_a(\mathcal{R}_a)$ to $1 \in \mathfrak{A}$ and obtain the following identity in right principal powers of a

$$a^n + \alpha_1 a^{n-1} + \ldots + \alpha_{n-1} a + \alpha_n 1 = 0 \quad ,$$

i.e. $f_a \in \mathbb{K}\langle X\rangle$ is an annihilating polynomial of a. If $\mathfrak{A}$ has no unit element we apply the zero transformation $f_a(\mathcal{R}_a)$ to a and obtain the following identity in right principal powers of a

$$a^{n+1} + \alpha_1 a^n + \ldots + \alpha_{n-1} a^2 + \alpha_n a = 0 \quad ,$$

i.e. $\tilde{f}_a(X) := f_a(X)X \in \mathbb{K}\langle X\rangle$ is an annihilating polynomial of $a \in \mathfrak{A}$. Hence the right minimal polynomial m_a of a is a divisor of f_a or $\tilde{f}_a$, respectively. The homogeneity of the coefficients as functions of the coordinates of a is obvious from the fact that the polynomial coefficients except from their sign are the elementary symmetric functions of the zeros of the polynomial. □

<u>Remark.</u> Let $a \in \mathfrak{A}$ and $m_{\mathcal{R}_a} \in \mathbb{K}\langle X\rangle$ be minimal polynomial of the right transformation $\mathcal{R}_a \in \mathfrak{M}(\mathfrak{A})$.
If the algebra $\mathfrak{A}$ has a unit element 1, then the minimal polynomial $m_a \in \mathbb{K}\langle X\rangle$ coincides with $m_{\mathcal{R}_a}$, since on the one hand we have

$$0 = m_{\mathcal{R}_a}(\mathcal{R}_a)(1) = m_{\mathcal{R}_a}(a) \quad ,$$

hence $m_{\mathcal{R}_a}$ divides m_a, and on the other hand we have

$$0 = m_a(a) = m_a(\mathcal{R}_a)(1) \quad ,$$

hence $m_a(\mathcal{R}_a) = 0$, thus m_a divides $m_{\mathcal{R}_a}$.
One might think that if $\mathfrak{A}$ has no unit element, then $m_{\mathcal{R}_a}(X)X$ would be the minimal polynomial of a. But this is not true in general, as we can demonstrate by example 1.1.
Let $\mathfrak{A} = \langle a_1, \ldots, a_n\rangle_{\mathbb{R}}$ with multiplication table given by $a_i a_j = \frac{1}{2}(a_i + a_j)$, $i,j = 1,\ldots,n$. Choose $a = a_1 \in \mathfrak{A}$, then the right transformation $\mathcal{R}_a$ is represented by the matrix

$$R_a = \begin{pmatrix} 1 & 1/2 & \cdots & 1/2 \\ & 1/2 & \cdots & 1/2 \\ & & \ddots & \\ & 0 & & 1/2 \end{pmatrix}$$

One can easily check that the minimal polynomial of $\mathcal{R}_a$ is

$$m_{\mathcal{R}_a}(X) = (X-1)(X - \tfrac{1}{2}) \; .$$

Furthermore the minimal polynomial of a is

$$m_a(X) = (X-1)X \quad ,$$

which is a proper divisor of $m_{\mathcal{R}_a}(X)X$.

C. The rank equation of an algebra

Again let $\mathfrak{A}$ be an n-dimensional algebra over $\mathbb{K}$. Until now we have been interested only in the minimal polynomial of a single element $a \in \mathfrak{A}$. Now we look for a polynomial in right principal powers that annihilates all elements of $\mathfrak{A}$. For this purpose we use the following idea which can be found in a book [76] by Jacobson. For the given field $\mathbb{K}$ let $\mathbb{F} = \mathbb{K}(\xi_1,\ldots,\xi_n)$ be the field of rational functions in n (associative and commutative) indeterminates over $\mathbb{K}$ and let $\mathfrak{A}_{\mathbb{F}} = \mathfrak{A} \otimes_{\mathbb{K}} \mathbb{F}$ be the algebra over the extension $\mathbb{F}$ of the field $\mathbb{K}$. Now choose a fixed basis $a_1,\ldots,a_n$ in $\mathfrak{A}$. If x is an element running over the whole algebra $\mathfrak{A}$, then in

$x = \sum_{i=1}^{n} \xi_i a_i$ the vector $(\xi_1,\ldots,\xi_n)^T$ runs over $\mathbb{K}^n$ and thus

$x = \sum_{i=1}^{n} \xi_i a_i$ is one element of $\mathfrak{A}_{\mathbb{F}}$. In this context Jacobson [76] calls x a generic element of $\mathfrak{A}$. We apply the ideas of the preceding section with $\mathbb{F}$ instead of $\mathbb{K}$. Then we find that there is exactly one normalized polynomial $m_x \in \mathbb{F}\langle X\rangle$ which annihilates $x \in \mathfrak{A}_{\mathbb{F}}$ and has minimal degree with respect to this property

$$m_x(X) = X^s + \beta_1(\xi_1,\ldots,\xi_n)X^{s-1} + \ldots + \beta_{s-1}(\xi_1,\ldots,\xi_n)X + \beta_s(\xi_1,\ldots,\xi_n).$$

By theorem 2.16 the coefficients β_i of m_x are homogeneous polynomials of degree i, i=1,...,s, in the coordinates $\xi_1,\ldots,\xi_n$ of x with respect to the basis $a_1,\ldots,a_n$ of $\mathfrak{A}$. (This is of course valid for any basis of $\mathfrak{A}$).

Following Jacobson [75] we call m_x the right minimal polynomial of the algebra $\mathfrak{A}$. Dickson [22] uses the term right rank equation of the algebra $\mathfrak{A}$. We follow the notation of Etherington [27] and we do not distinguish the cases where the algebra has a unit element or not. The normalized polynomial in $\mathbb{F}\langle X\rangle$ of minimal degree without absolute term that annihilates $x \in \mathfrak{A}_{\mathbb{F}}$ is called the right rank polynomial of $\mathfrak{A}$ (cf. the notation in 2.16). Consequently the right rank polynomial of an algebra is either the right minimal polynomial m_x of the algebra in the above sense or $m_x(X)X$ depending on whether the algebra has no unit element or has a unit element.

<u>Definition 2.17.</u> Let $x = \sum_{i=1}^{n} \xi_i a_i \in \mathfrak{A}_{\mathbb{F}}$. The normalized polynomial

$$q_x(X) = X^r + \beta_1(\xi_1,\ldots,\xi_n)X^{r-1} + \ldots + \beta_{r-1}(\xi_1,\ldots,\xi_n)X \in \mathbb{F}\langle X\rangle \quad ,$$

$\mathbb{F} = \mathbb{K}(\xi_1,\ldots,\xi_n)$, without absolute term which annihilates x and has minimal degree is called the <u>right rank polynomial</u> of $\mathfrak{A}$. The degree of q_x is called the <u>right rank</u> of the algebra $\mathfrak{A}$.

In view of our earlier remark the coefficient β_i of q_x is homogeneous of degree i in the coefficients $\xi_1,\ldots,\xi_n$ of x, $i=1,\ldots,r-1$. As an abbreviation we introduce

$$\beta_i(x) := \beta_i(\xi_1,\ldots,\xi_n), \quad i=1,\ldots,r-1.$$

Then the right rank polynomial reads

$$q_x(X) = X^r + \beta_1(x)X^{r-1} + \ldots + \beta_{r-1}(x)(X) \quad . \tag{2.27}$$

By definition we have the equation

$$E_x q_x(X) = q_x(x) = 0 \quad ,$$

which is called the <u>right rank equation</u> of the algebra $\mathfrak{A}$.

We have given an exact definition of the right minimal polynomial and the right rank polynomial for a non associative algebra which may not have a unit element. From now on we take the liberty as usual in algebra to write the right rank polynomial (also the right minimal polynomial etc.) for a generic element $x \in \mathfrak{A}$. With this convention the right rank polynomial reads

$$q_x(x) = x^r + \beta_1(x)x^{r-1} + \ldots + \beta_{r-1}(x)x \; . \tag{2.28}$$

3. TRAIN ALGEBRAS, GENETIC ALGEBRAS AND SPECIAL TRAIN ALGEBRAS

In the following we assume that $\mathfrak{A}$ is an n-dimensional commutative, not necessarily associative algebra over a field $\mathbb{K}$. Then the aforementioned right and left properties coincide and we can neglect such attributes.

A. Train algebras

In many algebras which arise in connection with genetical problems the coefficients of the rank polynomial do not depend explicitely on all n coordinates of a generic element but only on a $\mathbb{K}$-valued function of these coordinates. Etherington [27] gave the following definition.

Definition 3.1. Let $\mathfrak{A}$ be a baric algebra with weight homomorphism ω. $\mathfrak{A}$ is called a train algebra if the coefficients $\beta_i(x)$ of the rank polynomial (2.28) of $\mathfrak{A}$ are only functions of $\omega(x)$.

The coefficients $\beta_i(x)$ of the rank polynomial of an algebra are homogeneous of degree i in the coordinates of x and thus in a train algebra with weight homomorphism ω these coefficients are constant multiples of the i-th power $\omega^i(x)$ of the weight $\omega(x)$, i.e. the rank equation of $\mathfrak{A}$ has the form

$$q_x(x) = x^r + \gamma_1 \omega(x) x^{r-1} + \gamma_2 \omega^2(x) x^{r-2} + \dots + \gamma_{r-1} \omega^{r-1}(x) x = 0. \quad (3.1)$$

In a train algebra $\mathfrak{A}$ of rank r all elements $x \in \mathfrak{A}$ of weight 1 satisfy the equation

$$x^r + \gamma_1 x^{r-1} + \gamma_2 x^{r-2} + \dots + \gamma_{r-1} x = 0 \quad . \quad (3.2)$$

If we apply the weight homomorphism ω to this equation we obtain

$$1 + \gamma_1 + \dots + \gamma_{r-1} = 0 \quad .$$

Hence $\lambda_o = 1 \in \mathbb{K}$ is a zero of the polynomial

$$p(X) := X^r + \gamma_1 X^{r-1} + \dots + \gamma_{r-1} X \in \mathbb{K}\langle X\rangle \quad . \quad (3.3)$$

In any extension field $\mathbb{L}$ of $\mathbb{K}$ which contains the splitting field of p the polynomial p splits into linear factors

$$p(x) = X(X-1)(X-\lambda_1)\dots(X-\lambda_{r-2}) \quad . \quad (3.4)$$

Definition 3.2. Let $\mathfrak{A}$ be a train algebra with rank equation (3.1) and let the polynomial p split into linear factors in $\mathbb{L} \supseteq \mathbb{K}$ as in (3.4). Then the elements $\lambda_0 := 1, \lambda_1, \ldots, \lambda_{r-2}$ of $\mathbb{L}$ are called the principal train roots of $\mathfrak{A}$.

Then the rank equation (3.1) can be factorized according to the product (2.26) as

$$q_x(x) = x(x - \omega(x))(x - \lambda_1\omega(x)) \ldots (x - \lambda_{r-2}\omega(x)) = 0 . \quad (3.5)$$

From (3.1) we obtain immediately the following result of Etherington [27].

Lemma 3.3. Let $\mathfrak{A}$ be a train algebra of rank r with weight homomorphism $\omega : \mathfrak{A} \longrightarrow \mathbb{K}$. Then every element in $\mathfrak{N} := \ker\omega$ is nilpotent of an index not greater than r.

Abraham [2] showed that the maximal index of nilpotency of the elements of $\mathfrak{N}$ may be less than the rank of a train algebra. It is clear that every nilpotent element of a train algebra is contained in $\mathfrak{N}$, since from $y^l = 0$ for an $l \in N$ follows $0 = \omega(y^l) = \omega^l(y)$ and thus $\omega(y) = 0$, cf. Schafer [101] .

Until now we have not considered whether the weight homomorphism of a baric algebra is uniquely determined. In fact, in general this is not the case as can be seen from the following example.

Example 3.4. Let $\mathfrak{A} = \langle a_1, a_2, a_3 \rangle_{\mathbb{R}}$ be a commutative 3-dimensional algebra with multiplication table

	a_1	a_2	a_3
a_1	a_1+a_2	a_2	a_2
a_2		a_2	a_2
a_3			a_2+a_3

Then $\mathfrak{N}_1 := \langle a_1, a_2 \rangle$ and $\mathfrak{N}_2 := \langle a_2, a_3 \rangle$ are ideals of $\mathfrak{A}$ which do not contain $\mathfrak{A}^2$. From lemma 1.11 follows that $\mathfrak{A}$ has two different weight homomorphisms. □

The following assertion is a special case of a result of Holgate [67].

Lemma 3.5. Let $\mathfrak{A}$ be a baric algebra with weight homomorphism ω. If $\mathfrak{N} := \ker\omega$ is nil, then ω is uniquely determined.

Proof. Let $\varphi : \mathfrak{A} \longrightarrow \mathbb{K}$ be any non-trivial homomorphism and let $x \in \mathfrak{N}$. Then there is an $l \in \mathbb{N}$ with $x^l = 0$. Since φ is a homomorphism into $\mathbb{K}$ it follows $\varphi(x) = 0$. Let $x \in \mathfrak{A} \setminus \mathfrak{N}$, thus $\omega(x) \neq 0$. Then we have $\omega(\frac{x^2}{\omega(x)} - x) = 0$, hence $\frac{x^2}{\omega(x)} - x \in \mathfrak{N}$. Consequently $\varphi(\frac{x^2}{\omega(x)} - x) = 0$, i.e. $\varphi(x)(\frac{\varphi(x)}{\omega(x)} - 1) = 0$, thus $\varphi(x) = 0$ or $\varphi(x) = \omega(x)$. Since φ has been assumed to be non-trivial we have $\varphi = \omega$. □

From lemma 3.3, with assistance of lemma 3.5 follows

Corollary 3.6. A train algebra has exactly one weight homomorphism.

Let $\mathfrak{A}$ be a train algebra and assume that the elements $x \in \mathfrak{A}$ of weight 1 satisfy (3.2). If we multiply equation (3.2) l times successively with the element x from the right (or from the left) we obtain

$$x^{r+l} + \gamma_1 x^{r+l-1} + \ldots + \gamma_{r-1} x^{l+1} = 0, \quad l=0,1,\ldots, \qquad (3.6)$$

this says that the principal powers of each element $x \in \mathfrak{A}$ of weight 1 satisfy a linear difference equation in $\mathfrak{A}$ with constant coefficients. This fact provides a simple method to compute principal powers. Suppose the principal train roots $\lambda_1,\ldots,\lambda_{r-2}$ of the algebra $\mathfrak{A}$ are known. If one has computed the first r-1 principal powers $x = x^1, x^2,\ldots,x^{r-1}$ then the following principal powers x^{r+l}, $l \in \mathbb{N}_0$, can be computed without using the multiplication constants of the algebra $\mathfrak{A}$. Furthermore we can apply the same idea to arbitrary elements $x \neq 0$ of the algebra which do not necessarily satisfy $\omega(x) = 1$, because

$$x^l = \omega^l(x)\tilde{x}^l, \qquad l \in \mathbb{N},$$

where

$$\tilde{x} := \frac{x}{\omega(x)} \quad .$$

If a fixed basis $a_1,\ldots,a_n$ of $\mathfrak{A}$ is given, then the linear difference equation (3.6) for $x \in \mathfrak{A}$ with $\omega(x) = 1$ is a system of n linear uncoupled difference equations for the coordinates of the principal powers $x^j = \sum_{i=1}^{n} \xi_i^{(j)} a_i$ of x:

$$\xi_i^{(r+l)} + \gamma_1\xi_i^{(r+l-1)} + \ldots + \gamma_{r-1}\xi_i^{(l+1)} = 0 \ , \ l \in \mathbb{N}_o, \ i=1,\ldots,n. \qquad (3.7)$$

The solutions of the difference equations (3.7) can be presented in the form

$$\xi_i^{(j+1)} = \sum_{k=0}^{r-2} \beta_{ki}(j)\lambda_k^j \ , \qquad j \in \mathbb{N}, \quad i=1,\ldots,n, \qquad (3.8)$$

where $j \longmapsto \beta_{ki}(j)$, $k=0,\ldots,r-2$, are polynomials in j. The coefficients of these polynomials can be computed from the initial data $\xi_i^{(1)},\ldots,\xi_i^{(r-1)}$, $i=1,\ldots,n$. For the principal powers of x we obtain the representation

$$x^j = \sum_{k=0}^{r-2} b_k(j)\lambda_k^j \ , \qquad j \in \mathbb{N}, \qquad (3.9)$$

with

$$b_k(j) := \sum_{i=1}^{n} \beta_{ki}(j)a_i \ , \qquad k=0,1,\ldots,r-2.$$

Introducing in $\mathfrak{A}$ the natural topology of an n-dimensional space $\mathbb{R}^n$ or $\mathbb{C}^n$, respectively, we have as a simple consequence from matrix theory

<u>Theorem 3.7.</u> Let $\mathfrak{A}$ be a train algebra of rank r over $\mathbb{R}$ or $\mathbb{C}$. Assume the principal train roots $\lambda_o,\lambda_1,\ldots,\lambda_{r-2}$ satisfy $|\lambda_i| < 1$, $i=1,\ldots,r-2$. Then the sequence $\{x^j\}$ of principal powers converges for every element of weight 1.

These facts are important in particular with respect to applications in genetics. Assume that the inheritance of a population can be described by a gametic algebra $\mathfrak{A}$. Let $x \in \mathfrak{A}$ represent the distribution of gametic types in the population. With random mating within the population the filial population is represented by x^2. If the filial population is randomly mated with the initial population we obtain $x^2x = x^3$ and so on. The generation sequence which is created by random mating with the initial population corresponds to the sequence of principal powers of x. Hence, if the gametic algebra $\mathfrak{A}$ is a train algebra and the principal train roots of $\mathfrak{A}$ are known, we can compute the distribution of gametes in the (r+l)th generation from those in the first r-1 generations, l=0,1,2,... , and we can apply the convergence result, cf. also secton 4C.

As well as sequences of principal powers, sequences of plenary powers $x^{[1]} = x$, $x^{[2]}$, $x^{[3]}$,... (cf.definition 2.1), $x \in \mathfrak{A}$, are of major interest in genetics, because they include all sequences describing the generations of a population with random mating within itself. In an n-dimensional algebra $\mathfrak{A}$ over $\mathbb{K}$ the plenary powers $x^{[1]},\ldots,x^{[n]}$, $x^{[n+1]}$

of every $x \in \mathfrak{A}$ are linearly dependent. Hence there exists a normalized equation

$$x^{[k]} + \sigma_1(x)x^{[k-1]} + \ldots + \sigma_{k-1}(x)x^{[1]} = 0 \qquad (3.10)$$

of minimal degree over the field $\mathbb{F} = \mathbb{K}(\xi_1,\ldots,\xi_n)$ of rational functions in n indeterminates over $\mathbb{K}$ which is satisfied for every $x \in \mathfrak{A}$, and this equation is uniquely determined.

Now assume that the algebra $\mathfrak{A}$ is a train algebra with weight homomorphism ω. Etherington [27] has posed the problem under which conditions on $\mathfrak{A}$ the coefficients $\sigma_i(x)$, $i=1,\ldots,k-1$, of (3.10) only depend on $\omega(x)$, i.e. (3.10) assumes the form

$$x^{[k]} + \tau_1(\omega(x))x^{[k-1]} + \ldots + \tau_{k-1}(\omega(x))x^{[1]} = 0. \qquad (3.11)$$

In this case we say that the plenary powers form a train and call equation (3.11) the <u>train equation</u> for <u>plenary powers</u>. (Here we deviate from Etherington's notation who has applied this term only to equation (3.11) for elements of weight 1). Similarly the rank equation (3.1) of a train algebra is sometimes called the train equation for principal powers.

If the plenary powers form a train, then the train equation for plenary powers of elements of weight 1 has constant coefficients, $\tau_i := \tau_i(1)$, $i=1,\ldots,k-1$,

$$x^{[k]} + \tau_1 x^{[k-1]} + \ldots + \tau_{k-1}x^{[1]} = 0 \quad . \qquad (3.12)$$

If we replace in this equation $x^{[1]}$ by $x^{[l]}$, $l \in \mathbb{N}$, and thus $x^{[i]}$ by $x^{[i+l-1]}$, $i=1,\ldots,k$, then we obtain a linear difference equation with constant coefficients for the plenary powers of elements of weight 1

$$x^{[k+l]} + \tau_1 x^{[k+l-1]} + \ldots + \tau_{k-1}x^{[l]} = 0, \; l \in \mathbb{N}. \qquad (3.13)$$

This difference equation can be solved if the plenary train roots, i.e. the characteristic roots except 0, of the characteristic polynomial corresponding to (3.12) and the initial data are known.

Etherington [28] has shown that in a train algebra of rank 2 or 3 the plenary powers always form a train and he gives an explicit construction for the train equation for plenary powers from the train equation for principal powers, cf. section 4C. Later Reiersøl [98] and Holgate [61] have shown that in gametic algebras for coupled and partially coupled loci the plenary powers form a train and they have given formulas for the computation of the plenary train roots. Subsequently Holgate [62] and Heuch [52] have succeeded in proving that in certain classes of train algebras the plenary powers form a train, cf. also section 4C. For completeness we mention that Etherington [27], [28] has investigated further sequences of elements in baric algebras which represent mating schemes. In Etherington's paper [32] the concept of trains for such general situations is still somewhat vague. Heuch [52] has made the concept of a train and of a train equation in a baric algebra precise and he has applied these ideas to further mating schemes, cf. section 7D.

B. Genetic algebras

In general those algebras which arise in connection with genetical problems are more special than train algebras. For this reason Etherington [27] defined special train algebras. In a special train algebra the kernel of the weight homomorphism is nilpotent and the principal powers of the kernel are ideals of the algebra, cf. definition 3.20. Schafer [101] found that not all algebras from genetics are of this type. The copular algebra for simple Mendelian segregation already fails to have this property, cf. the proof of theorem 3.23. He defined the mathematical term "genetic algebra", cf. definition 3.7. Genetic algebras are a natural subclass of train algebras.

For a particular algebra it is in general difficult to show that it is genetic in the sense of Schafer. For a long time it has been known that the special train algebras over a suitable extension field have a basis with respect to which all left transformations $\mathcal{L}_x$ correspond to lower triangular matrices with eigenvalues of the form $\omega(x)$, $\omega(x)\lambda_1, \dots, \omega(x)\lambda_{n-1}$, where $\lambda_1, \dots, \lambda_{n-1}$ are constants independent of x. However only in 1971 Gonshor [47] succeeded in proving that an algebra has this property if and only if it is genetic in the sense of Schafer, cf. Theorem 3.13.

For algebras arising from genetical models one can rather often find a basis with this special property, hence such algebras are genetic algebras, cf. section 7. For genetic algebras the convergence theorems of section 4C can be exploited.

In the following we shall reverse the historical development and first

consider genetic algebras and later give some comments on special train algebras.
The following definition is due to Schafer [101].

Definition 3.8. A baric algebra $\mathfrak{A}$ with weight homomorphism ω is called Schafer genetic if the coefficients of the characteristic polynomial of any transformation $\mathcal{T} \in \mathfrak{T}(\mathfrak{A})$

$$\mathcal{T} = f(\mathcal{L}_{x_1}, \dots, \mathcal{L}_{x_s})$$

are functions of $\omega(x_1), \dots, \omega(x_s)$ only.

Holgate [67] has indicated that this definition is equivalent to the following property

Theorem 3.9. Let the algebra $\mathfrak{A}$ have a weight homomorphism ω. The following properties are equivalent.

(1) $\mathfrak{A}$ is Schafer genetic.

(2) Any two transformations $\mathcal{T}_1, \mathcal{T}_2 \in \mathfrak{T}(\mathfrak{A})$ with representations

$$\mathcal{T}_1 = f(\mathcal{L}_{x_1}, \dots, \mathcal{L}_{x_s}), \quad \mathcal{T}_2 = f(\mathcal{L}_{y_1}, \dots, \mathcal{L}_{y_s})$$

for elements $x_1, \dots, x_s, y_1, \dots, y_s \in \mathfrak{A}$ with $\omega(x_i) = \omega(y_i)$, $i=1,\dots,s$, have the same characteristic polynomial.

Since the rank polynomial (2.27) of an algebra is a divisor of $f_x(X)X$, cf. theorem 2.16, where f_x is the characteristic polynomial of the right transformation $\mathcal{R}_x$ which is equal to the left transformation $\mathcal{L}_x$, it follows immediately that

Theorem 3.10. Let $\mathfrak{A}$ be Schafer genetic. Then $\mathfrak{A}$ is a train algebra.

With assistance of corollary 3.6 we obtain

Corollary 3.11. The weight homomorphism of a Schafer genetic algebra is uniquely determined.

It is appropriate to put

$$m := n-1$$

and to number the basis elements of an $(m+1)$-dimensional algebra from 0 to m.

Definition 3.12. Let $\mathfrak{A}$ be an (m+1)-dimensional algebra over $\mathbb{K}$. $\mathfrak{A}$ is called Gonshor genetic if the extended algebra $\mathfrak{A}_{\mathbb{L}}$, where $\mathbb{L}$ is a suitable algebraic extension of $\mathbb{K}$, admits a basis $c_o, c_1, \ldots, c_m$, $c_o \in \mathfrak{A}$, where the multiplication constants λ_{ijk} defined by

$$c_i c_j = \sum_{k=0}^{m} \lambda_{ijk} c_m, \qquad i,j=0,1,\ldots,m,$$

have the following properties

(i) $\lambda_{000} = 1$

(ii) $\lambda_{0jk} = 0$ for $k < j$

(iii) $\lambda_{ijk} = 0$ for $k \leq \max(i,j)$ if $i,j > 0$.

Such a basis $c_o, \ldots, c_m$ is called a canonical basis.

In order to underline that $\mathfrak{A}$ has a canonical basis over $\mathbb{L}$, $\mathbb{L} \supseteq \mathbb{K}$, we say

" $\mathfrak{A}$ is Gonshor genetic over $\mathbb{L}$"

or

" $\mathfrak{A}_{\mathbb{L}}$ is Gonshor genetic".

The following is trivial: If $\mathfrak{A}$ is Gonshor genetic over $\mathbb{L}$, then $\mathfrak{A}$ is Gonshor genetic over any extension field of $\mathbb{L}$.

Remark. In Gonshor's terminology an algebra $\mathfrak{A}$ is genetic if it admits a canonical basis over $\mathbb{K}$. But with this restriction not all Schafer genetic algebras will be Gonshor genetic. The implication would then be true only for algebraically closed fields.
For this reason we have included the transition to a suitable extension field in the definition.

Gonshor [47] has proved the following result

Theorem 3.13. Let $\mathfrak{A}$ be an (m+1)-dimensional algebra over $\mathbb{K}$. $\mathfrak{A}$ is Gonshor genetic if and only if $\mathfrak{A}$ is Schafer genetic.

Proof. I. Let $\mathfrak{A}$ be Gonshor genetic. Then $\mathfrak{A}_{\mathbb{L}}$ has a canonical basis $c_o, \ldots, c_m$ with respect to which the multiplication constants λ_{ijk}, $i,j,k=0,\ldots,m$, satisfy axioms (i) - (iii) from definition 3.12. We define a mapping $\omega_{\mathbb{L}}: \mathfrak{A}_{\mathbb{L}} \longrightarrow \mathbb{L}$ by

$$c_o \longmapsto 1, \quad c_i \longmapsto 0, \quad i=1,\dots,m,$$

and linear extension onto $\mathfrak{A}_{\mathbb{L}}$.
In view of $\lambda_{ooo} = 1$, $\lambda_{ojo} = 0$ for $j > 0$ and $\lambda_{ijo} = 0$ for $i,j > 0$ the mapping $\omega_{\mathbb{L}}$ is an algebra homomorphism of $\mathfrak{A}_{\mathbb{L}}$ onto $\mathbb{L}$. The kernel $\mathfrak{N}_{\mathbb{L}} := \langle c_1,\dots,c_m\rangle_{\mathbb{L}}$ is an ideal of $\mathfrak{A}_{\mathbb{L}}$ in view of (ii) and (iii). Consequently $\mathfrak{N} := \mathfrak{A} \cap \mathfrak{N}_{\mathbb{L}}$ is an ideal of the $\mathbb{K}$-algebra $\mathfrak{A}$. Since $\omega_{\mathbb{L}|\mathfrak{N}_{\mathbb{L}}} \equiv 0$ we have $\omega_{\mathbb{L}|\mathfrak{N}} = 0$. Consequently $\omega_{\mathbb{L}|\mathfrak{N}}$ can be interpreted as a trivial algebra homomorphism of $\mathfrak{N}$ into $\mathbb{K}$. Since $c_o \in \mathfrak{A}$ and $\omega_{\mathbb{L}}(c_o) = 1 \in \mathbb{K}$ the mapping $\omega_{\mathbb{L}|\mathfrak{A}} =: \omega$ is a non-trivial algebra homomorphism of $\mathfrak{A}$ into $\mathbb{K}$. Hence $\mathfrak{A}$ is a baric algebra.
Now let $\mathcal{T} = f(\ell_{x_1}, \ell_{x_2}, \dots, \ell_{x_s}) \in \mathcal{T}(\mathfrak{A})$ be a transformation of $\mathfrak{A}$. With respect to the canonical basis $c_o,\dots,c_m$ of $\mathfrak{A}_{\mathbb{L}}$ to each ℓ_x with $x = \sum_{i=0}^{m} \xi_i c_i \in \mathfrak{A}$ there corresponds the matrix

$$L_x = \sum_{i=0}^{m} \xi_i \Lambda_i$$

with

$$\Lambda_o := \begin{pmatrix} 1 & & & \\ \lambda_{oo1} & \lambda_{o11} & & 0 \\ \cdot & \cdot & \cdot & \\ \cdot & \cdot & \cdot & \\ \cdot & \cdot & & \cdot \\ \lambda_{oom} & \lambda_{i1m} & \cdots\cdots & \lambda_{omm} \end{pmatrix}$$

and

$$\Lambda_i := \begin{pmatrix} 0 & & & & \\ \cdot & \cdot & & & \\ \cdot & & \cdot & & \\ \cdot & & \cdot & & 0 \\ \lambda_{ioi} & 0 \;.\;.\; 0 & & & \\ \lambda_{io,i+1} & \dots & \lambda_{ii,i+1} & 0 & \\ \cdot & & \cdot & & \cdot \\ \cdot & & \cdot & & \cdot \\ \lambda_{iom} & \dots & \lambda_{iim} & \dots \lambda_{i,m-1,m} & 0 \end{pmatrix} \qquad i=1,\dots,m.$$

Thus L_x is a lower triangular matrix with diagonal elements

$$\xi_o \cdot 1 = \omega(x)\cdot 1, \; \xi_o\lambda_{oii} = \omega(x)\lambda_{oii}, \quad i=1,\dots,m.$$

The set of lower triangular matrices is closed with respect to addition and multiplication. Thus also to $\mathcal{T} = f(\mathcal{L}_{x_1}, \ldots, \mathcal{L}_{x_s})$ there corresponds a lower triangular matrix T with diagonal elements

$$f(\omega(x_1), \ldots, \omega(x_s)), \quad f(\omega(x_1)\lambda_{oii}, \ldots, \omega(x_s)\lambda_{oii}), \; i=1,\ldots,m.$$

Hence the coefficients of the characteristic polynomial of $\mathcal{T}$, namely

$$\det(\lambda I - T) = \prod_{i=0}^{m} (\lambda - f(\omega(x_1)\lambda_{oii}, \ldots, \omega(x_s)\lambda_{oii})),$$

are functions of $\omega(x_j)$, $j=1,\ldots,s$, only.

II. On the other hand let $\mathfrak{A}$ be Schafer genetic with the weight homomorphism ω. We choose a basis $c_o, b_1, \ldots, b_m$ in $\mathfrak{A}$ such that $\omega(c_o) = 1$ and $\omega(b_i) = 0$ for $i=1,\ldots,m$. Let $\mathcal{L}_{c_o}, \mathcal{L}_{b_1}, \ldots, \mathcal{L}_{b_m}$ be the corresponding left transformations.

Let $\mathfrak{h}$ be the multiplicative subsemigroup of $\mathfrak{T}(\mathfrak{A})$ generated by the elements $\mathcal{L}_{c_o}^k \mathcal{L}_{b_i}$, $i=1,\ldots,m$, $k=0,1,2,\ldots$. Each element of this semigroup contains at least one factor of the form $\mathcal{L}_{b_i}$. Since $\mathfrak{A}$ is Schafer genetic by assumption and $\omega(b_i) = 0$, $i=1,\ldots,m$, each element $\mathfrak{X} \in \mathfrak{h}$ has the characteristic polynomial $p(\lambda) = \lambda^{m+1}$, i. e. for each $\mathfrak{X} \in \mathfrak{h}$ the characteristic roots are $(m+1)$-times 0, i.e. every $\mathfrak{X} \in \mathfrak{h}$ is nilpotent. From a theorem quoted in Jacobson's book [76] chapter II, section 2, we infer that an algebra generated by a multiplicative semigroup of nilpotent endomorphisms of a finite-dimensional vector space is itself nilpotent. Let $\mathfrak{U} := \langle \mathfrak{h} \rangle$ the subalgebra of $\mathfrak{T}(\mathfrak{A})$ generated by $\mathfrak{h}$.

We define

$$\mathfrak{N}_1 := \ker \omega \;, \quad \mathfrak{N}_{l+1} := \mathfrak{U}\mathfrak{N}_l, \qquad L = 1,2,\ldots .$$

Since $\mathfrak{N}_1$ is an ideal of $\mathfrak{A}$, the sets $\mathfrak{N}_1, \mathfrak{N}_2, \ldots$ form a descending chain of subalgebras of $\mathfrak{A}$. For some $(r+1) \in N$ this chain becomes stationary at $\langle 0 \rangle$ since $\mathfrak{U}$ is nilpotent, thus

$$\mathfrak{N}_1 > \mathfrak{N}_2 > \ldots \mathfrak{N}_r > \mathfrak{N}_{r+1} = \langle 0 \rangle .$$

Now we consider the factor spaces $\mathfrak{N}_l/\mathfrak{N}_{l+1}$, $l=1,\ldots,r$.

In view of $\mathcal{L}_{b_i} \in \mathfrak{U}$ we have $\mathcal{L}_{b_i}\mathfrak{N}_l \subseteq \mathfrak{U}\mathfrak{N}_l = \mathfrak{N}_{l+1}$, thus the linear mapping $\mathcal{L}_{b_i}^{(l)}$ induced by $\mathcal{L}_{b_i}$ in $\mathfrak{N}_l/\mathfrak{N}_{l+1}$ is the zero mapping, $l=1,\ldots,r$.

From $\mathcal{L}_{c_o}\mathfrak{U} \subseteq \mathfrak{U}$ it follows that $\mathcal{L}_{c_o}\mathfrak{N}_l \subseteq \mathfrak{N}_l$. Thus $\mathcal{L}_{c_o}$ induces a linear mapping $\mathcal{L}_{c_o}^{(l)}$ on $\mathfrak{N}_l/\mathfrak{N}_{l+1}$, $l=1,\ldots,r$.

Let $\mathbb{L}$ be an extension field of $\mathbb{K}$ which contains the splitting field of the characteristic polynomial of $\mathcal{L}_{c_o}$. We choose a basis $c_1^{(l)} + \mathfrak{N}_{l+1}, \ldots, c_{k_l}^{(l)} + \mathfrak{N}_{l+1}$ of $\mathfrak{N}_l/\mathfrak{N}_{l+1}$ over $\mathbb{L}$, $l=1,\ldots,r$, such that to $\mathcal{L}_{c_o}^{(l)}$ there corresponds a lower triangular matrix. Then we show that the elements

$$c_o,\ c_1^{(1)},\ldots,\ c_{k_1}^{(1)},\quad c_1^{(2)},\ldots,\ c_{k_2}^{(2)},\ldots,\ c_1^{(r)},\ldots,\ c_{k_r}^{(r)}$$

form a canonical basis of $\mathfrak{A}_{\mathbb{L}}$. With this aim we enumerate these basis elements in the given order from 0 to m. The multiplication table of $\mathfrak{A}_{\mathbb{L}}$ with respect to this basis is

$$c_i c_j = \sum_{k=0}^{m} \lambda_{ijk} c_k, \qquad i,j=0,\ldots,m.$$

Since ω is an algebra homomorphism and $\omega(c_o) = 1$ and $\omega(c_i) = 0$, $i=1,\ldots,m$, we have $\lambda_{ooo} = 1$. With respect to the new basis the linear transformation $\mathcal{L}_{c_o}$ is represented by the matrix

$$\Lambda_o := \begin{pmatrix} \lambda_{ooo} & \lambda_{o1o} & \cdots & \lambda_{omo} \\ \vdots & \vdots & & \vdots \\ \lambda_{oom} & \lambda_{o1m} & \cdots & \lambda_{omm} \end{pmatrix}$$

By construction the mapping $\mathcal{L}_{c_o}^{(l)}$ induced by $\mathcal{L}_{c_o}$ on $\mathfrak{N}_l/\mathfrak{N}_{l+1}$ is respresented by a lower triangular matrix, $l=1,\ldots,r$.
Thus Λ_o must be a lower triangular matrix, i.e.

$$\lambda_{ojk} = 0 \qquad \text{for } k < j .$$

Each c_i, $i=1,\ldots,m$, is a linear combination of $b_1,\ldots,b_m$, $c_i = \sum_{h=1}^{m} \sigma_{hi} b_h$.
For $i,j=1,\ldots,m$ we have

$$c_i c_j = \sum_{k=0}^{m} \lambda_{ijk} c_k$$

and also

$$c_i c_j = \sum_{h=1}^{m} \sigma_{hi} b_h c_j = \sum_{h=1}^{m} \sigma_{hi} \mathcal{L}_{b_h} c_j .$$

Let $c_j \in \mathfrak{N}_l$ but $c_j \notin \mathfrak{N}_{l+1}$. Then $\mathcal{L}_{b_h} c_j \in \mathfrak{N}\mathfrak{N}_l = \mathfrak{N}_{l+1}$. If we compare coefficients we obtain $\lambda_{ijk} = 0$ for $k \leq j$. Since $\mathfrak{A}$ is commutative we have also $\lambda_{ijk} = 0$ for $k \leq j$. With this, 3.12 (iii) is proved. Therefore $c_o, c_1, \ldots, c_m$ form a canonical basis of $\mathfrak{A}_{\mathbb{L}}$. □

Remark. From the proof of theorem 3.13 we infer that every genetic algebra possesses over a suitable extension field a basis with respect to which to all left transformations correspond lower triangular matrices. But this latter property is not sufficient for an algebra to be genetic as becomes obvious from the following example.

Example 3.14. Let $\mathfrak{A} = \langle c_o, c_1 \rangle_{\mathbb{R}}$ have the multiplication table

	c_o	c_1
c_o	$c_o + c_1$	$2c_1$
c_1		$-c_1$

Then the left transformations $\mathcal{L}_{c_o}$, $\mathcal{L}_{c_1}$ are represented by the matrices

$$L_{c_o} = \begin{pmatrix} 1 & 0 \\ 1 & 2 \end{pmatrix} \quad , \quad L_{c_1} = \begin{pmatrix} 0 & 0 \\ 2 & -1 \end{pmatrix} .$$

$\mathfrak{N} := \langle c_1 \rangle$ is an ideal of $\mathfrak{A}$ and $\mathfrak{A}^2 \subseteq \mathfrak{N}$, thus $\mathfrak{A}$ is baric with weight homomorphism $\omega : \mathfrak{A} \longrightarrow \mathbb{R}$, $\omega(c_o) := 1$, $\omega(c_1) := 0$.
But $c_1 \in \mathfrak{N}$ is not nilpotent as it would be in a genetic algebra by theorem 3.10 and lemma 3.3. Hence the algebra $\mathfrak{A}$ is not genetic. □

After we have shown that Schafer's definition 3.4 and the modification 3.12 of Gonshor's definition are equivalent in the following we shall only use the term "genetic" algebra. As we have seen the construction of a canonical basis requires in general the transition to an extension field $\mathbb{L}$ of $\mathbb{K}$. We shall denote the natural extension of the weight homomorphism $\omega : \mathfrak{A} \longrightarrow \mathbb{K}$ onto $\mathfrak{A}_{\mathbb{L}}$ again by ω.

A canonical basis of a genetic algebra is not uniquely determined as lemma 3.16 shows. On the other hand the multiplication constants $\lambda_{ooo} = 1, \lambda_{o11}, \ldots, \lambda_{omm}$ are uniquely determined since they are the characteristic roots of any left transformation $\mathcal{L}_x$ of $\mathfrak{A}_{\mathbb{L}}$ with $\omega(x) = 1$.

Definition 3.15. Let $\mathfrak{A}$ be a genetic algebra with canonical basis $c_o, c_1,\dots,c_m$ over $\mathbb{L} \supseteq \mathbb{K}$ and multiplication table $c_i c_j = \sum_{k=0}^{m} \lambda_{ijk} c_k$, $i,j=0,1,\dots,m$. Then $\lambda_{ooo} = 1, \lambda_{o11}, \dots, \lambda_{omm}$ are called the train roots of $\mathfrak{A}$.

If we call $\lambda_o = 1, \lambda_1, \dots, \lambda_m$ the train roots of a genetic algebra then we understand that the enumeration is chosen in such a way that there is a basis $c_o, c_1,\dots,c_m$ which satisfies the axioms 3.12 (i) - (iii).

Lemma 3.16. Let $\mathfrak{A}$ be a genetic algebra with a canonical basis c_o, $c_1,\dots,c_m$ over $\mathbb{L} \supseteq \mathbb{K}$. Then for any $a \in \mathfrak{A}_{\mathbb{L}}$ with $\omega(a) = 1$ the elements $a, c_1,\dots,c_m$ form a canonical basis of $\mathfrak{A}_{\mathbb{L}}$.

Proof. Let $a \in \mathfrak{A}_{\mathbb{L}}$ with $\omega(a) = 1$. This element can be represented as a linear combination

$$a = \sum_{i=0}^{m} \sigma_i c_i \quad \text{with} \quad \sigma_o = 1 .$$

In view of 3.12 (i) - (iii) we have

$$\begin{aligned} a^2 &= \Big(\sum_{i=0}^{m} \sigma_i c_i\Big)\Big(\sum_{j=0}^{m} \sigma_j c_j\Big) = \sum_{k=0}^{m} \Big(\sum_{i,j=0}^{m} \sigma_i \lambda_{ijk} \sigma_j\Big) c_k \\ &= \sum_{i,j=0}^{m} \sigma_i \lambda_{ijo} \sigma_j c_o + \sum_{k=1}^{m} \Big(\sum_{i,j=0}^{m} \sigma_i \lambda_{ijk} \sigma_j\Big) c_k \\ &= \qquad c_o \qquad + \sum_{k=1}^{m} \Big(\sum_{i,j=0}^{m} \sigma_i \lambda_{ijk} \sigma_j\Big) c_k, \end{aligned}$$

and for every $j > 0$ we obtain

$$\begin{aligned} ac_j &= \Big(\sum_{i=0}^{m} \sigma_i c_i\Big) c_j = \sigma_o c_o c_j + \sum_{i=1}^{m} \sigma_i c_i c_j \\ &= \sum_{k=j}^{m} \lambda_{ojk} c_k + \sum_{i=1}^{m} \sigma_i \sum_{k=\max(i,j)+1}^{m} \lambda_{ijk} c_k \\ &= \sum_{k=j}^{m} \lambda_{ojk} c_k + \sum_{k=j+1}^{m} \Big(\sum_{i=1}^{m} \sigma_i \lambda_{ijk}\Big) c_k. \end{aligned}$$

Thus we have shown that 3.12(i) and (ii) are satisfied for the multiplication constants of $\mathfrak{A}_{\mathbb{L}}$ with respect to $a, c_1,\dots,c_m$. The axiom 3.12(iii) is satisfied because it refers only to $c_1,\dots,c_m$. □

Thus in the definition of a canonical basis the condition on c_o is unnecessary.

In terms of matrices a genetic algebra can be characterized as follows:

Theorem 3.17. Let $\mathfrak{A}$ be an $(m+1)$-dimensional commutative algebra over $\mathbb{K}$ with weight homomorphism ω. Then the following assertions are equivalent.

(1) $\mathfrak{A}$ is genetic.

(2) For any given basis $a_o, a_1,\dots,a_m$ of $\mathfrak{A}$ with $\omega(a_i) = 1$, $i=1,\dots,m$, and multiplication constants γ_{ijk}, $i,j,k=0,1,\dots,m$, there is a nonsingular matrix $S = (\sigma_{ij}) \in \mathfrak{M}_{m+1}(\mathbb{L})$, $\mathbb{L} \supseteq \mathbb{K}$, with $\sigma_{oj} = 1$, $j=0,1,\dots,m$, such that all matrices $S\Gamma_i S^{-1}$, $\Gamma_i := (\gamma_{ijk})^T$, $i=0,1,\dots,m$, are lower triangular matrices with the same diagonal $(1, \lambda_1,\dots,\lambda_m)$.

Proof. I. Let $\mathfrak{A}$ be genetic with canonical basis $c_o, c_1,\dots,c_m$ over $\mathbb{L} \supseteq \mathbb{K}$. Let $\mathfrak{A}_{\mathbb{L}}$ have multiplication constants λ_{ijk}, $i,j,k=0,1,\dots,m$, with respect to this basis. Then the matrices $\Lambda_p := (\lambda_{pjk})^T_{\substack{j=0,\dots,m\\ k=0,\dots,m}}$ are lower triangular matrices whereby $\Lambda_1,\dots,\Lambda_m$ are nilpotent, i.e. the diagonals of these matrices contain only zeros.
Let $a_o, a_1,\dots,a_n$ be any basis of $\mathfrak{A}$ with $\omega(a_i) = 1$, $i=0,1,\dots,m$.
Let $S = (\sigma_{ij})$ be the matrix carrying the canonical basis into this basis

$$a_j = \sum_{i=0}^{m} \sigma_{ij} c_i, \quad j=0,1,\dots,m.$$

In view of $\omega(c_o) = 1$ and $\omega(c_i) = 0$, $i=1,\dots,m$, we have $\sigma_{io} = 1$ for all $i=0,\dots,m$. From (2.10) it follows that the matrices Γ_i and Λ_p, $i,p=0,\dots,m$, are related by the following equations

$$S\Gamma_i S^{-1} = \sum_{p=0}^{m} \sigma_{pi} \Lambda_p, \quad i=0,\dots,m. \tag{3.14}$$

Thus all matrices $S\Gamma_i S^{-1}$, $i=0,\dots,m$, are lower triangular matrices with the same diagonal $(1, \lambda_{o11},\dots,\lambda_{omm})$.

II. Assume (2) is valid. Let $T := S^{-1}$, $T = (\tau_{ij})$.
From $ST^{-1} = I$ and $\sigma_{oo} = \sigma_{1o} = \dots = \sigma_{mo} = 1$ it follows that

$$\sum_{j=0}^{m} \tau_{jo} = 1 \quad \text{and} \quad \sum_{j=0}^{m} \tau_{jk} = 0 \qquad \text{for } k=1,\dots,m.$$

We show that the basis

$$c_j := \sum_{i=0}^{m} \tau_{ij} a_i, \qquad j=0,\dots,m,$$

of $\mathfrak{A}_{\mathbb{L}}$ satisfies axioms 3.11(i) - (iii).
Let λ_{ijk}, $i,j,k=0,\dots,m$, be the multiplication constants of $\mathfrak{A}_{\mathbb{L}}$ with respect to the basis $c_0, c_1, \dots, c_m$. Because of (3.14) these multiplication constants satisfy a linear system of n^3 equations

$$\sum_{p=0}^{m} \sigma_{pi} \Lambda_p = B_i, \qquad i=0,\dots,m, \tag{3.15}$$

where by assumption the matrices $B_i =: (\beta_{ijk})^T_{\substack{i=0,\dots,m\\k=0,\dots,m}}$ are lower triangular matrices with the same diagonal $(\lambda_0, \lambda_1, \dots, \lambda_m)$, $\lambda_0 = 1$, i.e. $\lambda_{ijk} = 0$ for $k < j$ and $\beta_{ijj} = \lambda_j$ for all $i=0,\dots,m$. We introduce vectors

$\lambda_{.jk} := (\lambda_{ojk}, \dots, \lambda_{mjk})$ and $\beta_{.jk} := (\beta_{ojk}, \dots, \beta_{mjk})$, $j,k=0,\dots,m$, and rewrite (3.15) as

$$\lambda_{.jk} S = \beta_{.jk} \quad , \qquad i,k=0,\dots,m \quad . \tag{3.16}$$

Equations (3.16) have a unique solution given by

$$\lambda_{.jk} = T\beta_{.jk}, \quad , \qquad j,k=0,\dots,m.$$

Let $\varepsilon := (1,\dots,1) \in \mathbb{K}^{m+1}$. From $\beta_{.jj} = \lambda_j \varepsilon$, $j=0,\dots,m$, and the special properties of T it follows that

$$\lambda_{.jj} = (\lambda_j, 0, \dots, 0), \quad j=0,\dots,m,$$

i.e. $\lambda_{ooo} = \lambda_o = 1$, $\lambda_{ojj} = \lambda_j$, and for every $i \in \{1,\dots,m\}$ we have $\lambda_{ijj} = 0$ for all $j=0,\dots,m$.
For $k < j$ we have $\beta_{.jk} = 0$, whence $\lambda_{.jk} = 0$.
The λ_{ijk} and the β_{ijk} reflect the commutativity of $\mathfrak{A}$, thus $\lambda_{ijk} = 0$ for $k \leq \max(i,j)$ if $i,j > 0$. Hence the multiplication constants λ_{ijk} satisfy axioms 3.12 (i) - (iii). □

A genetic algebra can be characterized by a chain of subalgebras with some additional properties

Theorem 3.18. Let $\mathfrak{A}$ be an (m+1)-dimensional algebra over $\mathbb{K}$.
The following assertions are equivalent.
(1) $\mathfrak{A}$ is genetic.

(2) $\mathfrak{A}$ has an m-dimensional ideal $\mathfrak{N}$ with $\mathfrak{A}^2 \nsubseteq \mathfrak{N}$ and $\mathfrak{A}_{\mathbb{L}}$ has a descending chain of ideals

$$\mathfrak{A}_{\mathbb{L}} > \mathfrak{N}_{\mathbb{L}} = \mathfrak{N}_1 > \ldots > \mathfrak{N}_m > \mathfrak{N}_{m+1} = \langle o \rangle$$

over an appropriate extension field $\mathbb{L}$ of $\mathbb{K}$, where $\dim \mathfrak{N}_i = m+1-i$, $i=1,\ldots,m$, and

$$\mathfrak{N}_i \mathfrak{N}_j \subseteq \mathfrak{N}_l, \quad l = \max(i,j)+1, \; i,j=1,\ldots,m.$$

Proof. Let $\mathfrak{A}$ be genetic. Then $\mathfrak{A}$ is baric by definition 3.8. From lemma 1.9 it follows that $\mathfrak{A}$ has an m-dimensional ideal $\mathfrak{N}$ with $\mathfrak{A}^2 \nsubseteq \mathfrak{N}$. By theorem 3.13. there is an extension field $\mathbb{L}$ of $\mathbb{K}$ such that $\mathfrak{A}_{\mathbb{L}}$ has a canonical basis $c_o,\ldots,c_m$.
Put $\mathfrak{N}_i := \langle c_i, c_{i+1},\ldots,c_m \rangle_{\mathbb{L}}$, $i=1,\ldots,m$, then $\dim_{\mathbb{L}} \mathfrak{N}_i = m+1-i$, $i=1,\ldots,m$. From 3.12(iii) it follows immediately that $\mathfrak{N}_i \mathfrak{N}_j \subseteq \mathfrak{N}_l$, $l = \max(i,j)+1$, $i,j=1,\ldots,m$.
II. Suppose (2) holds. From lemma 1.11 it follows that $\mathfrak{A}$ has a non trivial algebra homomorphism ω with kernel $\mathfrak{N}$. Choose a basis $c_1,\ldots,c_m$ of $\mathfrak{N}_1$ such that $c_j \in \mathfrak{N}_j$, $c_j \notin \mathfrak{N}_{j+1}$, $j=1,\ldots,m$, and complete it by $c_o \in \mathfrak{A}$ with $\omega(c_o) = 1$ to obtain a basis of $\mathfrak{A}_{\mathbb{L}}$. Since ω is an algebra homomorphism we have 3.12(i). Since each $\mathfrak{N}_j$ is an ideal of $\mathfrak{A}_{\mathbb{L}}$ we have 3.12(ii), and from $\mathfrak{N}_i \mathfrak{N}_j \subseteq \mathfrak{N}_l$, $l = \max(i,j)+1$ it follows that 3.12(iii). □

Lemma 3.19. Let $\mathfrak{A}$ be a genetic algebra over $\mathbb{L} \geq \mathbb{K}$ with weight homomorphism ω. Then $\mathfrak{N} := \ker\omega$ is nilpotent.

Proof. By theorem 3.16 the algebra $\mathfrak{A}_{\mathbb{L}}$ has a sequence of ideals

$$\mathfrak{A}_{\mathbb{L}} > \mathfrak{N}_1 > \ldots \mathfrak{N}_m > \mathfrak{N}_{m+1} = \langle o \rangle,$$

where $\mathbb{L}$ is a suitable extension field of $\mathbb{K}$, and $\mathfrak{N}_i \mathfrak{N}_j \subseteq \mathfrak{N}_l$, $l = \max(i,j)+1$, $i,j=1,\ldots,m$.
Since $\mathfrak{N} = \mathfrak{N}_1 \cap \mathfrak{A}$ we obtain for the principal powers of $\mathfrak{N}$ by induction

$$\mathfrak{N}^i = \langle \mathfrak{N}\mathfrak{N}^{i-1} \rangle \subseteq \langle \mathfrak{N}_1 \mathfrak{N}_{i-1} \rangle \subseteq \mathfrak{N}_i, \quad i=1,\ldots,m+1,$$

thus $\mathfrak{N}^{m+1} = \langle o \rangle$, i.e. $\mathfrak{N}$ is nilpotent. □

Remark. This result has been proved by Schafer [101] with the help of a theorem of Albert [8] which says that an ideal $\mathfrak{b}$ of an algebra is nilpotent if and only if the subalgebra $\mathfrak{b}^*$ of the transformation algebra $\mathcal{T}(\mathfrak{A})$ which is generated by all $\mathcal{L}_b, \mathcal{R}_b$, $b \in \mathcal{B}$, is nilpotent. The direct proof using the definition of a (Gonshor) genetic algebra appears to be simpler.

In a baric algebra $\mathfrak{A}$ with weight homomorphism ω the property that $\ker\omega$ be nilpotent is necessary but not sufficient (cf. Lemma 3.19 and the following example).

Example 3.20. Let $\mathfrak{A} := \langle c_o, c_1, c_2\rangle_{\mathbb{R}}$ have the multiplication table

	c_o	c_1	c_2
c_o	c_o	c_1	c_1
c_1		c_2	0
c_2			0

Then $\mathfrak{N} = \langle c_1, c_2\rangle$ is an ideal of $\mathfrak{A}$ which does not contain $\mathfrak{A}^2$, hence $\mathfrak{A}$ is a baric algebra with weight homomorphism $\omega: \mathfrak{A} \longrightarrow \mathbb{R}$ defined by $\omega(c_o) = 1$, $\omega(c_1) = \omega(c_2) = 0$ and $\ker\omega = \mathfrak{N}$.
Furthermore $\mathfrak{N}^2 = \langle c_2\rangle$, $\mathfrak{N}^3 = \langle 0\rangle$, thus $\mathfrak{N}$ is nilpotent.
Now consider the transformation $\mathcal{T} = \mathcal{L}_x\mathcal{L}_y$ for elements $x = \alpha c_o$, $y = \beta c_o + \gamma c_1 \in \mathfrak{A}$ of weight α and β, respectively, $\alpha, \beta, \gamma \neq 0$.
This transformation is represented by the matrix

$$\begin{pmatrix} \alpha & 0 & 0 \\ 0 & \alpha & \alpha \\ 0 & 0 & 0 \end{pmatrix} \left(\begin{pmatrix} \beta & 0 & 0 \\ 0 & \beta & \beta \\ 0 & 0 & 0 \end{pmatrix} + \begin{pmatrix} 0 & 0 & 0 \\ \gamma & 0 & 0 \\ 0 & \gamma & 0 \end{pmatrix} \right) = \begin{pmatrix} \alpha\beta & 0 & 0 \\ \alpha\gamma & \alpha\beta + \alpha\gamma & 0 \\ 0 & 0 & 0 \end{pmatrix}$$

which has the characteristic polynomial

$$p(\lambda) = \lambda^3 - (2\alpha\beta + \alpha\gamma)\lambda^2 + (\alpha^2\beta^2 + \alpha^2\beta\gamma)\lambda .$$

Since this polynomial does not only depend on the weights α and β of x and y but also on γ, the algebra $\mathfrak{A}$ is not genetic, cf. definition 3.8. □

In view of lemma 3.3 and lemma 3.19 the question whether every train algebra is a genetic algebra is related to the question whether every non associative, commutative nilalgebra is nilpotent. The latter problem has been settled negatively by Suttles who gave a counter-example. Abraham [2] has used Suttles' example to construct a train algebra which is not a genetic algebra.

Theorem 3.21. There is a train algebra which is not a genetic algebra.

Proof. Suttles' algebra $\mathfrak{N} := \langle c_1, \dots c_5 \rangle_{\mathbb{R}}$ has the multiplication table

	c_1	c_2	c_3	c_4	c_5
c_1	0	c_3	c_4	0	$-c_3$
c_2		0	c_5	c_3	0
c_3			0	0	0
c_4				0	0
c_5					0

This algebra is nil but not nilpotent:

$$x = \sum_{i=1}^{5} \xi_i c_i$$

$$x^2 = 2(\xi_1\xi_2 - \xi_1\xi_5 + \xi_2\xi_4)c_3 + 2\xi_1\xi_3 c_4 + 2\xi_2\xi_3 c_5$$

$$x^3 = 2\xi_1(\xi_1\xi_2 - \xi_1\xi_5 + \xi_2\xi_4)c_4 + 2\xi_2(\xi_1\xi_2 - \xi_1\xi_5 + \xi_2\xi_4)c_5$$

$$x^4 = 0$$

but

$$\mathfrak{N}^2 = \langle c_3, c_4, c_5 \rangle = \mathfrak{N}^r \quad \text{for all } r \geq 2.$$

This example shows also the existence of a solvable algebra $\mathfrak{N}$, in view of $\mathfrak{N}^{[3]} = \mathfrak{N}^2 \mathfrak{N}^2 = \langle 0 \rangle$, which is not nilpotent.
Abraham [2] considers the algebra $\mathfrak{A} = \langle c_0, c_1, \dots, c_5 \rangle_{\mathbb{R}}$ with

$$c_0^2 = c_0 \quad \text{and } c_0 c_i = \tfrac{1}{2} c_i, \quad i=1,\dots,5.$$

$\mathfrak{A}$ is a baric algebra with weight homomorphism $\omega : \mathfrak{A} \longrightarrow \mathbb{R}$ defined by $\omega(c_0) = 1$, $\omega(c_i) = 0$, $i=1,\dots,5$.
Let $y = \xi_0 c_0 + x$, $x \in \mathfrak{N}$, be an arbitrary element of $\mathfrak{A}$, thus $\omega(y) = \xi_0$.
Then

$$y^2 - \xi_0 y = x^2$$

$$(y^2 - \xi_0 y)(y - \tfrac{1}{2}\xi_0) = x^3$$

$$(y^2 - \xi_0 y)(y - \tfrac{1}{2}\xi_0)^2 = x^4 = 0.$$

Thus $\mathfrak{A}$ is a train algebra of rank 4 with principal train roots 1, 1/2, 1/2. But $\mathfrak{A}$ is not a genetic algebra since $\mathfrak{N} = \ker \omega$ is not nilpotent as has been shown by Suttles. □

Remark. The first example for a train algebra which is not genetic given by Schafer [101] is not quite satisfactory since the underlying field is of characteristic 2.

A further characterization of genetic algebras has been given by Holgate [67] in terms of transformation algebras. We restate his result in a slightly weaked form.

Theorem 3.22. Let $\mathfrak{A}$ be an (m+1)-dimensional baric algebra over an algebraically closed field $\mathbb{K}$ of characteristic 0. Let ω be a weight homomorphism. Let $\mathfrak{T}^-$ be the Lie algebra associated with the transformation algebra $\mathfrak{T} = \mathfrak{T}(\mathfrak{A})$. Then the following statements are equivalent:
(1) $\mathfrak{A}$ is genetic.
(2) $\mathfrak{T}^-$ is solvable and $\mathfrak{N} := \ker\omega$ is nilpotent.

Proof. Instead of using $\mathfrak{T}$ ($\mathfrak{T}^-$) we can use the isomorphic matrix algebra $\mathfrak{T}_{m+1}(\mathbb{K})$ ($\mathfrak{T}_{m+1}(\mathbb{K})^-$) generated by all matrices corresponding to left transformations and the identity with respect to a fixed basis in $\mathfrak{A}$.
I. Let $\mathfrak{A}$ be genetic. Then by lemma 3.19 the ideal $\mathfrak{N}$ is nilpotent. Let $c_o,\dots,c_m$ be a canonical basis of $\mathfrak{A}$. Then to all left transformations of $\mathfrak{A}$ with respect to this basis correspond lower triangular matrices. Thus all elements of $\mathfrak{T}_{m+1}(\mathbb{K})$ are lower triangular matrices. Let $A,B\in\mathfrak{T}_{m+1}(\mathbb{K})$. Then the commutator $[A,B] = AB - BA$ of A and B is a nilpotent lower triangular matrix. Hence the second plenary power $(\mathfrak{T}_{m+1}(\mathbb{K})^-)^{[2]}$ contains only nilpotent lower triangular matrices and thus $(\mathfrak{T}_{m+1}(\mathbb{K})^-)^{[2]}$ and also $\mathfrak{T}_{m+1}(\mathbb{K})^-$ are solvable.
II. Suppose statement (2) holds. Then the algebra $\mathfrak{A}$ together with the Lie algebra $\mathfrak{T}^-$ of linear transformations of $\mathfrak{A}$ satisfy the hypothesis of a theorem of Lie, cf. Jacobson [76] , p.50. Thus there is a basis $a_o,a_1,\dots,a_n$ of $\mathfrak{A}$ with respect to which all elements of $\mathfrak{T}$ are represented by lower triangular matrices. Since the ideal $\mathfrak{N}$ of $\mathfrak{A}$ is nilpotent by assumption, $\mathfrak{N}^*$ is also nilpotent by a theorem of Albert [8]. Therefore to all elements of $\mathfrak{N}^*$ with respect to this basis correspond nilpotent lower triangular matrices.
Now let $x,y\in\mathfrak{A}$ with $\omega(x) = \omega(y)$. By linearity we have $\mathcal{L}_x - \mathcal{L}_y = \mathcal{L}_{x-y}\in\mathfrak{N}^*$. Thus L_{x-y} is a nilpotent triangular matrix and L_x and L_y have the same diagonal.
Suppose $\mathcal{F}, \mathcal{G}\in\mathfrak{T}$ have representations $\mathcal{F} = f(\mathcal{L}_{x_1},\dots,\mathcal{L}_{x_s})$, $f\in\mathbb{K}\{X_1,\dots,X_s\}$,

and $\mathcal{S} = f(\mathcal{L}_{y_1},\ldots,\mathcal{L}_{y_s})$ with $\omega(x_i) = \omega(y_i)$ for $i=1,\ldots,s$. Then to $\mathcal{T}$ and $\mathcal{S}$ correspond matrices $f(L_{x_1},\ldots, L_{x_s})$ and $f(L_{y_1},\ldots, L_{y_s})$ both of which are lower triangular matrices and they have the same diagonal. Therefore $\mathcal{T}$ and $\mathcal{S}$ have the same characteristic polynomial which says that $\mathfrak{A}$ is a genetic algebra. □

<u>Remark.</u> The result of Holgate [67], theorem 1 differs from theorem 3.22 as the condition " $\mathfrak{N}$ is nilpotent" is replaced by " $\mathfrak{N}$ is nil". In proving this theorem the author concludes that $\mathcal{L}_z$ is nilpotent if $z \in \mathfrak{N}$ is nilpotent under the condition that $\mathfrak{T}^-$ is solvable. We do not know whether this conclusion is always correct. However if we replace " $\mathfrak{N}$ is nilpotent" by " $\mathfrak{N}$ is nil of finite nilindex", i.e. that the indices of nilpotency of all elements of $\mathfrak{N}$ are bounded, then theorem 3.22 remains valid: On one hand it follows from (1) that $\mathfrak{N}$ is nilpotent and thus nil of finite nilindex, and on the other hand Gerstenhaber [43] has shown that in a commutative, not necessarily finite-dimensional algebra $\mathfrak{N}$ over a field of characteristic 0 which is nil of finite nilindex every left transformation is nilpotent.+)
Of course the problem would be settled if we knew that any finite-dimensional nilalgebra over a field of characteristic 0 has finite nilindex.

Furthermore Holgate [67] mentioned that the property "kerω is nilpotent" in theorem 3.19 can be replaced by the property that $\mathfrak{A}$ has no partial weight homomorphism essentially different from ω, i.e. there does not exist a weight homomorphism defined on a subalgebra $\mathfrak{B}$ of $\mathfrak{A}$ which is not identical with ω on this subalgebra. The proof of this result depends essentially on theorem 1 in [67] and therefore we refrain from further discussion.+)

+) Holgate [private communication] has indicated that the step in his proof of his theorem 1 in [67], where he stated that if $\mathfrak{A}$ is baric and $\mathfrak{T}^-$ is solvable, $x^n = 0$ implies $\mathcal{L}_x^n = 0$ for all $x \in \mathfrak{N}$, can be deduced by elementary transformations of the basis of $\mathfrak{A}$. He has also given a direct proof that under these conditions, if $\mathfrak{N}$ is nil it has finite nilindex and hence is nilpotent by Gerstenhaber's result. Therefore the strong version of theorem 3.22 with "$\mathfrak{N}$ is nilpotent" replaced by "$\mathfrak{N}$ is nil" is true as well as the further, above mentioned theorem of Holgate where "$\mathfrak{N}$ is nil(potent)" is replaced by "there does not exist a partial weight homomorphism essentially different from ω".

In the following we describe three further characterizations of genetic algebras given by Gonshor [48]. For this we require two additional kinds of powers of ideals of algebras.

Definition 3.23. Let $\mathfrak{B}$ be an ideal of an algebra $\mathfrak{A}$.
(1) $\mathfrak{B}^{\{k\}}$ is the set of all linear combinations of products of elements of $\mathfrak{A}$ with at least k factors from $\mathfrak{B}$.
(2) $\mathfrak{B}^{\langle k\rangle}$ is the set of all linear combinations of left products $(a_r \cdot \ldots \cdot (a_3(a_2a_1))\ldots)$ of elements of $\mathfrak{A}$ with at least k factors from $\mathfrak{B}$.

The following statements are immediate.

Lemma 3.24. Let $\mathfrak{B}$ be an ideal of $\mathfrak{A}$. Then
(1) $\mathfrak{B}^{\{k\}}$ and $\mathfrak{B}^{\langle k\rangle}$ are ideals of $\mathfrak{A}$ for any $k \in \mathbb{N}$.
(2) $\mathfrak{B}^{\langle k\rangle} \leq \mathfrak{B}^{\{k\}}$, $k \in \mathbb{N}$.
(3) $\mathfrak{B}^{\{2^{k-1}\}} \leq \mathfrak{B}^{\langle k\rangle}$, $k \in \mathbb{N}$.

Gonshor [48] uses these powers of the kernel of a weight homomorphism to characterize genetic algebras.

Theorem 3.25. Let $\mathfrak{A}$ be a baric (m+1)-dimensional algebra over $\mathbb{K}$ with weight homomorphism ω. Let $\mathfrak{N} := \ker\omega$. The following statements are equivalent:
(1) $\mathfrak{A}$ is genetic.
(2) There is an $s \in \mathbb{N}$ such that $\mathfrak{N}^{\{s+1\}} = \langle o\rangle$.
(3) There is an $r \in \mathbb{N}$ such that $\mathfrak{N}^{\langle r+1\rangle} = \langle o\rangle$.
(4) $\mathcal{L}_b \in \operatorname{rad} \mathfrak{T}(\mathfrak{A})$ for all $b \in \mathfrak{N}$.

Proof. Assume that (1) holds. We choose a canonical basis $c_o,\ldots,c_m$ of $\mathfrak{A}_{\mathbb{L}}$, where $\mathbb{L}$ is an algebraic extension field of $\mathbb{K}$. Let $\mathfrak{N}_k$ be the subspace of $\mathfrak{A}_{\mathbb{L}}$ spanned by $c_k,\ldots,c_m$, k=1,...,m, and $\mathfrak{N}_{m+1} := \langle o\rangle$. From 3.12 (iii) it follows that $\mathfrak{N}^{\{k\}}$ is contained in $\mathfrak{N}_k$. Hence $\mathfrak{N}^{\{m+1\}} = \langle o\rangle$.
II. Statement (3) follows from (2) by the preceding lemma.
III. We do not prove the implication (3) $\Longrightarrow$ (1) at this point since the proof is similar to the earlier proof by Etherington [31] for the existence of a canonical basis in a special train algebra (cf. part I of the proof of theorem 3.28). The essential difference is that the sequence of ideals of the principal powers of $\mathfrak{N}$ has to be replaced by the follow-

ing sequence of ideals

$$\mathfrak{A} > \mathfrak{N}^{\langle 1\rangle} > \mathfrak{N}^{\langle 2\rangle} > \ldots > \mathfrak{N}^{\langle r\rangle} > \mathfrak{N}^{\langle r+1\rangle} = \langle 0\rangle .$$

IV. Assume statement (3). Let $b \in \mathfrak{N}$. Then $\mathfrak{T}(\mathfrak{A})\mathcal{L}_b$ is a left ideal of $\mathfrak{T}(\mathfrak{A})$ and

$$(\mathfrak{T}(\mathfrak{A})\,\mathcal{L}_b)^{r+1}\,\mathfrak{A} \subseteq \mathfrak{N}^{\langle r+1\rangle} = \langle 0\rangle .$$

Therefore $\mathfrak{T}(\mathfrak{A})\mathcal{L}_b$ is a nilpotent left ideal of $\mathfrak{T}(\mathfrak{A})$ and consequently $\mathfrak{T}(\mathfrak{A})\mathcal{L}_b$ is in the radical of $\mathfrak{T}(\mathfrak{A})$. Since the transformation algebra $\mathfrak{T}(\mathfrak{A})$ contains the identical mapping $\mathfrak{J}$ we have $\mathcal{L}_b \in \operatorname{rad} \mathfrak{T}(\mathfrak{A})$.
V. Assume statement (4). Since the radical of $\mathfrak{T}(\mathfrak{A})$ is nilpotent of an index $r \in \mathbb{N}$ we have $\mathfrak{N}^{\langle r\rangle}\mathfrak{A} = \langle 0\rangle$ and thus $\mathfrak{N}^{\langle r+1\rangle} = \langle 0\rangle$. □

The following investigations of the structure of genetic algebras going back to Schafer [101] do not really fit into the frame-work of this section. However we mention them for reasons of completeness.
Since for every baric algebra $\mathfrak{A}$ with weight homomorphism ω the factor algebra $\mathfrak{A}/\mathfrak{N}$, $\mathfrak{N} := \ker\omega$, is isomorphic to the simple algebra $\mathbb{K}$, the algebra $\mathfrak{A}$ has a radical which is contained in $\mathfrak{N}$, cf. definition 2.5. On the other hand the radical contains every nilpotent ideal (theorem 2.6), together with Lemma 3.19 we have the following theorem

<u>Theorem 3.26.</u> Let $\mathfrak{A}$ be a genetic algebra with weight homomorphism ω. Then $\mathfrak{N} := \ker\omega$ is the radical of $\mathfrak{A}$.

In section 4C we shall investigate under which conditions a genetic algebra $\mathfrak{A}$ possesses a Levi decomposition, i.e. $\mathfrak{A}$ is the direct sum of its radical $\mathfrak{R}$ and a subalgebra isomorphic with the factor algebra $\mathfrak{A}/\mathfrak{R}$.

C. Special train algebras

<u>Definition 3.27.</u> Let $\mathfrak{A}$ be a baric algebra over $\mathbb{K}$ with weight homomorphism ω. $\mathfrak{A}$ is called a <u>special train algebra</u> if $\mathfrak{N} := \ker\omega$ is nilpotent and the principal powers $\mathfrak{N}^i$, $i \in \mathbb{N}$, are ideals of $\mathfrak{A}$.

(We underline that the latter property is trivial only for associative algebras).

Notice that in a special train algebra the principal powers of $\mathfrak{N}$ form a sequence of ideals of $\mathfrak{A}$ which becomes stationary at $\langle 0\rangle$ for an integer r+1,

$$\mathfrak{A} > \mathfrak{N} = \mathfrak{N}^1 > \mathfrak{N}^2 > \ldots\ldots > \mathfrak{N}^r > \mathfrak{N}^{r+1} = \langle 0 \rangle .$$

In order to prove that a special train algebra is a train algebra, Etherington [31] shows first that every special train algebra has a canonical basis (a "canonical form of the algebra" in Etherington's work), and further that the coefficients of the characteristic polynomial of any left transformation $\mathcal{L}_x$ are functions of $\omega(x)$ only. Conversely Etherington [31] first concludes that any algebra with a canonical basis is a special train algebra. Later in his corrigendum of [28] Etherington mentioned that not every algebra which possesses a canonical basis is a special train algebra. A further counterexample coming from genetics has been given by Schafer [101] , cf. theorem 3.30. If in addition to the existence of a canonical basis we require that the principal powers of the kernel of a weight homomorphism are ideals of the algebra, then the algebra is a special train algebra.
This fact is expressed in the following theorem

Theorem 3.28. Let $\mathfrak{A}$ be a baric algebra with weight homomorphism ω . Then the following statements are equivalent.

(1) $\mathfrak{A}$ is a special train algebra.
(2) $\mathfrak{A}$ is (Gonshor) genetic and all principal powers of $\ker\omega$ are ideals of $\mathfrak{A}$.

Proof. I. Assume (1) is valid. Following definition 3.27 $\mathfrak{N} := \ker\omega$ is nilpotent of an index r and the principal powers of $\mathfrak{N} = \mathfrak{N}^1$ are ideals of $\mathfrak{A}$. We show that $\mathfrak{A}$ has a canonical basis over a suitable extension field $\mathbb{L}$ of $\mathbb{K}$. Let $c_o \in \mathfrak{A}$ with $\omega(c_o)=1$. Furthermore let $b_1^{(1)},\ldots,b_{k_1}^{(1)},\ldots,b_1^{(r)},\ldots,b_{k_r}^{(r)}$ be a basis of $\mathfrak{N}^1$ such that

$b_1^{(l)},\ldots,b_{k_l}^{(l)} \in \mathfrak{N}^l$ but not in $\mathfrak{N}^{l+1}$, $l=1,\ldots,r-1$.

In view of $\mathfrak{N}\mathfrak{N}^l \subseteq \mathfrak{N}^{l+1}$ the left transformations $\mathcal{L}_{b_1^{(l)}},\ldots,\mathcal{L}_{b_k^{(l)}}$ induce the trivial mapping on the factor spaces $\mathfrak{N}^l/\mathfrak{N}^{l+1}$, $l=1,\ldots,r$. With respect to the basis c_o, $b_1^{(1)},\ldots,b_{k_r}^{(r)}$ the left transformations $\mathcal{L}_{b_1^{(l)}},\ldots,\mathcal{L}_{b_{k_l}^{(l)}}$ are represented by nilpotent lower triangular matrices of the form

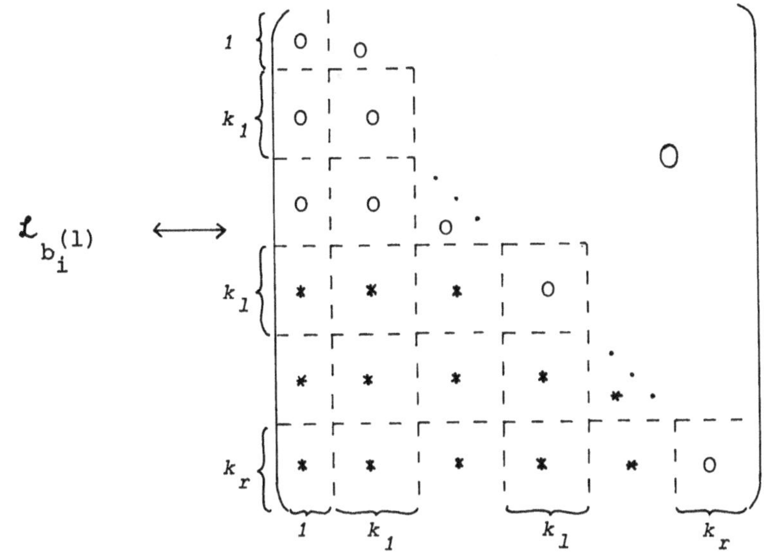

Since $\mathfrak{N}^1, \mathfrak{N}^2, \ldots, \mathfrak{N}^r$ are ideals of $\mathfrak{A}$, the left transformation $\mathcal{L}_{c_o}$ is represented by a lower block triangular matrix

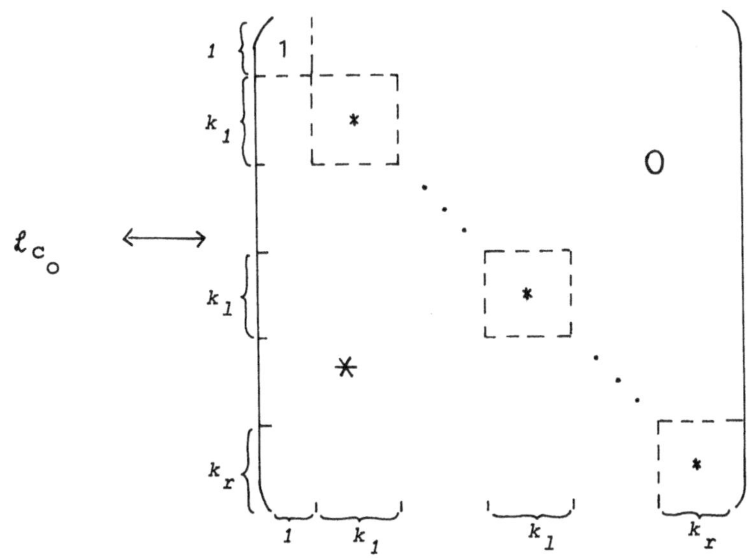

Now let $\mathbb{L}$ be an extension field of $\mathbb{K}$ containing the splitting field of the characteristic polynomial of $\mathcal{L}_{c_o}$. We choose the bases

$$c_1^{(l)} + \mathfrak{N}^{l+1}, \ldots, c_{k_l}^{(l)} + \mathfrak{N}^{l+1}$$

of $\mathfrak{N}^l / \mathfrak{N}^{l+1}$, $l=1,\ldots,r$, such that the linear mappings $\mathcal{L}_{c_o^{(l)}}$ induced by $\mathcal{L}_{c_o}$ in $\mathfrak{N}^l / \mathfrak{N}^{l+1}$ are represented by lower triangular matrices, $l=1,\ldots,r$. Then $c_o, c_1^{(1)}, \ldots, c_{k_1}^{(1)}, \ldots, c_1^{(r)}, \ldots, c_{k_r}^{(r)}$ form an $\mathbb{L}$-basis of $\mathfrak{A}_{\mathbb{L}}$ with respect to which $\mathcal{L}_{c_o}$ is represented by a lower triangular matrix and with respect to which the linear transformations $\mathcal{L}_{b_i^{(l)}}$, $i=1,\ldots,k_l$, $l=1,\ldots,r$, are still represented by nilpotent lower block triangular matrices.

Applying similar arguments as at the end of the proof of theorem 3.13 we find that

$$c_o, c_1^{(1)}, \ldots, c_{k_1}^{(1)}, \ldots, c_1^{(r)}, \ldots, c_{k_r}^{(r)}$$

form a canonical basis of $\mathfrak{A}_{\mathbb{L}}$. Thus $\mathfrak{A}$ is a genetic algebra.

II. Assume statement (2). Let $c_o,\ldots,c_m$ be a canonical basis of $\mathfrak{A}_{\mathbb{L}}$, $c_o \in \mathfrak{A}$. We know that $\mathfrak{N}_{\mathbb{L}} := \langle c_1,\ldots,c_m\rangle_{\mathbb{L}}$ is nilpotent, consequently the kernel of ω, namely $\mathfrak{N} = \mathfrak{N}_{\mathbb{L}} \cap \mathfrak{A}$, is nilpotent. Since the principal powers of $\mathfrak{N}$ are ideals of $\mathfrak{A}$ the algebra $\mathfrak{A}$ is a special train algebra. □

A trivial consequence of this theorem is

Theorem 3.29. Every special train algebra is genetic.

Theorem 3.30. There is a genetic algebra which is not a special train algebra.

Proof. Let $\mathfrak{A} := \langle c_o,c_1,c_2,c_3,c_4,c_5\rangle_{\mathbb{R}}$ have the following multiplication table

	c_o	c_1	c_2	c_3	c_4	c_5
c_o	c_o	$\frac{1}{2}c_1$	$\frac{1}{4}c_3$	0	0	0
c_1		$\frac{1}{4}c_2$	$\frac{1}{8}c_4$	0	0	0
c_2			$\frac{1}{16}c_5$	0	0	0
c_3				0	0	0
c_4					0	0
c_5						0

From the multiplication table it is immediately clear that $c_o,\ldots,c_5$ form a canonical basis of $\mathfrak{A}$ and thus $\mathfrak{A}$ is a genetic algebra. Clearly, $\mathfrak{A}$ is also baric where the weight homomorphism $\omega : \mathfrak{A} \longrightarrow \mathbb{R}$ is defined by $\omega(c_o) := 1$, $\omega(c_i) := 0$, $i=1,\ldots,5$.

The principal powers of $\mathfrak{N} = \ker\omega$ are

$$\mathfrak{N}^1 = \langle c_1,\ldots,c_5\rangle , \quad \mathfrak{N}^2 = \langle c_2,c_4,c_5\rangle , \quad \mathfrak{N}^3 = \langle c_4,c_5\rangle , \quad \mathfrak{N}^4 = \langle 0\rangle ,$$

Table 3.31. Let $\mathfrak{A}$ be an (m+1)-dimensional commutative algebra over $\mathbb{K}$

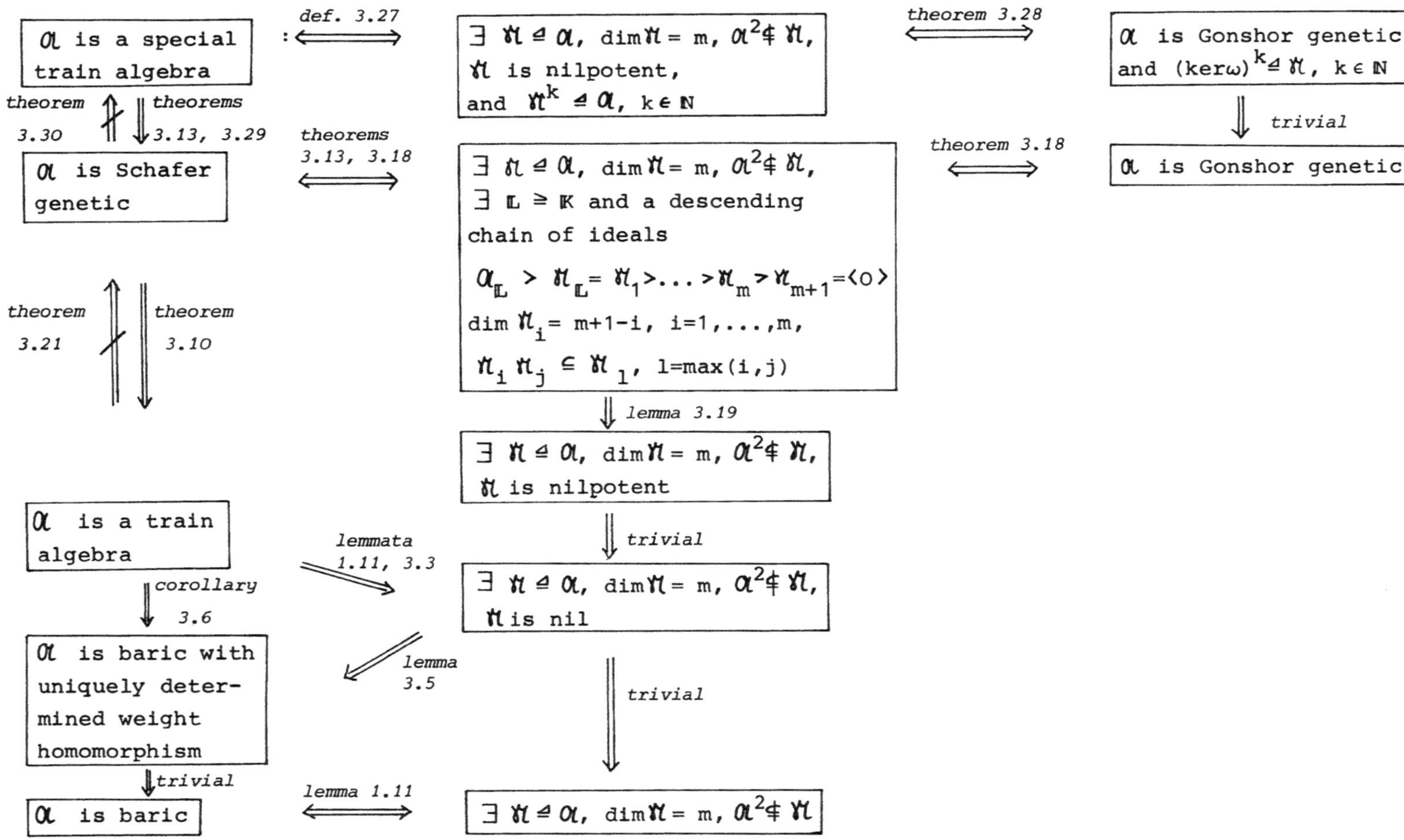

i.e. the principal powers form a descending sequence of subalgebras of $\mathfrak{A}$. However $c_o c_2 = \frac{1}{4} c_3 \notin \mathfrak{N}^2$. Thus $\mathfrak{N}^2$ is not an ideal of $\mathfrak{A}$ and $\mathfrak{A}$ is not a special train algebra.
Actually, this algebra is not just an ad hoc example. It is isomorphic with the copular algebra for simple Mendelian segregation with two alleles, cf. sections 6C and 7A. □

We collect essential statements of this section together with lemma 1.11 in table 3.31.

Etherington [28] has investigated the structure of commutative train algebras of ranks 2 and 3 and he has shown that every train algebra of rank 2 is a special train algebra. Furthermore for special train algebras of rank 3 he gave the "typical" multiplication table which however contained an error as explained in the corrigendum. From this multiplication table it would have been clear that a train algebra of rank 3 with train roots different from 1/2 is a special train algebra. Abraham [2] has shown this statement without any conditions on the train roots.

Theorem 3.32. A train algebra of rank not greater than 3 over a field $\mathbb{K}$, char $\mathbb{K} \neq 2$, is a special train algebra.

The proof of this theorem depends essentially on the following theorem.

Theorem 3.33. A finite-dimensional commutative nilalgebra of nilindex $k \leq 3$ over $\mathbb{K}$, char $\mathbb{K} \neq 2$, is nilpotent.

Proof. I. Let $k = 1$, then $\mathfrak{N} = \langle 0 \rangle$ and the assertion is trivially correct.
II. Let $k = 2$, then $\dim \mathfrak{N} \geq 1$. If $\dim \mathfrak{N} = 1$ then $\mathfrak{N}^2 = \langle 0 \rangle$. Let $\dim \mathfrak{N} = 2$. Then there exist two linearly independent elements c_1, c_2 in $\mathfrak{N}$. For every $y = \eta_1 c_1 + \eta_2 c_2$, $\eta_1, \eta_2 \in \mathbb{K}$, we have $0 = y^2 = 2\eta_1\eta_2 c_1 c_2$. In view of char $\mathbb{K} \neq 2$ we have $c_1 c_2 = 0$. Since this is true for any two elements of $\mathfrak{N}$, $\mathfrak{N}$ is nilpotent of index 2.
III. Let $k = 3$. We can assume the $\dim \mathfrak{N} \geq 3$. Since the assertion is trivial for $\dim \mathfrak{N} = 1$, and for $\dim \mathfrak{N} = 2$ we can embed $\mathfrak{N}$ into a 3-dimensional algebra $\tilde{\mathfrak{N}} := \langle \mathfrak{N}, c_3 \rangle$ with $c_3 \notin \mathfrak{N}$ and $c_3^2 = 0$, $\mathfrak{N} c_3 = \langle 0 \rangle$ which is a nil algebra of nilindex 3.
Thus let $\dim \mathfrak{N} \geq 3$. Let $c_1, c_2, c_3 \in \mathfrak{N}$ be linearly independent. With arbitrary $\eta_1, \eta_2, \eta_3 \in \mathbb{K}$ we form the linear combination

$$y := \eta_1 c_1 + \eta_2 c_2 + \eta_3 c_3 \quad .$$

For the principal powers we have

$$y^2 = \eta_1^2c_1^2 + \eta_2^2c_2^2 + \eta_3^2c_3^2 + 2\eta_1\eta_2c_1c_2 + 2\eta_1\eta_3c_1c_3 + 2\eta_2\eta_3c_2c_3$$

$$\begin{aligned} 0 = y^3 &= \eta_1^2\eta_2(c_1^2c_2 + 2(c_1c_2)c_1 + \eta_1\eta_2^2(c_1c_2^2 + 2(c_1c_2)c_2) \\ &+ \eta_1^2\eta_3(c_1^2c_3 + 2(c_1c_3)c_1) + \eta_1\eta_3^2(c_1c_3^2 + 2c_1c_3)c_3) \\ &+ \eta_2^2\eta_3(c_2^2c_3 + 2(c_2c_3)c_2) + \eta_2\eta_3^2(c_2c_3^2 + 2(c_2c_3)c_3) \\ &+ 2\eta_1\eta_2\eta_3((c_1c_2)c_3 + (c_2c_3)c_1 + (c_3c_1)c_2) . \end{aligned}$$

Since η_1, η_2, η_3 range over $\mathbb{K}$ the factors of $\eta_1^2\eta_2, \dots, \eta_2\eta_3^2$ and $\eta_1\eta_2\eta_3$ must vanish and thus

$$c_1^2c_j + 2(c_ic_j)c_i = 0 \qquad \text{for all } i,j=1,2,3, \quad i\neq j,$$

and

$$(c_ic_j)c_k + (c_jc_k)c_i + (c_kc_i)c_j = 0 \qquad \text{for all } i,j,k=1,2,3, i\neq j\neq k\neq i.$$

Since $\mathfrak{N}$ is nil of nilindex 3 we have $c_i^3 = 0$ and hence in the first equation we need not require $i \neq j$. Because the second equation reduces to the first, if two or three indices coincide, we need not require $i\neq j\neq k\neq i$ in the second equation. Hence these equations are satisfied for any $x,y,z \in \mathfrak{N}$, i.e.

$$x^2y + 2(xy)\ x = 0, \tag{3.17}$$

$$(xy)z + (yz)x + (zx)\ y = 0, \tag{3.18}$$

the last equation is the Jacobi identity (2.12).
The first equation (3.17) saying that x^2y and $(xy)x$ are linearly dependent is comparable to the identity (2.16) characterizing a commutative alternative algebra.
Using the left transformations $\mathcal{L}_x$ and $\mathcal{L}_y$ and the commutativity of $\mathfrak{A}$ we can rewrite equations (3.17) and (3.18) as

$$\mathcal{L}_{x^2} + 2(\mathcal{L}_x)^2 = 0 \tag{3.19}$$

$$\mathcal{L}_{xy} + \mathcal{L}_y\mathcal{L}_x + \mathcal{L}_x\mathcal{L}_y = 0. \tag{3.20}$$

In (3.20) we substitute $y = x^2$, then

$$\mathcal{L}_{x^3} + \mathcal{L}_{x^2}\mathcal{L}_x + \mathcal{L}_x\mathcal{L}_{x^2} = 0$$

and using again (3.19)

$$\mathcal{L}_{x^3} = 4(\mathcal{L}_x)^3 \quad . \tag{3.21}$$

Since $\mathfrak{N}$ has the nilindex 3 we have $x^3 = 0$ and thus $\mathcal{L}_{x^3} = 0$.
In view of char $\mathbb{K} \neq 2$, from (3.21) follows $(\mathcal{L}_x)^3 = 0$.
The rest of the proof is similar to the proof of the theorem of Schafer for alternative algebras ([102], Theorem 3.2):
To any subalgebra $\mathcal{C}$ of $\mathfrak{N}$ we associate $\mathcal{C}^*$, which is the subalgebra of $\mathcal{J}(\mathfrak{N})$ generated by all $\mathcal{L}_x$, $x \in \mathcal{C}$. We define

$$\Phi := \{ \mathcal{C} \leq \mathfrak{N} \mid \mathcal{C}^* \text{ is nilpotent} \} .$$

The family Φ is not empty because $\langle 0 \rangle \in \Phi$. Since $\mathfrak{N}$ has finite dimension there is a $\mathcal{B} \in \Phi$ of maximal dimension. Let k be the index of nilpotency of $\mathcal{B}^*$.
Suppose $\mathcal{B} \neq \mathfrak{N}$. We can construct an element $x \in \mathfrak{N} \setminus \mathcal{B}$ with $x\mathcal{B}^* \subseteq \mathcal{B}$.
(For any $y \in \mathfrak{N} \setminus \mathcal{B}$ we have either $y\mathcal{B}^* \subseteq \mathcal{B}$ or there is a $\mathcal{B}_1 \in \mathcal{B}^*$ with $y\mathcal{B}_1 \in \mathfrak{N} \setminus \mathcal{B}$.
In the first case we define $x := y$. In the second case we have either $(y\mathcal{B}_1)\mathcal{B}^* \subseteq \mathcal{B}$ or there is a $\mathcal{B}_2 \in \mathcal{B}^*$ with $y\mathcal{B}_1\mathcal{B}_2 \in \mathfrak{N} \setminus \mathcal{B}$.
We set $x := y\mathcal{B}_1$ or we continue.
This process comes to an end after finitely many steps at latest at $y\mathcal{B}_1 \ldots \mathcal{B}_k = 0 \in \mathcal{B}$, since $\mathcal{B}^*$ is nilpotent of index k.)
Now we consider the subalgebra $\mathcal{C}$ of $\mathfrak{N}$ generated by $\mathcal{B}$ and x. Since $x\mathcal{B}^* \subseteq \mathcal{B}$ we have $xy \in \mathcal{B}$ for all $y \in \mathcal{B}$. Since furthermore $\mathcal{L}_{x^2} = -2(\mathcal{L}_x)^2$ and $\mathcal{L}_{x^n} = 0$ for $n \geq 3$ every element $\mathcal{T} \in \mathcal{C}^*$ is a linear combination of products composed of factors $\mathcal{B} \in \mathcal{B}^*$ or $\mathcal{L}_x$.
From (3.20) we obtain for all $y \in \mathcal{B}$

$$\mathcal{L}_x\mathcal{L}_y = -\mathcal{L}_y\mathcal{L}_x - \mathcal{L}_{xy} \quad , \tag{3.22}$$

where $\mathcal{L}_{xy} \in \mathcal{B}^*$, i.e. $\mathcal{L}_x\mathcal{L}_y$ and $-\mathcal{L}_y\mathcal{L}_x$ differ only by a transformation $-\mathcal{L}_{xy} \in \mathcal{B}^*$.
It follows by induction that for every $\mathcal{B} \in \mathcal{B}^*$ there are uniquely determined elements $\mathcal{B}_1, \mathcal{B}_2, \mathcal{B}_3, \mathcal{B}_4, \mathcal{B}_5 \in \mathcal{B}^*$ such that

$$\mathcal{L}_x\mathcal{B} = \mathcal{B}_1\mathcal{L}_x + \mathcal{B}_2$$

$$\mathcal{L}_x^2\mathcal{B} = \mathcal{B}_3\mathcal{L}_x^2 + \mathcal{B}_4\mathcal{L}_x + \mathcal{B}_5 ,$$

hence

$$\mathcal{C}^* = \langle \mathcal{B}^*, \mathcal{B}^*\mathcal{L}_x, \mathcal{B}^*\mathcal{L}_x^2 \rangle .$$

Using these "commutation rules" and the identity $\mathcal{L}_x^3 = 0$ we obtain by induction

$$(\mathfrak{C}^*)^{3n} = \langle (\mathfrak{B}^*)^n \mathcal{L}_x^2,\ (\mathfrak{B}^*)^n \mathcal{L}_x, (\mathfrak{B}^*)^n \rangle \quad , \quad n \in \mathbb{N}.$$

Since $\mathfrak{B}^*$ is nilpotent of index k, the algebra $\mathfrak{C}^*$ is nilpotent of an index not greater than 3k. This is a contradiction to $\mathfrak{B}$ being maximal. Hence $\mathfrak{B} = \mathfrak{N}$ and $\mathfrak{N}^*$ is nilpotent. By a theorem of Albert ([8] , Lemma 5) this is equivalent to saying that $\mathfrak{N}$ is nilpotent itself. □

<u>Proof of theorem 3.32.</u> Now let $\mathfrak{A}$ be a train algebra of rank $r \leq 3$ with weight homomorphism ω. In any case all $x \in \mathfrak{A}$ satisfy an equation of the form

$$x(x - \omega(x))(x - \lambda\omega(x)) = 0, \tag{3.23}$$

(if the rank should be 2, one can multiply the rank equation by the factor $(x - \lambda\omega(x))$).

By lemma 3.3 $\mathfrak{N} := \ker\omega$ is nil of index ≤ 3 and thus nilpotent by theorem 3.32. It remains to prove that the principal powers $\mathfrak{N}^k$ of $\mathfrak{N}$ are ideals of $\mathfrak{A}$ for k=2,3,... .

Let dim $\mathfrak{N}$ = m. For m = 1 the assertion is trivial since $\mathfrak{N}^2 = \langle 0 \rangle$. Therefore assume $m \geq 2$. We choose $c_o \in \mathfrak{A}$ with $\omega(c_o) = 1$. Then there is $z_o \in \mathfrak{N}$ such that

$$c_o^2 = c_o + z_o$$

$$c_o^3 = c_o + z_o + c_o z_o \ .$$

Let $x_o = c_o + \eta z \in \mathfrak{A}$ be an arbitrary element of weight 1. For the principle powers of x we obtain

$$x^2 = c_o + z_o + 2\eta c_o z + \eta^2 z^2$$

$$x^3 = c_o + z_o + \eta c_o z + c_o z_o + \eta z_o z$$

$$+ 2\eta(c_o z)c_o + 2\eta^2(c_o z)z + \eta^2 c_o z^2 + \eta^3 z^3 \quad .$$

From (3.23) it follows that all elements $x \in \mathfrak{A}$ of weight 1 satisfy the equation

$$x^3 - (1+\lambda)x^2 + \lambda x = 0 \ .$$

We substitute the expressions for the principal powers of x and collect terms with equal powers of η, then we obtain

$$\eta^3 z^3 + \eta^2 [2(c_o z)z + c_o z^2 - (1+\lambda)z^2] + \eta[z_o z + 2(c_o z)c_o - (2\lambda+1)c_o z + \lambda z] + [c_o z_o + \lambda z_o] = 0.$$

Since $\eta \in \mathbb{K}$ was arbitrarily chosen the factors of η^3, η^2, η^1 and η^o vanish, in particular

$$2(c_o z)z + c_o z^2 - (1+\lambda)z^2 = 0 \tag{3.24}$$

holds.

$\mathfrak{N}$ has at least dimension 2. Choose any two linearly independent elements $z_1, z_2 \in \mathfrak{N}$ and consider

$$z = \eta_1 z_2 + \eta_2 z_2 \quad .$$

Substitute this expression for z in (3.24) and collect terms.
Since η_1, η_2 run over $\mathbb{K}$ the coefficients must vanish, in particular from the coefficient of $\eta_1 \eta_2$ we obtain

$$2(c_o z_1)z_2 + 2(c_o z_2)z_1 + 2c_o(z_1 z_2) - 2(1-\lambda)z_1 z_2 = 0 \ .$$

Since char $\mathbb{K} \neq 2$ we can solve for $c_o(z_1 z_2)$

$$c_o(z_1 z_2) = -(c_o z_1)z_2 - 2(c_o z_2)z_1 + (1+\lambda)z_1 z_2 \ . \tag{3.25}$$

In view of (3.24) this equation is also valid for dependent elements $z_1, z_2 \in \mathfrak{N}$. Thus it is universal in $\mathfrak{N}$. Since all terms on the right hand side are elements of $\mathfrak{N}^2$ we have $c_o(z_1 z_2) \in \mathfrak{N}^2$ for all $z_1, z_2 \in \mathfrak{N}$, thus we have shown that the subalgebra $\mathfrak{N}^2$ is an ideal of $\mathfrak{A}$.
Now we can show by induction that all higher principal powers of $\mathfrak{N}$ are ideals of $\mathfrak{A}$. Suppose $\mathfrak{N}^{k-1}$ is an ideal of $\mathfrak{A}$, assume $z \in \mathfrak{N}^{k-1}$, $z_2 \in \mathfrak{N}$. Then $(c_o z_1)z_2$, $(c_o z_2)z_1$, $z_1 z_2 \in \mathfrak{N}^k$ and from (3.25) follows that $\mathfrak{N}^k$ is an ideal of $\mathfrak{A}$. Hence we have shown that $\mathfrak{A}$ is a special train algebra. □

From theorem 2.32 we obtain that the train algebra which is not a genetic algebra, mentioned in the proof of theorem 3.21, has lowest possible rank (since every special train algebra is a genetic algebra).

4. IDEMPOTENTS AND SEQUENCES OF POWERS IN TRAIN ALGEBRAS, GENETIC ALGEBRAS AND ALGEBRAS WITH GENETIC REALIZATION

For the structure of an algebra the existence of idempotents, i.e. of elements $e \neq 0$ with $e^2 = e$, is of particular interest. But also from the biological aspect the existence of such elements is of interest because the equilibria of a population which can be described by an algebra correspond to idempotents of this algebra.

Again we make the general assumption that $\mathfrak{A}$ is a finite-dimensional, commutative algebra over a field $\mathbb{K}$.

A. Idempotents in train algebras and genetic algebras

Lemma 4.1. Let $\mathfrak{A}$ be a baric algebra with weight homomorphism ω. For every idempotent e of $\mathfrak{A}$ we have either

$$\omega(e) = 0 \quad \text{or} \quad \omega(e) = 1.$$

In a train algebra all elements of weight 0 are nilpotent (Lemma 3.2) and an idempotent can not be nilpotent. Thus we have

Lemma 4.2. Let $\mathfrak{A}$ be a train algebra with weight homomorphism ω. Then $\omega(e) = 1$ for every idempotent e of $\mathfrak{A}$.

Let $\mathfrak{A}$ be a baric algebra with weight homomorphism ω, and let $\mathfrak{N} := \ker \omega$. Then $\mathfrak{A}/\mathfrak{N} \cong \mathbb{K}$ by theorem 1.8. Thus $\mathfrak{A}/\mathfrak{N}$ is a 1-dimensional associative algebra with unit element and, of course, not nilpotent. Therefore there exists a Levi decomposition of $\mathfrak{A}$ with respect to $\mathfrak{N}$, i.e. there exists a subalgebra $\mathfrak{B} \leq \mathfrak{A}$ with $\mathfrak{B} \cong \mathfrak{A}/\mathfrak{N}$ such that the vector space $\mathfrak{A}$ is the direct sum of the vector spaces $\mathfrak{B}$ and $\mathfrak{N}$, if and only if $\mathfrak{A}$ possesses an idempotent, cf. Schafer [101].
Since by theorem 2.3 in a genetic algebra $\mathfrak{A}$ the kernel of the weight homomorphism ω is the radical of $\mathfrak{A}$, we have

Theorem 4.3. A genetic algebra $\mathfrak{A}$ is the direct sum of its radical $\mathfrak{R}$ and a subalgebra $\mathfrak{B}$ isomorphic with $\mathfrak{A}/\mathfrak{R}$ if and only if $\mathfrak{A}$ has an idempotent.

The following theorem was proved by Gonshor [45] initially for special train algebras. In a later publication [47] he indicates that the proof requires only the existence of a canonical basis whence the theorem is valid also for genetic algebras.

Theorem 4.4. Let $\mathfrak{A}$ be a genetic algebra of dimension m+1 over $\mathbb{K}$, char$\mathbb{K} \neq 2$. Let $\mathfrak{A}_{\mathbb{L}}$ with extension field $\mathbb{L}$ of $\mathbb{K}$ have a canonical basis $c_o,\dots,c_m$.
If all train roots of $\mathfrak{A}$ are different from 1/2 then $\mathfrak{A}_{\mathbb{L}}$ has exactly one idempotent.

Proof. Let the multiplication constants of $\mathfrak{A}_{\mathbb{L}}$ with respect to $c_o,\dots,c_m$ be λ_{ijk}, $i,j,k=0,\dots,m$. Suppose $0 \neq e = \sum_{i=0}^{m} \eta_i c_i$ is an idempotent of $\mathfrak{A}_{\mathbb{L}}$. Then $\eta_o = 1$ by lemma 4.2. The other coordinates of e satisfy the equations

$$\eta_k = \sum_{i,j=0}^{m} \eta_i \lambda_{ijk} \eta_j , \qquad k=1,\dots,m.$$

By way of 3.12(i)-(iii) these equations reduce to

$$\eta_k = 2\lambda_{okk}\eta_k + \sum_{i,j=0}^{k-1} \eta_i \lambda_{ijk} \eta_j , \qquad k=1,\dots,m. \tag{4.1}$$

If all train roots λ_{okk}, $k=1,\dots,m$, are different from 1/2 one can show by induction on k that this system has a unique solution.□

Many genetic algebras arising from concrete genetic situations have train roots equal 1/2, an example is the gametic algebra for simple Mendelian segregation where all train roots except 1 equal 1/2, cf. theorem 7.2. Therefore a generalization of this theorem is desirable. Gonshor [45] has indicated that similar theorems may be proved if several of the train roots equal 1/2. In particular if the train root $\lambda_1 = 1/2$ and all other train roots are different from 1/2 then there exists a 1-parametric family of idempotents if and only if for a canonical basis $c_o,\dots,c_m$ the product c_o^2 has no component in the direction of c_1.

Holgate [64] suggested that the following generalization of this result is true: If among the train roots $\lambda_o = 1, \lambda_1,\dots,\lambda_m$ exactly $\lambda_{t_1},\dots,\lambda_{t_r}$ equal 1/2 and c_o^2, $c_{t_i}c_{t_j}$, $c_o c_{t_i} - \frac{1}{2}c_{t_i}$, $i,j=1,\dots,r$, have no component in the direction of $c_{t_1},\dots,c_{t_r}$ then there is an r-parametric family of idempotents. Heuch [52] gave a counterexample to this conjecture. In fact this is an example that a genetic algebra may not have an idempotent at all.

Example 4.5. Let $\mathfrak{A} := \langle c_o, c_1, c_2 \rangle_R$ with multiplication table

	c_o	c_1	c_2
c_o	c_o+c_1	$\frac{1}{4}c_1+c_2$	$\frac{1}{2}c_2$
c_1		0	0
c_2			0

$\mathfrak{A}$ is a genetic algebra with c_o, c_1, c_2 as a canonical basis and the train roots are $\lambda_o = 1$, $\lambda_1 = 1/4$, $\lambda_2 = 1/2$. Although c_o^2, c_2^2 and $c_oc_2 - \frac{1}{2}c_2$ have no component in the direction of c_2, the genetic algebra $\mathfrak{A}$ has no idempotent: Suppose $e = c_o + \eta_1c_1 + \eta_2c_2$ is an idempotent, then following (4.1)

$$\eta_1 = \frac{1}{4}\eta_1 + 1$$
$$\eta_2 = \eta_2 + 2$$

these equations can not be satisfied. □

We can demonstrate the following generalization of theorem 4.4.

Theorem 4.6. Let $\mathfrak{A}$ be an (m+1)-dimensional genetic algebra with a canonical basis $c_o,\dots,c_m$ over $\mathbb{L}$, char $\mathbb{L} \neq 2$.
Assume that exactly r of the train roots equal 1/2, namely $\lambda_{t_1},\dots,\lambda_{t_r}$, $1 \leq t_1 < \dots < t_r \leq m$. If c_ic_j has no component with respect to c_{t_ρ} for $i,j=0,\dots,t_\rho-1$, $\rho=1,\dots,r$, then $\mathfrak{A}_{\mathbb{L}}$ has an r-parametric family of idempotents

$$\mathcal{E} = \left\{ c_o + \sum_{i=1}^{m} f_i(\eta_{t_1},\dots,\eta_{t_{\sigma(i)}})c_i \;\middle|\; \eta_{t_1},\dots,\eta_{t_r} \in \mathbb{L} \right\}, \qquad (4.2)$$

here f_i is a polynomial in $\eta_{t_1},\dots,\eta_{t_{\sigma(i)}}$, where $t_{\sigma(i)}$ denotes the greatest index t_j with $t_j \leq i$, $i=1,\dots,m$. Furthermore we have $f_{t_\rho}(\eta_{t_1},\dots,\eta_{t_\rho}) = \eta_{t_\rho}$, $\rho=1,\dots,r$, and $f_i(\eta_{t_1},\dots,\eta_{t_{\sigma(i)}}) = f_i(\eta_{t_1},\dots,\eta_{t_{\sigma(i-1)}})$ for $i \neq t_1,\dots,t_r$.

Proof. Let λ_{ijk}, $i,j,k=0,\dots,m$, be the multiplication constants with respect to $c_o,\dots,c_m$, where $\lambda_{oii} = \lambda_i$, $i=0,\dots,m$, are the train roots of $\mathfrak{A}$. The additional condition for the products of basis elements is equivalent with

$$\lambda_{ijt_\varrho} = 0, \qquad i,j=0,\dots,t_\varrho - 1,\ \varrho = 1,\dots,r. \tag{4.3}$$

The element $e = \sum_{l=0}^{m} \eta_l c_l \in \mathfrak{A}_{\mathbb{L}}$ is an idempotent if and only if $\eta_o = 1$ and $\eta_1,\dots,\eta_m$ satisfy equations (4.1). Assume that $\eta_o = 1$ and that

$$\eta_\varkappa = f_\varkappa(\eta_{t_1},\dots,\eta_{t_{\sigma(\varkappa)}}),\ \varkappa = 1,\dots,k-1,$$

satisfy the first k-1 equations of (4.1) for any choice $\eta_{t_1},\dots,\eta_{t_{\sigma(k-1)}} \in \mathbb{L}$, where the $f_\varkappa$, $\varkappa = 1,\dots,k-1$, are polynomials of the following form: if $\varkappa \in \{t_1,\dots,t_r\}$ then $f_\varkappa(\eta_1,\dots,\eta_\varkappa) = \eta_\varkappa$ and for $\varkappa \notin \{t_1,\dots,t_r\}$ the polynomial $f_\varkappa$ only depends on $\eta_{t_1},\dots,\eta_{t_{\sigma(\varkappa-1)}}$. Consider the k-th equation of (4.1). If $\lambda_k = 1/2$ then this equation is satisfied for all $\eta_k \in \mathbb{L}$ in view of (4.3). We put

$$f_\varkappa(\eta_{t_1},\dots,\eta_{t_{\sigma(k)}}) := \eta_k.$$

On the other hand if $\lambda_k \neq 1/2$ then the k-th equation has the solution

$$\eta_k = \frac{1}{1-2\lambda_k} \sum_{i,j=0}^{k-1} \eta_i \lambda_{ijk} \eta_j ,$$

where $\eta_1,\dots,\eta_{k-1}$ are polynomials depending only on $\eta_{t_1},\dots,\eta_{t_{\sigma(k-1)}}$, then also η_k is a polynomial in these variables.
Thus the assertion is proved. □

If a genetic algebra is given where exactly r train roots equal 1/2 and the additional hypothesis (4.3) is not satisfied then in general there will not be an r-parametric family of idempotents. On the other hand one can find genetic algebras where the additional hypothesis is not satisfied and which nevertheless have an r-parametric family of idempotents.

Example 4.7. Let $\mathfrak{A} = \langle c_o, c_1, c_2 \rangle_{\mathbb{R}}$ have the multiplication table

	c_o	c_1	c_2
c_o	$c_o+c_1-c_2$	$-c_1+c_2$	$\frac{1}{2}c_2$
c_1		$3c_2$	0
c_2			0

$\mathfrak{A}$ is a genetic algebra with a canonical basis c_o, c_1, c_2 and the train roots $\lambda_o = 1$, $\lambda_1 = -1$, $\lambda_2 = 1/2$.
The set of all idempotents is

$$\mathcal{E} = \left\{ c_o + \frac{1}{3}c_1 + \eta_2 c_2 \,\middle|\, \eta_2 \in \mathbb{R} \right\} .$$

The principal powers of the ideal $\mathfrak{N} := \langle c_1, c_2 \rangle$ of $\mathfrak{A}$ are $\mathfrak{N}^2 = \langle c_2 \rangle$ and $\mathfrak{N}^3 = \langle 0 \rangle$. Hence on the one hand side every canonical basis d_o, d_1, d_2 of $\mathfrak{A}$ satisfies

$$d_o \in \mathfrak{A},\ \omega(d_o) = 1, \quad d_1 \in \mathfrak{N} \setminus \mathfrak{N}^2, \quad d_2 \in \mathfrak{N}^2,$$

and on the other hand every set of three elements $d_o, d_1, d_2 \in \mathfrak{A}$ with these properties forms a canonical basis of $\mathfrak{A}$, i.e. the canonical bases of $\mathfrak{A}$ are exactly of the form

$$\begin{aligned} d_o &= c_o + \alpha_{o1} c_1 + \alpha_{o2} c_2 \\ d_1 &= \qquad \alpha_{11} c_1 + \alpha_{12} c_2 \\ d_2 &= \qquad\qquad\qquad \alpha_{22} c_2 \end{aligned}$$

with $\alpha_{ij} \in \mathbb{R}$, $i \leq j$, $i,j = 0,1,2$, and $\alpha_{11}, \alpha_{22} \neq 0$. One can choose a canonical basis d_o, d_1, d_2 such that d_o^2 and $d_o d_1$ have no component in the direction of d_2 (e.g. $d_o = c_o + \frac{1}{3}c_1$, $d_1 = c_1 - 4c_2$, $d_3 = \alpha_{22}c_2$, $\alpha_{22} \neq 0$) but in every case d_1^2 is a scalar multiple of d_2.
Hence there does not exist a canonical basis satisfying the assumptions of theorem 4.6. Therefore these assumptions are not necessary for the existence of an r-parametric family of idempotents. □

In many applications the train roots which equal 1/2 are direct successors of the train root $\lambda_o = 1$. For this reason we mention the following more detailed theorem which follows immediately from theorem 4.6 and from (4.1) by induction.

Theorem 4.8. Let $\mathfrak{A}$ be an (m+1)-dimensional genetic algebra with canonical basis $c_o,\dots,c_m$ over $\mathbb{L}$, char $\mathbb{L} \neq 2$. For the train roots $\lambda_o,\dots,\lambda_m$ assume $\lambda_1 = \lambda_2 = \dots = \lambda_r = 1/2$, $\lambda_i \neq 1/2$, $i=r+1,\dots,m$.
If $c_i c_j$ has no components with respect to c_ρ, $i,j=0,\dots,\rho-1$, $\rho = 1,\dots,r$, then $\mathfrak{A}_{\mathbb{L}}$ has an r-parametric family of idempotents

$$\mathcal{E} = \left\{ c_o + \sum_{i=1}^{r} \eta_i c_i + \sum_{j=r+1}^{m} f_j(\eta_1,\dots,\eta_r) c_j \;\middle|\; \eta_1,\dots,\eta_r \in \mathbb{L} \right\} \quad (4.4)$$

where f_j is a polynomial at most of degree 2^{j-r} in the coordinates $\eta_1,\dots,\eta_r$, $j=r+1,\dots,m$.

After having derived these sufficient conditions for the existence of idempotents in genetic algebras we add one general result.

Theorem 4.9. Let $\mathfrak{A}$ be a genetic algebra over $\mathbb{L}$, char $\mathbb{L} \neq 2$. If all train roots are different from 1/2 then $\mathfrak{A}_{\mathbb{L}}$ has exactly one idempotent. If some train roots equal 1/2 then $\mathfrak{A}_{\mathbb{L}}$ has either no idempotents or at least a 1-parametric family of idempotents.

Proof. Let $c_o,\dots,c_m$ be a canonical basis of $\mathfrak{A}$ with multiplication constants λ_{ijk}, $i,j,k=0,\dots,m$, and train roots $\lambda_i = \lambda_{oii}$, $i=0,\dots,m$. From theorem 4.4 we know that if all train roots are different from 1/2 then $\mathfrak{A}_{\mathbb{L}}$ has exactly one idempotent. Now we assume that exactly r train roots $\lambda_{t_1},\dots,\lambda_{t_r}$, $1 \leq t_1 < \dots < t_r \leq m$ equal 1/2.
The element $e = \sum_{i=0}^{m} \eta_i c_i \in \mathfrak{A}_{\mathbb{L}}$ is an idempotent of $\mathfrak{A}_{\mathbb{L}}$ if and only if $\eta_o = 1$ and the other coordinates satisfy (4.1). Either the first $t_r - 1$ equations of (4.1) cannot be satisfied for any t_r - tuple $(1,\eta_1,\dots,\eta_{t_r-1}) \in \mathbb{L}^r$ or there is at least one solution $(1,\eta_1^*,\dots,\eta_{t_r-1}^*)$.
In the first case $\mathfrak{A}_{\mathbb{L}}$ has no idempotent.
In the second case we have to check whether this t_r - tuple also satisfies the t_r - th equation which reads

$$0 = \sum_{i,j=0}^{t_r-1} \eta_i \lambda_{ijk} \eta_j \;,$$

in view of $\lambda_{t_r} = 1/2$. If the t_r - tuple $(1,\eta_1^*,\dots,\eta_{t_r-1}^*)$ does not satisfy this equation, then $\mathfrak{A}_{\mathbb{L}}$ has no idempotent. However if this equation is satisfied by $\eta_o = 1$, $\eta_1 = \eta_1^*, \dots, \eta_{t_r-1} = \eta_{t_r-1}^*$ then $\eta_{t_r} \in \mathbb{L}$ can be

chosen arbitrarily and the further $\eta_{t_r+1},\ldots,\eta_m$ are uniquely determined as polynomials in η_{t_r} by the last $m-t_r$ equations of (4.1), since by assumption all train roots $\lambda_{t+1},\ldots,\lambda_m$ are different from 1/2. In this case $\mathfrak{A}_{\mathbb{L}}$ has a family of idempotents parametrized by η_{t_r}.
Furthermore it follows immediately: if the first t_r-1 equations of (4.1) admit an l-parametric family of solutions, then $\mathfrak{A}_{\mathbb{L}}$ has an (l+1)-parametric family of idempotents. □

From this theorem it follows immediately that the condition given by Gonshor [45] is not just sufficient but also necessary for the existence of exactly one idempotent, i.e. we have

Theorem 4.10. Let $\mathfrak{A}$ be a genetic algebra over $\mathbb{L}$, char $\mathbb{L} \neq 2$. $\mathfrak{A}_{\mathbb{L}}$ has exactly one idempotent if and only if all train roots are different from 1/2.

B. Idempotents in algebras with genetic realization

We recall definition 1.5 of an algebra with a genetic realization over $\mathbb{R}$ or $\mathbb{C}$. Such an algebra admits a basis $a_0,a_1,\ldots,a_m$ over $\mathbb{R}$ with multiplication constants γ_{ijk} satisfying the relations

$$0 \leq \gamma_{ijk} \leq 1, \quad i,j,k=0,\ldots,m, \quad \sum_{k=0}^{m} \gamma_{ijk} = 1, \quad i,j=0,\ldots,m.$$

Since the m-dimensional simplex

$$\mathcal{S}_m := \left\{ \sum_{i=0}^{m} \alpha_i a_i \quad 0 \leq \alpha_i \leq 1, \quad i=0,\ldots,m, \quad \sum_{i=0}^{m} \alpha_i = 1 \right\} \qquad (4.5)$$

spanned by $a_0,\ldots,a_m$ is mapped into itself by the mapping $x \longmapsto x^2$ and this mapping is continuous with respect to the usual topology of $\mathbb{R}^{m+1}$ Brouwer's fixed point theorem yields the following result.

Theorem 4.11. Let $\mathfrak{A}$ be an (m+1)-dimensional algebra over $\mathbb{R}$, $\mathbb{C}$ respectively, which has a genetic realization. Then $\mathfrak{A}$ has at least one idempotent and this idempotent is in $\mathcal{S}_m$.

Many authors in the field of algebras in genetics have expressed the view that the algebras occuring in applications do usually contain an idempotent.

However with the exception of [112] , [113], it usually has not been recognized that for an algebra with a genetic realization the existence of an idempotent is guaranteed. In this connection we mention the work of Ljubič [84] who indicated the existence of equilibria which, in the notation of algebras, correspond to idempotents. Ljubič poses the problem of finding all fixed points of a quadratic mapping of $\mathfrak{S}_m$ into itself. This is just the problem of finding all idempotents of an algebra with genetic realization which lie in $\mathfrak{S}_m$.
Using the preceding theorem and theorem 4.4 we obtain the following theorem.

Theorem 4.12. Let $\mathfrak{A}$ be an (m+1)-dimensional genetic algebra over $\mathbb{R}$, $\mathbb{C}$ respectively, which has a genetic realization. If all train roots of $\mathfrak{A}$ are different from 1/2 then $\mathfrak{A}_{\mathbb{C}}$ has exactly one idempotent and this idempotent is in $\mathfrak{S}_m$.

Collecting our result we have the following:
For an algebra with a genetic realization (therefore for any algebra arising in a genetical application) the existence of an idempotent is guaranteed. If, in addition, the algebra is genetic one can find a canonical basis such that the first element is an idempotent. The multiplication constants with respect to such a canonical basis satisfy, besides 3.12 (i)-(iii), the condition

$$\lambda_{ook} = 0 \qquad \text{for } k=1,\dots,m. \tag{4.6}$$

In the following we give a variety of examples of genetic algebras with their manifolds of idempotents. In addition we give the rank equations which can be easily verified.

Example 4.13. Let $\mathfrak{A}$ be a 3-dimensional genetic algebra over $\mathbb{K}$, char$\mathbb{K} \neq 2$, with idempotent c_o and canonical basis c_o, c_1, c_2. The multiplication table of $\mathfrak{A}$ has the form

	c_o	c_1	c_2
c_o	c_o	$\lambda_1 c_1 + \alpha c_2$	$\lambda_2 c_2$
c_1		βc_2	0
c_2			0

where 1, λ_1, λ_2 are the train roots of $\mathfrak{A}$. In the following table for each choice of $\lambda_1, \lambda_2, \alpha$ and β we give the manifold of idempotents and the rank equation:

λ_1, λ_2	α, β	idempotents	rank equation for x $\omega(x)=1$
$\lambda_1, \lambda_2 \neq \frac{1}{2}$ $\lambda_1 \neq \lambda_2$	arbitrary	c_o	$x(x-1)(x-\lambda_1)(x-\lambda_2)=0$
$\lambda_1 = \lambda_2 \neq \frac{1}{2}$	$(\alpha,\beta) \neq (0,0)$		$x(x-1)(x-\lambda_1)^2=0$
	$(\alpha,\beta)=(0,0)$		$x(x-1)(x-\lambda_1)=0$
$\lambda_1 = \frac{1}{2}$ $\lambda_2 \neq \frac{1}{2}$	$\alpha=0,\ \beta=0$	$\ell_1=\{c_o+\eta_1 c_1 \mid \eta_1 \in \mathbb{L}\}$	$x(x-1)(x-\lambda_2) = 0$
	$\alpha \neq 0,\ \beta=0$	$\ell_1^*=\{c_o+\eta_1 c_1 + \eta_2 c_2 \mid \eta_1 \in \mathbb{L},\ \eta_2 = \frac{2\alpha}{1-2\lambda_2}\eta_1\}$	
	α arbitrary $\beta \neq 0$	$\mathfrak{h}=\{c_o+\eta_1 c_1+\eta_2 c_2 \mid \eta_1 \in \mathbb{L}, \eta_2 = \frac{2\alpha\eta_1+\beta\eta_1^2}{1-2\lambda_2}\}$	
$\lambda_1 = \frac{1}{2}$ $\lambda_2 = \frac{1}{2}$	$\alpha=0,\ \beta=0$	$\ell=\{c_o+\eta_1 c_1+\eta_2 c_2 \mid \eta_1, \eta_2 \in \mathbb{L}\}$	$x(x-1) = 0$
	$\alpha \neq 0,\ \beta=0$ $\alpha=0,\ \beta \neq 0$	$\ell_2=\{c_o+\eta_2 c_2 \mid \eta_2 \in \mathbb{L}\}$	$x(x-1)(x-\frac{1}{2}) = 0$
	$\alpha \neq 0,\ \beta \neq 0$	$\ell_2=\{c_o+\eta_2 c_2 \mid \eta_2 \in \mathbb{L}$ $\ell_2^*=\{c_o - \frac{2\alpha}{\beta}c_1+\eta_2 c_2 \mid \eta_2 \in \mathbb{L}\}$	
$\lambda_1 \neq \frac{1}{2}$ $\lambda_2 = \frac{1}{2}$	α arbitrary $\beta \neq 0$	$\ell_2=\{c_o+\eta_2 c_2 \mid \eta_2 \in \mathbb{L}\}$	$x(x-1)(x-\lambda_1)(x-\frac{1}{2})=0$
	α arbitrary $\beta = 0$		$x(x-1)(x-\lambda_1)=0$

Since by now we know that every algebra with a genetic realization has an idempotent in γ_m we can prove the following theorem.

<u>Theorem 4.14.</u> Let $\mathfrak{A}$ be a genetic algebra with a genetic realization. Then with the exception of $\lambda_o = 1$ the absolute values of the train roots do not exceed 1/2.

Proof. Let $a_o,\dots,a_m$ be a natural basis of $\mathfrak{A}$ over $\mathbb{R}$ in the sense of definition 1.4. By theorem 1.9 $\mathfrak{A}$ is a Banach algebra with respect to the l^1-norm defined by

$$\|x\|_a := \sum_{i=0}^{m} |\xi_i| \qquad \text{for} \quad x = \sum_{i=0}^{m} \xi_i a_i \in \mathfrak{A} .$$

By theorem 4.13 the algebra $\mathfrak{A}$ has at least one idempotent e in γ_m. Now let $c_o = e, c_1,\dots,c_m$ be a canonical basis of $\mathfrak{A}$ ($\mathfrak{A}_{\mathbb{C}}$). Let λ_{ijk}, $i,j,k=0,\dots,m$, be the multiplication constants of $\mathfrak{A}$($\mathfrak{A}_{\mathbb{C}}$) with respect to this basis. In addition to 3.12(i)-(iii) we have (4.6), since $c_o = e$. Let $\|\cdot\|_c$ be the l^1-norm with respect to this canonical basis. Since all norms on $\mathfrak{A}$ are topologically equivalent there is a constant $\beta > 0$ such that

$$\|x\|_c \leq \beta \|x\|_a \qquad \text{for all } x \in \mathfrak{A} \ (\mathfrak{A}_{\mathbb{C}}) .$$

Take $x = \sum_{i=0}^{m} \xi_i c_i \in \gamma_m$. Let the plenary powers $x^{[r]}$ of x (cf. definition 2.1) with respect to $c_o,\dots,c_m$ have the representation

$$x^{[r]} = \sum_{i=0}^{m} \xi_i^{[r]} c_i , \qquad r \in \mathbb{N} .$$

Because of the special properties of the multiplication constants we have the following difference equations for the coordinates

$$\xi_o^{[r+1]} = 1, \qquad r \in \mathbb{N} ,$$

$$\xi_k^{[r+1]} = 2\lambda_{okk}\xi_k^{[r]} + \sum_{i,j=0}^{k-1} \xi_i^{[r]}\lambda_{ijk}\xi_j^{[r]} , \quad k=1,\dots,m, \quad r \in \mathbb{N}. \tag{4.7}$$

Let $k \in \{1,\dots,m\}$. Choose $x \in \gamma_m$ such that the first k coordinates $\xi_o,\dots,\xi_{k-1}$ coincide with those of c_o, but the (k+1)-th is different, i.e.

$$\xi_o = 1, \quad \xi_1 = \dots = \xi_{k-1} = 0, \quad \xi_k \neq 0 .$$

Then the difference equation for $\xi_k^{[r]}$ reduces to

$$\xi_k^{[r+1]} = 2\lambda_{okk}\xi_k^{[r]} , \qquad r \in \mathbb{N},$$

which has the solution

$$\xi_k^{[r+1]} = (2\lambda_{okk})^r \xi_k^{[1]} .$$

In view of $x^{[r]} \in \gamma_m$ we have $\|x^{[r]}\|_a = 1$ and thus $\|x^{[r]}\|_c = \sum_{i=0}^{m} |\xi_i^{[r]}| \leq \beta$

whence $|\xi_i^{[r]}| \leq \beta$ for all $i=0,\ldots,m$ and all $r \in \mathbb{N}$.

Since $\xi_k^{[1]} \neq 0$, we have $|2\lambda_{okk}| \leq 1$. □

Theorem 4.14 gives a necessary condition for genetic algebras to have a genetic realization. Together with the existence of an idempotent this property is also sufficient if the dimension is $m + 1 = 2$.
Since a 2-dimensional algebra with a genetic realization has necessarily a canonical basis over $\mathbb{R}$ we assume that the underlying field is $\mathbb{R}$.

Theorem 4.15. Let $\mathfrak{A}$ be a 2-dimensional algebra over $\mathbb{R}$ with train roots $\lambda_o = 1$ and λ_1. Then the following assertions are equivalent.
(1) $\mathfrak{A}$ has a genetic realization.
(2) $\mathfrak{A}$ has an idempotent and $-1/2 \leq \lambda_1 \leq 1/2$.

Proof.I. (2) follows from (1) by theorems 4.11 and 4.14.
II. Suppose (2). Then we can choose the first element of a canonical basis c_o, c_1 to be an idempotent. Then the multiplication table of $\mathfrak{A}$ assumes the form

$$c_o^2 = c_o, \quad c_o c_1 = \lambda_1 c_1, \quad c_1^2 = 0 \quad .$$

Let $\alpha, \beta \in \mathbb{R}$, $\alpha \neq \beta$. Then

$$a_o := c_o + \alpha c_1, \quad a_1 := c_o + \beta c_1$$

form a new basis of $\mathfrak{A}$ with $\omega(a_i) = 1$, $i=0,1$.
The multiplication table of $\mathfrak{A}$ with respect to this basis is

$$a_o^2 = \frac{1}{\alpha-\beta}\left[(2\lambda_1\alpha - \beta)a_o + \alpha(1-2\lambda_1)a_1\right]$$

$$a_o a_1 = \frac{1}{\alpha-\beta}\left[(\lambda_1(\alpha+\beta)-\beta)a_o + (\alpha-\lambda_1(\alpha+\beta))a_1\right]$$

$$a_1^2 = \frac{1}{\alpha-\beta}\left[\beta(2\lambda_1-1)a_o + (\alpha-2\lambda_1\beta)a_1\right] \quad .$$

The algebra $\mathfrak{A}$ has a genetic realization if and only if there are $\alpha, \beta \in \mathbb{R}$, $\alpha \neq \beta$, with $\alpha - \beta > 0$, say, such that the following inequalities are satisfied

$$2\lambda_1\alpha - \beta \geq 0 \quad , \quad \alpha(1-1\lambda_1) \geq 0 \quad ,$$

$$\alpha - 2\lambda_1\beta \geq 0 \quad , \quad \beta(2\lambda_1 - 1) \geq 0 \quad ,$$

$$\lambda_1(\alpha+\beta) - \beta \geq 0 \quad , \quad \alpha - \lambda_1(\alpha+\beta) \geq 0 \quad .$$

An equivalent set of conditions is:
$\alpha, \beta \in \mathbb{R}, \alpha > \beta$ with

$$\beta \leq 2\lambda_1 \alpha \leq \alpha \quad , \qquad \beta \leq 2\lambda_1 \beta \leq \alpha . \tag{4.8}$$

Because of $-1/2 \leq \lambda_1 \leq 1/2$ one can choose $\alpha = 1$, $\beta = -1$ to satisfy (4.8). □

Remark. The proof shows that if $\lambda_1 \neq -1/2$, then α and β can be chosen in continuously many different ways such that the algebra $\mathfrak{A}$ has more than countably many genetic realizations.

Theorem 4.14 can be interpreted in matrix terms. If an algebra $\mathfrak{A}$ has a genetic realization then with respect to a natural basis $a_o, \dots, a_m$ to the left transformations $\mathcal{L}_{a_o}, \dots, \mathcal{L}_{a_m}$ there correspond column stochastic matrices

$$\Gamma_i := (\gamma_{ijk})^T_{j=0,\dots,m} \quad , \quad i=0,\dots,m \quad .$$

The eigenvalues of these matrices do not exceed 1 in absolute value. By theorem 3.17 $\mathfrak{A}$ is genetic if and only if there is a non singular matrix $S = (\sigma_{ij}) \in \mathfrak{M}_{m+1}(\mathbb{C})$ with $\sigma_{oj} = 1$, $j=0,\dots,m$, such that all matrices $S\Gamma_i S^{-1}$, $i=0,\dots,m$, are lower triangular matrices with the same diagonal. Using theorem 4.14 we obtain the following bound for the eigenvalues of certain stochastic matrices.

Theorem 4.16. Let $\Gamma_o, \Gamma_1, \dots, \Gamma_m$ be column stochastic matrices with the property that the i-th column of the j-th matrix equals the j-th column of the i-th matrix. Suppose there is a non singular matrix $S = (\sigma_{ij}) \in \mathfrak{M}_{m+1}(\mathbb{C})$ with $\sigma_{oj} = 1$, $j=0,\dots,m$, such that all matrices $S\Gamma_i S^{-1}$, $i=0,\dots,m$, are lower (upper, respectively) triangular matrices with the same diagonal, then all matrices Γ_i, $i=0,\dots,m$, have the same eigenvalues $\lambda_o = 1, \lambda_1, \dots, \lambda_m$, and $|\lambda_i| \leq 1/2$, $i=1,\dots,m$.

We return to the problem of when a genetic algebra has a genetic realization. Let $\mathfrak{A}$ be a genetic algebra over $\mathbb{R}$ or $\mathbb{C}$ with train roots $\lambda_o = 1$, $\lambda_1, \dots, \lambda_m$.
If $\mathfrak{A}$ has a genetic realization then $1, \lambda_1, \dots, \lambda_m$ are the eigenvalues of a stochastic matrix Γ_o. Therefore a necessary condition is that the train roots are real or pairwise conjugate. Furthermore the train roots except for $\lambda_o = 1$ are not only bounded by 1 in absolute value but by 1/2.

A necessary condition is

$$1 + \sum_{i=1}^{m} \lambda_i \geqq 0 \quad , \tag{4.9}$$

since the sum of the eigenvalues is equal to the trace of the matrix. The question of conditions on the multiplication constants with respect to a canonical basis which ensure that a genetic algebra has a genetic realization is related to the problem of when a given set of m+1 complex numbers $\lambda_o = 1, \lambda_1, \ldots, \lambda_m$ are eigenvalues of a stochastic $(m+1)\times(m+1)$-matrix.

This problem has been dealt with by Kolmogoroff, Dmitriev, Dynkin, Suleĭmanova, Perfect, Oliveira. We quote some results. Suleimanova [108] shows by a geometrical method that m+1 real numbers are eigenvalues of a positive stochastic diagonalizable matrix if the sum of the moduli of the negative λ_i is less than 1.

Perfect [93] shows for m=2 and m=3 that real numbers $1, \lambda_1, \ldots, \lambda_m$ are eigenvalues of a positive diagonalizable stochastic matrix if and only if $1 + \sum_{i=1}^{m} \lambda_i > 0$.

Furthermore Perfect [94] shows that m+1 real numbers $1, \lambda_1, \ldots, \lambda_m$ with $0 \leqq \lambda_i < 1$, $i=1,\ldots,m$, are eigenvalues of a positive diagonalizable stochastic matrix.

Finally the following result of Perfect [94] is worth mentioning. In this context: m+1 real numbers $\lambda_o = 1, \lambda_1, \ldots, \lambda_m$ are the eigenvalues of a stochastic matrix if the set can be split into subsets such that each subset contains at least one positive number and the sum of moduli of negative does not exceed this positive number.

Oliveira [91] shows that this condition is not necessary. He constructs a stochastic matrix with eigenvalues

$$\lambda_o = 1,\ \lambda_1 = -\frac{1}{4},\ \lambda_2 = -\frac{1}{10},\ \lambda_3 = -\frac{1}{2},\ \lambda_4 = -\frac{1}{5},\ \lambda_5 = \frac{1}{20} \quad .$$

In our opinion these attempts indicate that the general problem for genetic algebras will be rather difficult to solve. The necessary conditions given in theorem 4.14 fits into the general results of Dmitriev and Dynkin [23] , [24] and Karpelevič [80] .

These authors have solved a problem posed by Kolmogoroff, namely to find for any given $n \in \mathbb{N}$, the subset of the unit disk consisting of all eigenvalues of all stochastic n×n-matrices. Trivially $M_n \subseteq M_{n+1}$, $n \in \mathbb{N}$, and $M_1 = \{1\}$, $M_2 = \{-1,1\}$, and also $M_3 = \{-1,1\} \cup D_3$ where D_3 is the equilateral triangle with corners at the cube roots of unity. Thus

M_2 contains the interval $[-1/2, 1/2]$ for the second train root of a 2-dimensional genetic algebra with a genetic realization, and each set M_n, n=3, contains the disk $\{\lambda \in \mathbb{C} \mid |\lambda| \leq 1/2\}$.

C. Sequences of principal powers and plenary powers in genetic algebras

If in some population the inheritance can be described by an algebra $\mathfrak{A}$, that is an algebra with a genetic realization, then the generation sequence arising from random mating with the initial population or from random mating within a generation is represented by the sequence of principal powers x^r or by the sequence of plenary powers $x^{[r]}$, respectively, where $x \in \mathfrak{A}$ represents the initial population. If $\mathfrak{A}$ is a train algebra then by definition x satisfies the rank equation (3.1) of the algebra from which it follows that the principal powers of x, where x is of weight 1, satisfy a linear difference equation (3.6) with constant coefficients. Then the j-th principal power of x can be explicitly computed from the principal train roots, cf. (3.9).
If in addition the algebra $\mathfrak{A}$ is genetic then by theorem 4.14 the train roots of $\mathfrak{A}$, with the exception of 1, do not exceed 1/2 in absolute value. Since the train roots contain the principal train roots from theorem 3.6 there follows

Theorem 4.17. Let $\mathfrak{A}$ be a genetic algebra over $\mathbb{R}$ or $\mathbb{C}$. If $\mathfrak{A}$ has a genetic interpretation, then the sequence of principal powers of any element $x \in \mathfrak{A}$ of weight 1 converges.

In order to gain insight into the behavior of sequences of plenary powers one can proceed in a similar manner starting from Etherington's question, cf. [27] , of under which conditions the plenary powers satisfy a train equation and possibly solve the related difference equation explicitly, cf. Heuch [58] .
In genetic algebras one can derive the convergence of sequences of plenary powers directly from the train roots.
The following theorem has been given by Gonshor [45] for special train algebras. It also holds for genetic algebras as he remarked in [47] .

Theorem 4.18. Let $\mathfrak{A}$ be a genetic algebra over $\mathbb{R}$ or $\mathbb{C}$. If all train roots different from $\lambda_0 = 1$ are less than 1/2 in absolute value then the sequence of plenary powers converges towards the idempotent of $\mathfrak{A}$ ($\mathfrak{A}_{\mathbb{C}}$) for any element $x \in \mathfrak{A}$ of weight 1.

Proof. Let $\mathfrak{A}$ have the dimension m+1. Since all train roots of $\mathfrak{A}$ are different from 1/2 the algebra $\mathfrak{A}$ ($\mathfrak{A}_{\mathbb{C}}$) has exactly one idempotent by theorem 4.4. This can be chosen as the first element of a canonical basis $c_o,\dots,c_m$. Let λ_{ijk}, $i,j,k=0,\dots,m$, be the multiplication constants with respect to this basis. Since c_o is an idempotent we have $\lambda_{oo1} = \dots = \lambda_{oom} = 0$. The train roots of $\mathfrak{A}$ are $\lambda_i := \lambda_{oii}$, $i=0,\dots,m$.

For every

$$x = \sum_{i=0}^{m} \xi_i c_i$$

with $\xi_o = 1$ the coordinates of the plenary powers

$$x^{[l]} = \sum_{i=0}^{m} \xi_i^{[l]} c_i$$

satisfy the difference equations

$$\xi_o^{[l+1]} = 1$$

$$\xi_k^{[l+1]} = 2\lambda_k \xi_k^{[l]} + \sum_{i,j=0}^{k-1} \xi_i^{[l]} \lambda_{ijk} \xi_j^{[l]} \quad , \ k=1,\dots,m. \qquad (4.10)$$

We show by induction that each sequence of coordinates $\xi_k^{[l]}$ converges to 0, $k=1,\dots,m$, and thus $\lim_{l\to\infty} x^{[l]} = c_o$.

The first equation of (4.10) reduces to

$$\xi_1^{[l+1]} = 2\lambda_1 \xi_1^{[l]}$$

and the solution is

$$\xi_1^{[l+1]} = (2\lambda_1)^l \xi_1^{[1]} .$$

In view of $|\lambda_1| < 1/2$ we have $\lim_{l\to\infty} \xi_1^{[l]} = 0$.

Assume that all sequences of coordinates $\xi_1^{[l]}, \dots, \xi_{k-1}^{[l]}$ converge to 0 and that the k-th sequence of coordinates satisfies the difference equation (4.10) for the index k. Since all sequences $\{\xi_1^{[l]}\},\dots,\{\xi_{k-1}^{[l]}\}$ converge to zero and $|\lambda_k| < 1/2$, the sequence $\{\xi_k^{[l]}\}$ also converges to 0 by well known arguments on linear difference equations. Thus the assertion is valid. □

Gonshor [45] has indicated some generalizations of this theorem for the case when some train roots equal 1/2, in particular if $\lambda_1 = 1/2$ is the only train root which equals 1/2 and in addition $\lambda_{0001} = 0$ holds. We show a generalization under the hypothesis of theorem 4.6.

Theorem 4.19. Let $\mathfrak{A}$ be an (m+1)-dimensional genetic algebra with a canonical basis $c_0,\dots,c_m$ over $\mathbb{R}$ or $\mathbb{C}$. Assume that among the train roots $\lambda_0 = 1, \lambda_1,\dots,\lambda_m$ exactly $\lambda_{t_1} = \dots = \lambda_{t_r} = 1/2$, $t_1<t_2<\dots<t_r$, and assume that the other train roots except λ_0 satisfy $|\lambda_i| < 1/2$. If c_ic_j has no component in the direction of c_{t_ϱ}, $i,j=0,\dots,t_\varrho-1$, $\varrho=1,\dots,r$, then the sequence of plenary powers of any element $x = \sum_{i=0}^{m}\xi_i c_i$ converges towards the idempotent of $\mathfrak{A}$ ($\mathfrak{A}_{\mathbb{C}}$) determined by the coordinates $\xi_{t_1},\dots,\xi_{t_r}$.

Proof. By theorem 4.6 the algebra $\mathfrak{A}(\mathfrak{A}_{\mathbb{C}})$ has an r-parametric family of idempotents

$$\mathcal{E} = \left\{ c_0 + \sum_{i=1}^{m} f_i(\eta_{t_1},\dots,\eta_{t_{\sigma(i)}})c_i \;\middle|\; \eta_1,\dots,\eta_r \in \mathbb{R}\ (\mathbb{C}) \right\}$$

where

$$f_{t_\varrho}(\eta_{t_1},\dots,\eta_{t_\varrho}) = \eta_{t_\varrho} \quad , \qquad \varrho = 1,\dots,r.$$

The coordinates $\xi_0^{[l]} = 1, \xi_1^{[l]},\dots,\xi_m^{[l]}$ of the l-th plenary power $x^{[l]}$ of x satisfy the difference equations (4.10). Assume $\{\xi_\varkappa^{[l]}\}$ converges towards

$$f_\varkappa(\xi_{t_1},\dots,\xi_{t_{\sigma(\varkappa)}}) \quad \text{for } \varkappa = 1,\dots,k-1.$$

We consider the k-th difference equation of (4.10).
Either $\lambda_k = 1/2$ or $|\lambda_k| < 1/2$. In the first case the index k is one of the indices $t_1,\dots,t_r$ and by assumption $\lambda_{ijk} = 0$ for $i,j,k=0,\dots,k-1$. Then the difference equation for $\xi_k^{[l]}$ reduces to

$$\xi_k^{[l+1]} = \xi_k^{[l]} \quad , \qquad l \in \mathbb{N},$$

with the initial condition $\xi_k^{[1]} = \xi_k$. The solution is $\xi_k^{[l]} = \xi_k$ which can be represented by $\xi_k = f_k(\xi_{t_1},\dots,\xi_k)$.

In the latter case we have the difference equation

$$\xi_k^{[l+1]} = 2\lambda_k\xi_k^{[l]} + \sum_{i,j=0}^{k-1} \xi_i^{[l]}\lambda_{ijk}\xi_j^{[l]} .$$

since $\xi_0^{[1]} = 1$ and by our induction hypothesis $\lim_{l\to\infty} \xi_\varkappa^{[l]} = f_\varkappa(\xi_{t_1},\ldots,\xi_{t_{\sigma(\varkappa)}})$ for $\varkappa = 1,\ldots,k-1$, and $|\lambda_k|<1/2$, we find that the sequence $\{\xi_k^{[l]}\}$ converges. The limit is the solution

$$f_k(\xi_{t_1},\ldots,\xi_{t_{\sigma(k)}}) := \frac{1}{1-2\lambda_k}\sum_{i,j=0}^{k-1} f_i(\xi_{t_1},\ldots,\xi_{t_{\sigma(i)}})\lambda_{ijk} f_j(\xi_{t_1},\ldots,\xi_{t_{\sigma(j)}})$$

of the k-th equation of (4.1) with ξ instead of η and $f_0 := 1$. □

Finally we discuss the problem which has been mentioned several times of whether the plenary powers satisfy a train equation. Etherington [28] has shown that such an assertion is valid in train algebras of rank 2 or 3. In a train algebra of rank 2 the train equations for principal powers and plenary powers coincide and have the form

$$x^2 - \omega(x)x = 0 \quad . \tag{4.11}$$

In a train algebra of rank 3 with rank equation

$$x^3 - (1+\lambda_1)\,\omega(x)x^2 + \lambda_1\omega^2(x)x = 0 \tag{4.12}$$

the train equation for plenary powers is

$$x^{[3]} - (1+2\lambda_1)\,\omega^2(x)x^{[2]} + 2\lambda_1\,\omega^3(x)x^{[1]} = 0 \quad . \tag{4.13}$$

Later Reiersøl [98] showed that in a gametic algebra for r coupled loci the plenary powers satisfy a train equation and he has shown how the plenary train roots can be computed recursively with respect to the dimension of $\mathfrak{A}$.

Furthermore Holgate [62] has shown by linearizing the operator $\mathfrak{E}: x \longmapsto x^2$ that the plenary powers in a genetic algebra with idempotent satisfy a train equation.

Within the frame-work of this monograph we can only sketch the principle of linearization and apply it to one example.

Let $\mathfrak{A}$ be an (m+1)-dimensional genetic algebra over $\mathbb{K}$ with an idempotent. One can assume that the idempotent is the first element of a canonical basis $c_0,\ldots,c_m$. Let λ_{ijk}, $i,j,k=0,\ldots,m$, be the multiplication constants with respect to this basis.

Then to the elements $x = \sum_{i=0}^{m} \xi_i c_i \in \mathfrak{A}$ of weight 1 there correspond the vectors $(1,\xi_1,\ldots,\xi_m)^T \in \mathbb{K}^{m+1}$. Using relations (4.10) one can embed the linear manifold

$$\mathfrak{U} := \left\{(1,\xi_1,\ldots,\xi_m)^T \,\middle|\, \xi_i \in \mathbb{K},\ i=1,\ldots,m\right\}$$

by a one-to-one map φ into a vector space $\mathcal{B}(\mathfrak{A})$ of higher dimension such that to the quadratic operator $\tilde{\mathcal{E}} : \mathfrak{U} \longrightarrow \mathfrak{U}$ with

$$\tilde{\mathcal{E}} : \begin{pmatrix} 1 \\ \xi_1 \\ \xi_2 \\ \cdot \\ \cdot \\ \cdot \\ \xi_m \end{pmatrix} \longmapsto \begin{pmatrix} 1 \\ 2\lambda_{011}\xi_1 \\ 2\lambda_{022}\xi_2 + \sum_{i,j=0}^{1} \xi_i\lambda_{ij2}\xi_j \\ \cdot \\ \cdot \\ \cdot \\ 2\lambda_{omm}\xi_m + \sum_{i,j=0}^{m-1} \xi_i\lambda_{ijm}\xi_j \end{pmatrix}$$

there corresponds a linear mapping $\hat{\mathcal{E}} : \mathcal{B}(\mathfrak{A}) \longrightarrow \mathcal{B}(\mathfrak{A})$ and the diagram is commutative.

$$\begin{array}{ccc} \mathfrak{U} & \xrightarrow{\ \tilde{\mathcal{E}}\ } & \mathfrak{U} \\ \varphi \big\downarrow & & \big\downarrow \varphi \\ \mathcal{B}(\mathfrak{A}) & \xrightarrow{\ \hat{\mathcal{E}}\ } & \mathcal{B}(\mathfrak{A}) \end{array}$$

The extension is accomplished by adding certain monomials in $\xi_1,\ldots,\xi_m$. The proof is performed by induction with respect to the dimension of the algebra.

Furthermore Holgate defines recursively a set which contains all eigenvalues of $\hat{\mathcal{E}}$ and hence all plenary train roots of $\mathfrak{A}$. Let $\hat{f}$ denote the characteristic polynomial of $\hat{\mathcal{E}}$. Then we have $\hat{f}(\hat{\mathcal{E}}) = 0$ and hence $\hat{f}(\hat{\mathcal{E}})(x) = 0$ which implies that the plenary powers of x satisfy a train equation of the form (3.10).

<u>Example 4.20.</u> Let $\mathfrak{A} = \langle c_o, c_1, c_2\rangle_{\mathbb{K}}$ be a 3-dimensional genetic algebra with idempotent c_o. Let the multiplication table be given by

	c_o	c_1	c_2
c_o	c_o	$\lambda_{o11}c_1 + \lambda_{o12}c_2$	$\lambda_{o22}c_2$
c_1		$\lambda_{112}c_2$	0
c_2			0

Let $\omega : \mathfrak{A} \longrightarrow \mathbb{K}$ be defined by $\omega(c_o) = 1$, $\omega(c_1) = \omega(c_2) = 0$. The mapping $\xi : x \longmapsto x^2$ induces the mapping

$$\tilde{\xi}: (1,\xi_1,\xi_2)^T \longmapsto (1,\ 2\lambda_{o11}\xi_1,\ 2\lambda_{o22}\xi_2 + 2\lambda_{o12}\xi_1 + \lambda_{112}\xi_1^2)^T$$

on the linear manifold

$$\mathfrak{U} = \left\{(1,\xi_1,\xi_2)^T \,\middle|\, \xi_1,\xi_2 \in \mathbb{K}\right\} .$$

We define the mapping $\varphi : \mathfrak{U} \longrightarrow \mathbb{K}^4$ by

$$(1,\xi_1,\xi_2)^T \longmapsto (1,\xi_1,\xi_1^2,\xi_2)^T \quad .$$

Then to the mapping $\tilde{\xi}$ there corresponds a linear mapping $\hat{\xi} : \mathbb{K}^4 \longrightarrow \mathbb{K}^4$ given by the matrix

$$\begin{pmatrix} 1 & 0 & 0 & 0 \\ 0 & 2\lambda_{o11} & 0 & 2\lambda_{o12} \\ 0 & 0 & 4\lambda_{o11}^2 & \lambda_{112} \\ 0 & 0 & 0 & 2\lambda_{o22} \end{pmatrix}$$

Let

$$\hat{f}(\lambda) = \lambda^4 + \alpha_1\lambda^3 + \alpha_2\lambda^2 + \alpha_3\lambda + \alpha_4$$

be the characteristic polynomial of $\hat{\xi}$. Then

$$\hat{f}(\hat{\xi})(x) = x^{[5]} - \alpha_1 x^{[4]} + \alpha_2 x^{[3]} + \alpha_3 x^{[2]} + \alpha_4 x^{[1]} = 0$$

for all $x \in \mathfrak{A}$ with $\omega(x) = 1$. Hence the plenary powers of any element $x \in \mathfrak{A}$ satisfy a train equation for which the plenary train roots are contained in $\{1,\ 2\lambda_{o11},(2\lambda_{o11})^2,\ 2\lambda_{o22}\}$. □

Abraham [1], [3] has extended Holgate's investigations on the linearization of the quadratic operator $\xi : x \longmapsto x^2$ and has found a formula for the dimension $\mathfrak{L}(\mathfrak{A})$ as a function of dim $\mathfrak{A}$.

Now we follow Heuch's proof, cf. [52], for the fact that in a genetic algebra with idempotent the plenary powers satisfy a train equation. Let $x = \sum_{i=0}^{m} \xi_i c_i, \xi_o = 1$. Then the coordinates of the l-th plenary power $x^{[l]} = \sum_{i=0}^{m} \xi_i^{[l]} c_i$ satisfy the difference equation (4.10) with

$\xi_o^{[l]} = 1$, $l \in \mathbb{N}$, and $\lambda_{ook} = 0$, $k=1,\ldots,m$.
By induction with respect to k it follows that the solutions have the form

$$\xi_k^{[l+1]} = \sum_{p=1}^{s_k} \beta_{pk}(l)\mu_{pk}^l \quad , \qquad k=1,\ldots,m, \tag{4.14}$$

in which $\beta_{pk}(l)$, $p=1,\ldots,s_k$, are polynomials in l, and the roots μ_{pk}, $p=1,\ldots,s_k$, are equal to $2\lambda_{okk}$ or are products of roots $\mu_{q,k-1}$ and $\mu_{r,k-1}$ of $\xi_{k-1}^{[l]}$ if $\lambda_{qrk} \neq 0$.
After a suitable renumeration of the roots μ_{pk} one obtains a representation

$$x^{[l+1]} = \sum_{i=0}^{v} b_i(l)\mu_i^l \quad , \qquad l \in \mathbb{N}, \tag{4.15}$$

where $b_i(l) \in \mathfrak{A}$, $i=1,\ldots,v$, are polynomials in l. From this explicit representation it follows that the plenary powers for any $x \in \mathfrak{A}$ of weight 1 satisfy a linear difference equation of the form (3.13).
Thus we have the following theorem.

Theorem 4.21. In a genetic algebra $\mathfrak{A}$ with idempotent the plenary powers of every element $x \in \mathfrak{A}$ satisfy a train equation (3.11).

Heuch [58] also provides formulas for the recursive computation of the roots and polynomials in (4.14) and he obtains in this way an explicit formula for the plenary powers of x.

5. THE NON COMMUTATIVE CASE

Non commutative algebras occur in genetic applications if not all individuals of a population mate randomly, e.g. in populations with a sex linked character or in partially selfing populations, cf. section 8. But also from a theoretical point of view it may be of interest to establish how far the theory can be carried over to non commutative algebras.

The definition 1.7 of a baric algebra is not restricted to the commutive case as already indicated in section 1. The same is true for algebras with a genetic realization, cf. definition 1.8. On the other hand all definitions in section 3 require commutativity. Some generalizations of these definitions for the non commutative case have been given by Etherington [27] , [31] and Holgate [70]. Etherington [32] has also computed equations satisfied by the plenary powers and other sequences of powers in a non commutative train algebra (cf. definition 5.1) with left and right rank equal to 2 or 3. But we do not know of any further investigations of the non commutative case.

In the following we shall see that all definitions of section 3 can be carried over to non commutative algebras as one-sided properties and as two-sided properties as well. However not all implications remain valid, and the non commutative case becomes somewhat involved. For this reason we have preferred to deal with the commutative case separately and postpone the non commutative case to this section.

A. Train algebras, genetic algebras and special train algebras

In the following let $\mathfrak{A}$ be an (m+1)-dimensional, non associative, non commutative algebra over $\mathbb{K}$.

<u>Definition 5.1.</u> Let $\mathfrak{A}$ be a baric algebra with weight homomorphism ω.

(1) $\mathfrak{A}$ is called a <u>left (right) train algebra</u> if the coefficients of the left (right) train equation of $\mathfrak{A}$ are functions of $\omega(x)$, $x \in \mathfrak{A}$, only. The degree of the left (right) train equation is called the <u>left (right) rank</u> of the algebra. The roots with the exception of $\lambda_{-1} = 0$ of the left (right) rank equation are called the <u>left (right) principal train roots</u> of the algebra $\mathfrak{A}$.

(2) $\mathfrak{A}$ is called a <u>train algebra</u> if $\mathfrak{A}$ is both a left and a right train algebra.

If one applies the weight homomorphism of a left (right) train algebra to the left (right) rank equation then one sees that $\lambda_o = 1$ is always a left (right) principal train root. If $\mathfrak{A}$ is a train algebra, then the left and right rank need not be equal, and even if these numbers coincide then the left and right train roots, except $\lambda_o = 1$, need not coincide.

The following assertions with the exception of definitions are formulated only for left properties. Similar assertions are valid for right properties. We recall definitions 2.1 and 2.2 of left principal powers and in particular the properties left nil and left nilpotent.

<u>Lemma 5.2.</u> Let $\mathfrak{A}$ be a baric algebra with weight homomorphism ω. Let $\mathfrak{N} := \ker\omega$ be left nil. Then ω is unique.

The proof proceeds exactly as the proof of lemma 3.4 in which commutativity has not been used.
Similarly the assertions 3.2 and 3.5 can be carried over.

<u>Lemma 5.3.</u> Let $\mathfrak{A}$ be a left train algebra of rank r with weight homomorphism $\omega : \mathfrak{A} \longrightarrow \mathbb{R}$, $\mathfrak{N} := \ker\omega$. Then

(1) Every element of $\mathfrak{N}$ is left nilpotent of index not greater than r.

(2) The weight homomorphism ω is unique.

<u>Definition 5.4.</u> Let $\mathfrak{A}$ be a baric algebra with weight homomorhphism ω, and let $\mathfrak{N} := \ker\omega$.

(1) $\mathfrak{A}$ is called a <u>weakly left (right) special train algebra</u> if $\mathfrak{N}$ is left (right) nilpotent and the left (right) principal powers of $\mathfrak{N}$ are left (right) ideals of $\mathfrak{A}$.

(2) $\mathfrak{A}$ is called a <u>left (right) special train algebra</u> if $\mathfrak{N}$ is nilpotent and the principal powers of $\mathfrak{N}$ are left (right) ideals of $\mathfrak{A}$.

(3) $\mathfrak{A}$ is called a <u>special train algebra</u> if $\mathfrak{N}$ is nilpotent and the principal powers of $\mathfrak{N}$ are ideals of $\mathfrak{A}$.

Trivially we have the implications (3) $\Longrightarrow$ (2) $\Longrightarrow$ (1). On the other hand the converse implications are not valid as it is shown by the following examples.

Example 5.5. (1) Let $\mathfrak{A} = \langle c_0, c_1, c_2 \rangle_{\mathbb{R}}$ have the following multiplication table:

	c_0	c_1	c_2
c_0	c_0	c_1	c_2
c_1	c_1	c_2	0
c_2	c_2	c_2	0

$\mathfrak{N} = \langle c_1, c_2 \rangle$ is an ideal of $\mathfrak{A}$ which does not contain $\mathfrak{A}^2$. For the left principal powers of $\mathfrak{N}$ we have ${}^2\mathfrak{N} = \langle c_2 \rangle$, ${}^3\mathfrak{N} = \langle 0 \rangle$. Furthermore ${}^2\mathfrak{N}$ is a left ideal of $\mathfrak{A}$, but $\mathfrak{N}$ is not nilpotent because $\mathfrak{N}^{(r)} = \langle c_2 \rangle$ for all $r \geqq 2$. Thus $\mathfrak{A}$ is a weakly left special train algebra but not a left special train algebra.

(2) Let $\mathfrak{A} = \langle c_0, c_1, c_2 \rangle_{\mathbb{R}}$ with multiplication table:

	c_0	c_1	c_2
c_0	c_0	c_1	c_2
c_1	c_1	c_2	0
c_2	c_1	0	0

$\mathfrak{N} = \langle c_1, c_2 \rangle$ is an ideal of $\mathfrak{A}$ which does not contain $\mathfrak{A}^2$. For the principal powers of $\mathfrak{N}$ we have $\mathfrak{N}^{(2)} = \langle c_2 \rangle$, $\mathfrak{N}^{(3)} = \langle 0 \rangle$, and $\mathfrak{N}^{(2)}$ is a left ideal but not a right ideal of $\mathfrak{A}$. Thus $\mathfrak{A}$ is a left special train algebra but not a special train algebra.

Definition 5.6. Let $\mathfrak{A}$ be a baric algebra with weight homomorphism ω.

(1) $\mathfrak{A}$ is left (right) Schafer genetic if for every transformation $\mathcal{T} \in \mathfrak{T}(\mathfrak{A})$ with the representation $\mathcal{T} = f(\mathcal{L}_{x_1}, \ldots, \mathcal{L}_{x_s})$ $(\mathcal{T} = f(\mathcal{R}_{x_1}, \ldots, \mathcal{R}_{x_s}))$ in left (right) transformations the coefficients of the characteristic polynomial are functions of $\omega(x_1), \ldots, \omega(x_s)$ only.

(2) $\mathfrak{A}$ is Schafer genetic if for every transformation $\mathcal{T} \in \mathfrak{T}(\mathfrak{A})$, $\mathcal{T} = f(\mathcal{L}_{x_1}, \mathcal{R}_{x_2}, \ldots, \mathcal{R}_{x_s})$ the coefficients of the characteristic polynomial are functions of $\omega(x_1), \omega(x_2), \ldots, \omega(x_s)$ only.

Trivially (2) $\Longrightarrow$ (1).

Let x be a generic element of $\mathfrak{A}$. Then the left and right rank polynomials of the algebra are divisors of $xg_x(x)$ and $f_x(x)x$, respectively, where g_x and f_x are the characteristic polynomials of $\mathcal{L}_x$ and $\mathcal{R}_x$, respectively. Hence we have the following theorem.

Theorem 5.7. If $\mathfrak{A}$ is left Schafer genetic (Schafer genetic) then $\mathfrak{A}$ is a left train algebra (train algebra).

Theorem 5.8. Every special train algebra is Schafer genetic.

Proof. Let $\mathfrak{A}$ be a special train algebra over $\mathbb{K}$. Then by definition $\mathfrak{A}$ has a weight homomorphism $\omega : \mathfrak{A} \longrightarrow \mathbb{K}$. Let $\mathfrak{N} := \ker\omega$. By definition 5.4(3) $\mathfrak{N}$ is nilpotent of an index r and the principal powers $\mathfrak{N}^{(1)} = \mathfrak{N}$, $\mathfrak{N}^{(2)}, \dots, \mathfrak{N}^{(r)}, \mathfrak{N}^{(r+1)} = \langle 0 \rangle$ of $\mathfrak{N}$ are ideals of $\mathfrak{A}$.
We choose $c_o \in \mathfrak{A}$ with $\omega(c_o) = 1$ and a basis $b_1^{(1)}, \dots, b_{k_1}^{(1)}, \dots, b_1^{(r)}, \dots, b_{k_r}^{(r)}$ of $\mathfrak{N}$ such that $b_1^{(l)}, \dots, b_{k_l}^{(l)}$ are in $\mathfrak{N}^{(l)}$ but not in $\mathfrak{N}^{(l+1)}$ for $l=1,\dots,r$.
In view of $\mathfrak{N}\mathfrak{N}^{(l)}, \mathfrak{N}^{(l)}\mathfrak{N} \subseteq \mathfrak{N}^{(l+1)}$ the transformations

$$\mathcal{L}_{b_1^{(l)}}, \dots, \mathcal{L}_{b_l^{(l)}}, \mathcal{R}_{b_1^{(l)}}, \dots, \mathcal{R}_{b_{k_l}^{(l)}}$$

induce zero mappings on the factor spaces $\mathfrak{N}^{(l)}/\mathfrak{N}^{(l+1)}$ for $l=1,\dots,r$.
Furthermore $\mathfrak{N}^{(l)}$ is an ideal of $\mathfrak{A}$ and

$$\mathcal{L}_{b_i^{(l)}} c_o = b_i^{(l)} c_o, \ \mathcal{R}_{b_i^{(l)}} c_o = c_o b_i^{(l)} \in \mathfrak{N}^{(l)}, \qquad i=1,\dots,k_l, \ l=1,\dots,r.$$

Therefore the left and right transformations $\mathcal{L}_{b_i^{(l)}}$ and $\mathcal{R}_{b_i^{(l)}}$ are represented by nilpotent block matrices $L_{b_i^{(l)}}$ and $R_{b_i^{(l)}}$, with respect to the basis $c_o, b_1^{(1)}, \dots, b_{k_r}^{(r)}$.
These block matrices $L_{b_i^{(l)}}$ and $R_{b_i^{(l)}}$, $i=1,\dots,k_l$, have the form

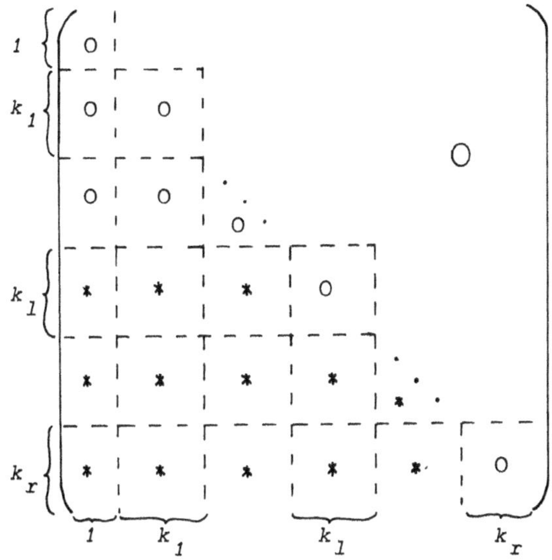

for $l=1,\dots,r$.

Since the principal powers of $\mathfrak{N}$ are ideals of $\mathfrak{A}$ the linear mappings $\mathcal{L}_{c_o}$ and $\mathcal{R}_{c_o}$ leave the factor spaces $\mathfrak{N}^{(l)}/\mathfrak{N}^{(l+1)}$, $l=1,\dots,r$, invariant. Since furthermore $\omega(c_o) = \omega(c_o^2) = 1$ there correspond to $\mathcal{L}_{c_o}$ and $\mathcal{R}_{c_o}$ block matrices L_{c_o} and R_{c_o} of the form

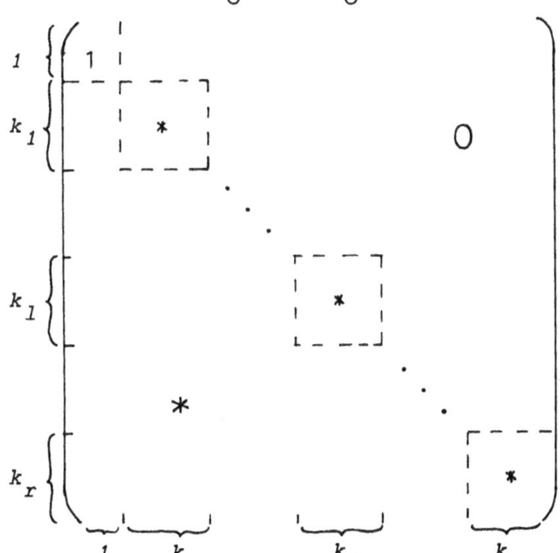

For a fixed block pattern the lower block triangular matrices (lower generalized triangular matrices in Gantmacher's terminology cf. [37]) form an algebra.

Therefore to every transformation $\mathcal{T} \in \mathfrak{T}(\mathfrak{A})$ there corresponds a lower block triangular matrix with respect to this basis. Every $\mathcal{T} \in \mathfrak{T}(\mathfrak{A})$ can be represented as $\mathcal{T} = f(\mathcal{L}_{x_1}, \mathcal{R}_{x_2}, \dots, \mathcal{R}_{x_s})$ for a certain

$f \in \mathbb{K}\{X_1,\ldots,X_s\}$. That is, every $\mathcal{T}$ is a linear combination of the identity on $\mathfrak{A}$ and of certain monomials in $\mathcal{L}_{x_1}, \mathcal{R}_{x_2}, \ldots, \mathcal{R}_{x_s}$. Let $\mathcal{L}_{x_i}\mathcal{R}_{x_j}$ be such a monomial.

Since $\mathcal{L}_{x_i}$ and $\mathcal{R}_{x_j}$ correspond to lower block matrices where the diagonal blocks are the same as in $\omega(x_i)L_{c_0}$ and $\omega(x_j)R_{c_0}$, respectively, to the transformation $\mathcal{L}_{x_i}\mathcal{R}_{x_j}$ there corresponds a lower block triangular matrix with diagonal blocks depending only on $\omega(x_i)$ and $\omega(x_j)$. This is true for any monomial in $\mathcal{L}_{x_1}, \mathcal{R}_{x_2}, \ldots, \mathcal{R}_{x_s}$ and thus for any transformation $\mathcal{T} \in \mathfrak{T}(\mathfrak{A})$.

For a lower block triangular matrix the determinant is just the product of the determinants of the diagonal blocks. Hence the characteristic polynomial of a $\mathcal{T} \in \mathfrak{T}(\mathfrak{A})$ with $\mathcal{T} = f(\mathcal{L}_{x_1}, \mathcal{R}_{x_2}, \ldots, \mathcal{R}_{x_s})$ depends only on $\omega(x_1), \omega(x_2), \ldots, \omega(x_s)$. □

In a similar manner one can show or directly derive from theorems 5.10 and 5.14 that every weakly left special train algebra and thus every left special train algebra is left Schafer genetic.

In section 3 we have seen that a commutative Schafer genetic algebra can be characterized as having a basis over an extension field with respect to which the multiplication constants form a certain triangular pattern. Here we ask for a similar characterization in the non commutative case. We start with the following definition.

<u>Definition 5.9.</u> Let $\mathfrak{A}$ be an algebra and let $c_0 \in \mathfrak{A}$, $c_1,\ldots,c_m$ be a basis of $\mathfrak{A}_{\mathbb{L}}$, where $\mathbb{L}$ is an extension field of $\mathbb{K}$. Let the multiplication table be given by $c_i c_j = \sum_{k=0}^{m} \lambda_{ijk} c_k$, $i,j=0,\ldots,m$.

(1) The algebra is called <u>weakly left (right) Gonshor genetic</u> if the field and the basis can be chosen such that

$$\lambda_{000} = 1, \quad \lambda_{0jk} = 0, \quad k < j, \quad \lambda_{ijk} = 0, \quad k \leq j \quad \text{if } i > 0$$

$$(\lambda_{000} = 1, \quad \lambda_{i0k} = 0, \quad k < i, \quad \lambda_{ijk} = 0, \quad k \leq i \quad \text{if } j > 0).$$

(2) The algebra is called <u>left (right) Gonshor genetic</u> if it is weakly left(right) Gonshor genetic and if in addition the basis can be chosen such that

$$\lambda_{ijk} = 0 \quad \text{for } k \leq \max(i,j) \text{ if } i,j > 0 \quad .$$

(3) The algebra is called Gonshor genetic if $\mathfrak{A}$ is left and right Gonshor genetic with respect to the same basis.

If $\mathfrak{A}$ has one of the above properties then c , $c_1,\ldots,c_m$ is called a canonical basis of $\mathfrak{A}$ and $\lambda_{ooo} = 1, \lambda_{oii}$ $(\lambda_{ooo}, \lambda_{ioi})$, $i=1,\ldots,m$, are the left (right) train roots of $\mathfrak{A}$.

In contrast to this definition Holgate [70] calls an algebra Schafer genetic if it satisfies definition 5.9(3)).

The implications (3) $\Longrightarrow$ (2) $\Longrightarrow$ (1) are obvious.
However the converse implications are not valid in general. In the following we shall see that between one-sided properties there are similar implications as in the commutative case.
However these do not carry over to the two-sided properties. In table 5.15 we have collected all general implications.

Theorem 5.10. A (weakly) left special train algebra is (weakly) left Gonshor genetic.

Proof. I. Let $\mathfrak{A}$ be a left special train algebra. We use the same notations as in the proof of theorem 5.8. In contrast to that situation the principal powers of $\mathfrak{N}$, except $\mathfrak{N}$ itself, need not be right ideals of $\mathfrak{A}$. But here also we can conclude that with respect to the basis $c_o, b_1^{(1)}, \ldots, b_{k_r}^{(r)}$ to the left and right transformations $\mathcal{L}_{b_1^{(1)}}, \ldots, \mathcal{L}_{b_{k_r}^{(r)}}$ $\mathcal{R}_{b_1^{(1)}}, \ldots, \mathcal{R}_{b_r^{(r)}}$ there correspond lower block triangular matrices with zero diagonal blocks, because $\mathcal{R}_{b_i^{(1)}} c_o = c_o b_i^{(1)}$, $i=1,\ldots,k_1$, $1=1,\ldots,r$.
Since the principal powers of $\mathfrak{N}$ are left ideals of $\mathfrak{A}$ the transformation $\mathcal{L}_{c_o}$ leaves the factor spaces $\mathfrak{N}^{(1)}/\mathfrak{N}^{(1+1)}$ invariant, $1=1,\ldots,r$.
In the 1-th factor space there can be constructed a basis $c_1^{(1)} + \mathfrak{N}^{(1)}, \ldots, c_{k_1}^{(1)} + \mathfrak{N}^{(1)}$ over a suitable extension field $\mathbb{L}_1$ of $\mathbb{K}$ such that to the linear mapping induced by $\mathcal{L}_{c_o}$ in this factor space there corresponds a lower triangular matrix, $1=1,\ldots,r$. Then

$$(c_o, c_1^{(1)}, \ldots, c_{k_1}^{(1)}, \ldots, c_1^{(r)}, \ldots, c_{k_r}^{(r)}) =: (c_o, c_1, \ldots, c_m)$$

is a basis of $\mathfrak{A}_{\mathbb{L}}$, $\mathbb{L} = \mathbb{L}_1, \ldots, \mathbb{L}_r$, with respect to which $\mathcal{L}_{c_o}$ is re-

presented by a lower triangular matrix and the transformations $\mathcal{L}_{c_1},\ldots,\mathcal{L}_{c_m},\mathcal{R}_{c_1},\ldots,\mathcal{R}_{c_m}$ are represented by nilpotent lower triangular matrices. Here we use the fact that basis transformations are performed within the factor spaces $\mathfrak{N}^{(l)}/\mathfrak{N}^{(l+1)}$, $l=1,\ldots,r$.
Let λ_{ijk}, $i,j,k=0,\ldots,m$, be the multiplication constants of $\mathfrak{A}$ with respect to this basis.
Since to all $\mathcal{L}_{c_1},\ldots,\mathcal{L}_{c_m},\mathcal{R}_{c_1},\ldots,\mathcal{R}_{c_m}$ there correspond nilpotent lower triangular matrices we have

$$\lambda_{ijk} = 0 \text{ for } k \leq \max(i,j) \quad \text{if } i,j > 0.$$

Since $\mathcal{L}_{c_o}$ is represented by a lower triangular matrix we have also

$$\lambda_{ojk} = 0 \text{ for } k < j,$$

and finally $\lambda_{ooo} = 1$ in view of $\omega(c_o) = 1$. Thus $\mathfrak{A}$ is left Gonshor genetic.
II. Let $\mathfrak{A}$ be a weakly left special train algebra. Now the principal powers are replaced by the left principal powers $^l\mathfrak{N}$ of $\mathfrak{N}$. We consider only left transformations. As in part I of the proof one shows that over a suitably chosen extension field $\mathbb{L}$ of $\mathbb{K}$ there is a basis $c_o, c_1,\ldots,c_m$ with respect to which the left transformations $\mathcal{L}_{c_o}$, $\mathcal{L}_{c_1},\ldots,\mathcal{L}_{c_m}$ are represented by lower triangular matrices, which, with the exception of the first one, are nilpotent. Thus the multiplication constants λ_{ijk}, $i,j,k=0,\ldots,m$, with respect to this basis satisfy

$$\lambda_{ojk} = 0 \qquad \text{for } k < j$$

$$\lambda_{ijk} = 0 \qquad \text{for } k \leq j \quad \text{if } i > 0,$$

and finally $\lambda_{ooo} = 1$. □

It is easy to find out that a special train algebra need not be Gonshor genetic. If we had required in part I of the preceding proof that $\mathfrak{A}$ be a special train algebra then we would have found a basis with respect to which $\mathcal{R}_{c_o}$ is represented by a lower triangular matrix. But then $\mathcal{L}_{c_o}$ would in general not be represented by a lower triangular matrix, cf. example 5.11.
The converse implications are not valid in general. Namely in the commutative case the various definitions of "genetic" coincide and the same is true for the different kinds of special train algebras. Thus the example of Schafer [101] , cf. the proof of theorem 3.27, gives an

example that (weakly left, left) Gonshor genetic algebras need not be (weakly left,left) special train algebras.

Example 5.11. Let $\mathfrak{A} = \langle c_o, c_1, c_2 \rangle_{\mathbb{R}}$ with multiplication table:

	c_o	c_1	c_2
c_o	c_o	$c_1 + c_2$	c_2
c_1	c_1	0	0
c_2	$c_1 + c_2$	0	0

$\mathfrak{N} := \langle c_1, c_2 \rangle$ is an ideal of $\mathfrak{A}$ which does not contain $\mathfrak{A}^2$. Thus $\mathfrak{A}$ is a baric algebra. Since $\mathfrak{N}^{(2)} = \langle 0 \rangle$, $\mathfrak{A}$ is a special train algebra. But $\mathfrak{A}$ does not have a canonical basis in the sense of definition 5.9(3) because the matrices

$$L_{c_o} = \begin{pmatrix} 1 & 0 & 0 \\ 0 & 1 & 0 \\ 0 & 1 & 1 \end{pmatrix} \quad \text{and} \quad R_{c_o} = \begin{pmatrix} 1 & 0 & 0 \\ 0 & 1 & 1 \\ 0 & 0 & 1 \end{pmatrix}$$

can not be simultaneously transformed into lower triangular form. □

By Theorem 5.8 every special train algebra is Schafer genetic. This proves that a Schafer genetic algebra need not be Gonshor genetic in the non commutative case. On the other hand we have

Theorem 5.12. Every Gonshor genetic algebra is Schafer genetic.

Proof. Let $\mathfrak{A}$ be Gonshor genetic with a canonical basis $c_o, \dots, c_m$ over an algebraic extension field $\mathbb{L}$ of $\mathbb{K}$ and let the left and right train roots be $\lambda_{ooo} = 1$, λ_{oii}, λ_{ioi}, $i=1,\dots,m$.
As we have seen in the proof of theorem 3.13 by $\omega(c_o) = 1$, $\omega(c_i) = 0$, $i=1,\dots,m$, and the linear extension onto $\mathfrak{A}_{\mathbb{L}}$ a non-trivial algebra homomorphism ω is defined and the restriction to $\mathfrak{A}$ gives a non-trivial algebra homomorphism of $\mathfrak{A}$ onto $\mathbb{K}$. To every left transformation $\mathcal{L}_{x_i}$ and every right transformation $\mathcal{R}_{x_j}$, $x_i, x_j \in \mathfrak{A}$, with respect to the canonical basis there correspond lower triangular matrices with diagonals

$$(\omega(x_i)\lambda_{ooo}, \ \omega(x_i)\lambda_{o11}, \dots, \omega(x_i)\lambda_{omm})$$

and

$$(\omega(x_j)\lambda_{ooo}, \ \omega(x_j)\lambda_{1o1}, \dots, \omega(x_j)\lambda_{mom}) ,$$

respectively. Consequently to every transformation $\mathcal{T} \in \mathcal{T}(\mathfrak{A})$ with representation $\mathcal{T} = f(\mathcal{L}_{x_1},\ldots,\mathcal{L}_{x_s})$, $f \in K\{X_1,\ldots,X_s\}$ there corresponds a lower triangular matrix with diagonal

$$(f(\omega(x_1)\lambda_{oii},\ldots,\omega(x_s)\lambda_{ioi}))\quad i=0,\ldots,m',$$

and the characteristic polynomial does depend only on $\omega(x_1),\ldots,\omega(x_s)$. □

Lemma 5.13. A Schafer genetic algebra is left Gonshor genetic.

Proof. Let $\mathfrak{A}$ be Schafer genetic. Then $\mathfrak{A}$ is in particular baric with weight homomorphism ω. We choose $c_o \in \mathfrak{A}$ with $\omega(c_o) = 1$ and a basis $b_1,\ldots,b_m$ of $\mathfrak{N} := \ker\omega$. Now let $\tilde{\mathfrak{h}}$ be the multiplicative subsemigroup of the transformation algebra $\mathcal{T}(\mathfrak{A})$ generated by $\mathcal{L}^k_{c_o}\mathcal{L}_{b_i}$, $\mathcal{L}_{c_o}\mathcal{R}_{b_i}$, $i=1,\ldots,m$, $k \in \mathbb{N}_o$. Every elements of $\tilde{\mathfrak{h}}$ contains at least one factor $\mathcal{L}_{b_i}$ or $\mathcal{R}_{b_i}$. Since $\mathfrak{A}$ is Schafer genetic every element of $\tilde{\mathfrak{h}}$ is nilpotent and from Jacobson [76], Theorem 1, p. 33 it follows that the subalgebra $\tilde{\mathfrak{U}}$ of $\mathcal{T}(\mathfrak{A})$ generated by $\tilde{\mathfrak{h}}$ is nilpotent. Therefore

$$\mathfrak{N}_1 := \mathfrak{N}\ ,\qquad \mathfrak{N}_{l+1} := \tilde{\mathfrak{U}}\mathfrak{N}_l\ ,\qquad l \in \mathbb{N},$$

form a descending chain of subalgebras of $\mathfrak{A}$ which ends after finitely many steps at $\langle 0\rangle$,

$$\mathfrak{N}_1 > \mathfrak{N}_2 > \ldots > \mathfrak{N}_r > \mathfrak{N}_{r+1} = \langle 0\rangle .$$

In view of $\mathcal{L}_{b_i}, \mathcal{R}_{b_i} \in \tilde{\mathfrak{U}}$, $i=1,\ldots,m$, these transformations generate the zero mapping on the factor spaces $\mathfrak{N}_l/\mathfrak{N}_{l+1}$, $l=1,\ldots,r$. Since $\mathcal{L}_{c_o}\tilde{\mathfrak{U}} \subseteq \tilde{\mathfrak{U}}$, $\mathcal{L}_{c_o}$ induces a linear mapping $\mathcal{L}^{(l)}_{c_o}$ on the factor space $\mathfrak{N}^{(l)}/\mathfrak{N}^{(l+1)}$, $l=1,\ldots,r$. Let $\mathbb{L}$ be an algebraic extension field of $\mathbb{K}$ such that all $\mathcal{L}^{(l)}_{c_o}$, $l=1,\ldots,r$, are represented by lower triangular matrices. As in the proof of theorem 3.12 it follows that

$$(c_o, c_1^{(1)},\ldots,c_{k_1}^{(1)},\ldots,c_1^{(r)}) =: (c_o,c_1,\ldots,c_m)$$

is a canonical basis of $\mathfrak{A}_{\mathbb{L}}$ in the sense of definition 5.9 (2). The argument of commutativity has to be replaced by the fact that c_ic_j is represented as $\mathcal{L}_{c_i}c_j$ and $\mathcal{R}_{c_j}c_i$. □

Table 5.15. Let $\mathfrak{A}$ be a finite dimensional algebra

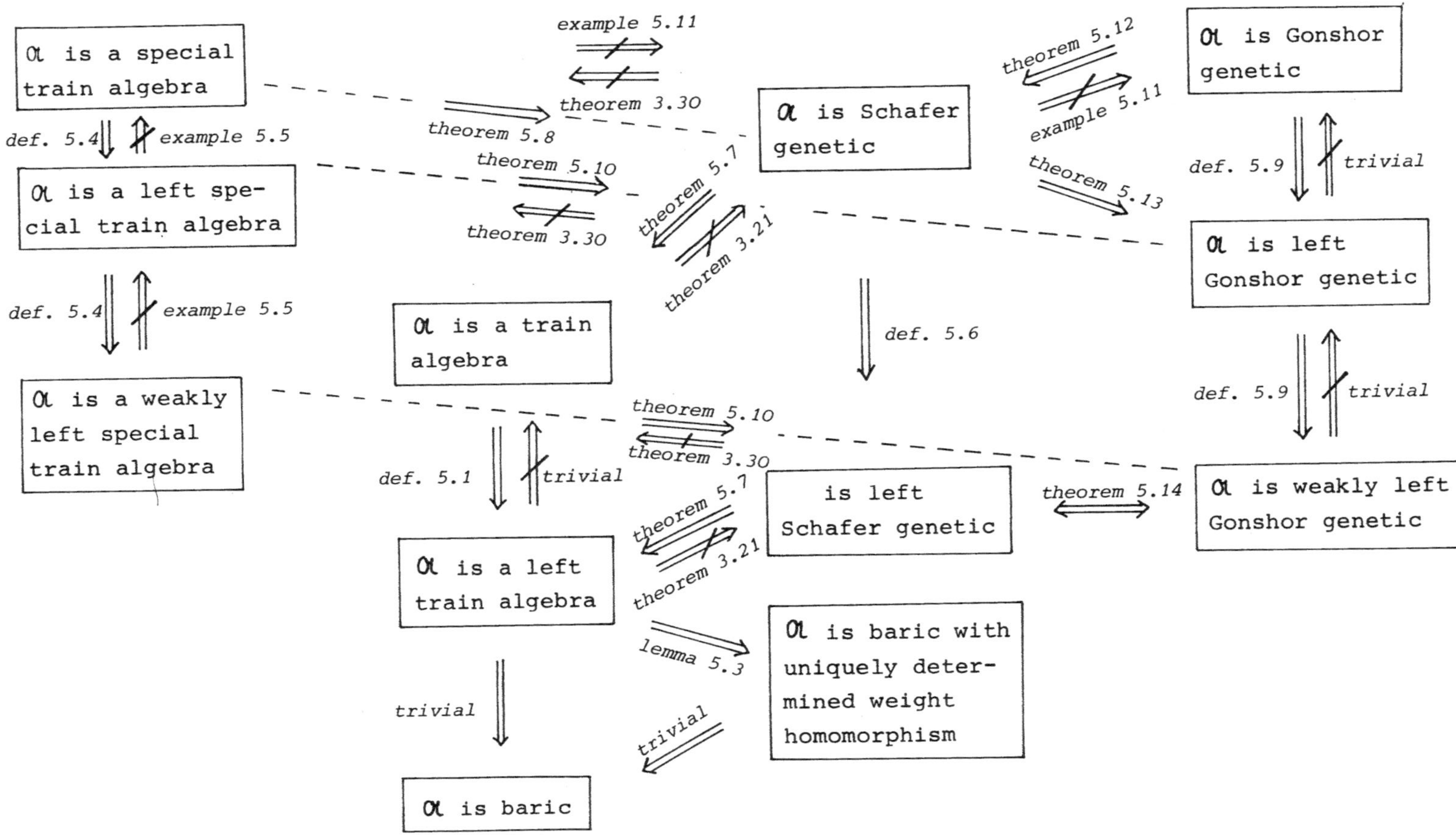

Theorem 5.14. An algebra $\mathfrak{A}$ is weakly left Gonshor genetic if and only if it is weakly left Schafer genetic.

Proof. The proof is identical with the proof of theorem 3.13 with the exception of the next to last sentence which has to be cancelled.□

Many of the characterizations of commutative genetic algebras and special train algebras introduced in section 3 can be carried over to the non commutative case. We shall not discuss these extensions in detail. In table 5.15 we have collected some implications which hold for non commutative baric algebras.

Theorem 5.16. Let $\mathfrak{A}$ be a 2-dimensional baric algebra over $\mathbb{K}$. Then the definitions 5.1(1) and (2), 5.4(1)-(3), 5.6(1), (2), and 5.9(1)-(3) are equivalent.

Proof. Since $\mathfrak{A}$ is a 2-dimensional baric algebra we can choose a basis c_o, c_1 of $\mathfrak{A}$ such that the multiplication table of $\mathfrak{A}$ assumes the following form:

	c_o	c_1
c_o	$c_o + \alpha c_1$	$\beta_1 c_1$
c_1	$\beta_2 c_1$	γc_1

Then by $\omega : \mathfrak{A} \longrightarrow \mathbb{K}$ with $\omega(c_o) := 1$, $\omega(c_1) := 0$ and the linear extension a weight homomorphism is defined. In view of the implications which are true in general, cf. table 5.15, it is sufficient to show that a 2-dimensional left train algebra is always a special train algebra and a genetic algebra. A 2-dimensional train algebra has rank either 2 or 3. Thus the left train equation has either the form

$$^2x - \omega(x)x = 0$$

or

$$^3x - (1+\lambda_1)\,\omega(x)\,(^2x) + \lambda_1\omega^2(x)x = 0 \quad , \qquad (5.1)$$

where $\lambda_o = 1$ or $\lambda_o = 1$ and $\lambda_1 = \beta_1$, respectively, are the left principal train roots.

Since every $x = \xi_o c_o + \xi_1 c_1 \in \mathfrak{A}$ satisfies left rank equation we have in the first case

$$\alpha + \xi_1(\beta_1 + \beta_2 - 1) + \xi_1^2\gamma = 0$$

from which follows that $\alpha = 0$, $\beta_1 + \beta_2 = 1$ and $\gamma = 0$.
In the second case the left rank of $\mathfrak{A}$ exceeds the dimension of $\mathfrak{A}$ by exactly 1. Thus $xg_x(x) = 0$ is the left rank equation of $\mathfrak{A}$ where g_x is the characteristic polynomial of $\mathcal{L}_x$. Thus

$$^3x - (\xi_0 + \xi_0\beta_1 + \xi_1\gamma)(^2x) + (\xi_0^2\beta_1 + \xi_0\xi_1\gamma)x = 0 . \qquad (5.2)$$

Equations (5.2) and (5.1) are identical if and only if $\gamma = 0$. In either case we have $\gamma = 0$ and thus $\mathfrak{A}$ is a special train algebra and Gonshor genetic. □

In the 3-dimensional case the situation is not so simple, as has been shown in example 5.11.

In order to further clarify the structure of the algebras under investigation we show the following theorem.

<u>Theorem 5.17.</u> Let $\mathfrak{A}$ be a baric algebra with weight homomorphism ω. If $\mathfrak{A}$ is a left special train algebra or $\mathfrak{A}$ is left Gonshor genetic or left Schafer genetic then
(1) $\mathfrak{N} = \ker\omega$ is nilpotent
(2) $\mathfrak{N}$ is the radical of $\mathfrak{A}$.

<u>Proof.</u> On one hand the radical of $\mathfrak{A}$ is contained in $\mathfrak{N}$ since $\mathfrak{A}/\mathfrak{N} \cong \mathbb{K}$. On the other hand the radical contains all nilpotent ideals. Thus (2) follows directly from (1). We show (1):
I. If $\mathfrak{A}$ is a left special train algebra then by definition $\mathfrak{N}$ is nilpotent.
II. If $\mathfrak{A}$ is left Gonshor genetic, with a canonical basis $c_0, c_1, \ldots, c_m$ over $\mathbb{L}$, then the l-th principal power $\mathfrak{N}^{(l)}$ of $\mathfrak{N}$ is contained in the subalgebra $\langle c_l, \ldots, c_m \rangle_{\mathbb{L}}$ of $\mathfrak{A}_{\mathbb{L}}$, $l=1, \ldots, m$, and $\mathfrak{N}^{(m+1)} = \langle 0 \rangle$, i.e. $\mathfrak{N}$ is nilpotent of an index less or equal m+1.
III. If $\mathfrak{A}$ is left Schafer genetic we refer to the proof of Schafer [101] for the commutative case. In fact let $\mathfrak{N}^*$ be the subalgebra of $\mathfrak{E}(\mathfrak{A})$ spanned by all $\mathcal{L}_x, \mathcal{R}_y$, $x,y \in \mathfrak{N}$. $\mathfrak{N}^*$ consists of all linear combinations of products of elements $\mathcal{L}_x, \mathcal{R}_y$, $x,y \in \mathfrak{N}$. Since $\mathfrak{A}$ is Schafer genetic every element $\mathcal{T} \in \mathfrak{N}^*$ has the same characteristic polynomial as the zero mapping of $\mathfrak{A}$ into itself, i.e. $f(\lambda) = \lambda^{m+1}$.
Therefore each $\mathcal{T} \in \mathfrak{N}^*$ is nilpotent and thus, since $\mathfrak{N}^*$ is a finite dimensional associative algebra, $\mathfrak{N}^*$ is nilpotent itself. By a theorem of Albert [8] the ideal $\mathfrak{N}$ of $\mathfrak{A}$ is also nilpotent. □

B. Idempotents and sequences of powers

We have seen that the various properties introduced in section 3 can be carried over to the non commutative case in various ways. Of course some implications do not remain valid. For reasons which will become obvious immediately, we restrict our considerations to Gonshor genetic algebras. The essential results on idempotents and sequences of powers in genetic algebras in section 4 are not based on the hypothesis of commutativity. The essential property is that the algebra has a basis with respect to which "sufficiently many" multiplication constants vanish.

In the following let $\mathfrak{A}$ be an (m+1)-dimensional algebra which need not be commutative nor associative. Let $\mathfrak{A}$ be Gonshor genetic. Let c_o, $c_1,\ldots,c_m$ be a canonical basis over an extension $\mathbb{L}$ of $\mathbb{K}$. Let

$$c_i c_j = \sum_{k=0}^{m} \lambda_{ijk} c_k, \qquad i,j=0,\ldots,m,$$

be the multiplication table of $\mathfrak{A}_{\mathbb{L}}$. Since $\mathfrak{A}$ is Gonshor genetic we have

$$\lambda_{ooo} = 1 \qquad \lambda_{oik} = \lambda_{iok} = 0 \qquad \text{for } k < i$$

and

$$\lambda_{ijk} = 0 \qquad \text{for } k \leq \max(i,j) \qquad \text{if } i,j > 0.$$

The necessary and sufficient conditions for an element $e = \sum_{i=0}^{m} \eta_i c_i \in \mathfrak{A}_{\mathbb{L}}$ to be an idempotent assume the form

$$\eta_o = 1$$

$$\eta_k = (\lambda_{okk} + \lambda_{kok})\eta_k + \sum_{i,j=0}^{k-1} \eta_i \lambda_{ijk} \eta_j, \quad k=1,\ldots,m. \qquad (5.3)$$

We see that the 2-fold train roots in the commutative case (4.1) are now replaced by the sums $\lambda_{okk} + \lambda_{kok}$, k=1,...,m, of corresponding train roots. Therefore we obtain similar results as in the commutative case if we replace the conditions $\lambda_{oii} = 1/2$ or $\neq 1/2$ by the hypothesis $\lambda_{oii} + \lambda_{ioi} = 1$ or $\neq 1$ in the non commutative case. We formulate results similar to 4.4, 4.6 and 4.9 without giving proofs.

<u>Theorem 5.18.</u> Let $\mathfrak{A}$ be a Gonshor genetic algebra with train roots $\lambda_{ooo} = 1, \lambda_{oii}, \lambda_{ioi}$, i=1,...,m.
If $\lambda_{oii} + \lambda_{ioi} \neq 1$ for all i=1,...,m, then $\mathfrak{A}_{\mathbb{L}}$, where $\mathbb{L}$ is an algebraic extension field of $\mathbb{K}$, contains exactly one idempotent.

Theorem 5.19. Let $\mathfrak{A}$ be a Gonshor genetic algebra. Assume that the train roots satisfy the equation $\lambda_{oii} + \lambda_{ioi} = 1$ exactly for the indices $i=t_1,\dots,t_r$ with $1 \leq t_1 < \dots < t_r \leq m$. If the product $c_i c_j$ does not have a component in the direction of c_{t_ρ} for $i,j=0,1,\dots,t_\rho-1$, $\rho=1,\dots,r$, then $\mathfrak{A}_{\mathbb{L}}$ has an r-parametric family of idempotents.

Theorem 5.20. Let $\mathfrak{A}$ be a Gonshor genetic algebra with train roots $\lambda_{ooo} = 1$, λ_{oii}, λ_{ioi}, $i=1,\dots,m$.
If for all train roots the condition $\lambda_{oii} + \lambda_{ioi} \neq 1$ is satisfied then $\mathfrak{A}_{\mathbb{L}}$ has exactly one idempotent.
If for some train roots $\lambda_{oii} + \lambda_{ioi} = 1$ holds then $\mathfrak{A}_{\mathbb{L}}$ has either no idempotents or at least a 1-parametric family of idempotents.

Since the definition of an algebra with a genetic realization is not restricted to the commutative case we have similar results as in theorem 4.11 and 4.12.

Theorem 5.21. If $\mathfrak{A}$ has a genetic realization then $\mathfrak{A}$ has at least one idempotent and this idempotent is in the simplex $\mathfrak{T}_m$.

Theorem 5.22. Let $\mathfrak{A}$ be a Gonshor genetic algebra over $\mathbb{R}$ ($\mathbb{C}$) which has a genetic realization. If the left and right train roots $\lambda_{oii}, \lambda_{ioi}$, $i=0,\dots,m$, of $\mathfrak{A}$ satisfy the conditions $\lambda_{oii} + \lambda_{ioi} \neq 1$ then $\mathfrak{A}_{\mathbb{C}}$ has exactly one idempotent.
This idempotent is in $\mathfrak{T}_m$.

Theorem 4.14 can also be carried over. Modifying the proof of that theorem we obtain

Theorem 5.23. Let $\mathfrak{A}$ be a Gonshor genetic algebra with a genetic realization. Then the left and right train roots satisfy the conditions

$$|\lambda_{oii} + \lambda_{ioi}| \leq 1, \quad |\lambda_{oii} - \lambda_{ioi}| \leq 1, \quad i=1,\dots,m.$$

Proof. From theorem 5.22 we know that the simplex $\mathfrak{T}_m$ of $\mathfrak{A}$ ($\mathfrak{A}_{\mathbb{C}}$) contains an idempotent y. We can choose y as first element of a canonical basis $c_o, c_1, \dots, c_m$.
Since $\mathfrak{A}$ has a genetic realization, $\mathfrak{A}$ has a natural basis $a_o, a_1, \dots, a_m$. We denote the l_1-norms of $\mathfrak{A}$ with respect to these bases by $\|\cdot\|_c$ and $\|\cdot\|_a$, respectively. There is a constant $\beta > 0$ such that $\|x\|_c \leq \beta \|x\|_a$ for all $x \in \mathfrak{A}$.

Let $x = \sum_{i=0}^{m} \xi_i c_i$, $z = \sum_{i=0}^{m} \zeta_i c_i \in \mathcal{V}_m$, hence $\xi_o = \zeta_o = 1$. Instead of plenary powers of x we consider the sequences $(xz)^{[r]}$ and $(zx)^{[r]}$ defind by

$$(xz)^{[1]} := xz \qquad (zx)^{[1]} := zx$$

$$(xz)^{[r+1]} := (xz)^{[r]}(zx)^{[r]}, \quad (zx)^{[r+1]} := (zx)^{[r]}(xz)^{[r]}, \quad r \in \mathbb{N}. \tag{5.4}$$

For simplicity we denote the coordinates with respect to the canonical basis of the iterates in the same way as the coordinates of plenary powers. Thus let

$$(xz)^{[r]} = \sum_{i=0}^{m} \xi_i^{[r]} c_i \qquad (zx)^{[r]} = \sum_{i=0}^{m} \zeta_i^{[r]} c_i, \quad r \in \mathbb{N}.$$

These coordinates satisfy the difference equations

$$\begin{aligned}
\xi_o^{[r]} &= 1 \\
\zeta_o^{[r]} &= 1 \\
\xi_k^{[r+1]} &= \lambda_{okk}\xi_k^{[r]} + \lambda_{kok}\zeta_k^{[r]} + \sum_{i,j=1}^{k-1} \xi_i^{[r]} \lambda_{ijk} \zeta_j^{[r]}, \quad r \in \mathbb{N}, \quad k=1,\dots,m, \\
\zeta_k^{[r+1]} &= \lambda_{kok}\xi_k^{[r]} + \lambda_{okk}\zeta_k^{[r]} + \sum_{i,j=0}^{k-1} \zeta_i^{[r]} \lambda_{ijk} \xi_j^{[r]}, \quad r \in \mathbb{N}, \quad k=1,\dots,m.
\end{aligned} \tag{5.5}$$

Let $k \in \{1,\dots,m\}$. Choose $x,y \in \mathcal{V}_m$ such that $\xi_1 = \dots = \xi_{k-1} = 0$ and $\zeta_1 = \dots = \zeta_{k-1} = 0$. Then the difference equations become linear. In matrix notation

$$\begin{pmatrix} \xi_k^{[r+1]} \\ \zeta_k^{[r+1]} \end{pmatrix} = \begin{pmatrix} \lambda_{okk} & \lambda_{kok} \\ \lambda_{kok} & \lambda_{okk} \end{pmatrix} \begin{pmatrix} \xi_k^{[r]} \\ \zeta_k^{[r]} \end{pmatrix}, \quad r \in \mathbb{N}.$$

The solution of this matrix equations is

$$\xi_k^{[r+1]} = \tfrac{1}{2}(\xi_k + \zeta_k)(\lambda_{okk} + \lambda_{kok})^r + \tfrac{1}{2}(\xi_k - \zeta_k)(\lambda_{okk} - \lambda_{kok})^r$$

$$\zeta_k^{[r+1]} = \tfrac{1}{2}(\xi_k + \zeta_k)(\lambda_{okk} + \lambda_{kok})^r - \tfrac{1}{2}(\xi_k - \zeta_k)(\lambda_{okk} - \lambda_{kok})^r .$$

In view of $(xz)^{[r]}$, $(zx)^{[r]} \in \mathcal{V}_m$ we have

$$\|(xz)^{[r]}\|_a = 1 \quad \text{and} \quad \|(zx)^{[r]}\|_a = 1.$$

Thus

$$\|(xz)^{[r]}\|_c = \sum_{i=0}^{m} |\xi_i^{[r]}| \leq \beta \text{ and } \|(zx)^{[r]}\|_c = \sum_{i=0}^{m} |\zeta_i^{[r]}| \leq \beta,$$

from where $|\xi_i^{[r]}|, |\zeta_i^{[r]}| \leq \beta$ for i=0,...,m and $n \in \mathbb{N}$. Choose ξ_k and ζ_k such that $\xi_k + \zeta_k \neq 0$ and $\xi_k - \zeta_k \neq 0$. Then $|\lambda_{okk} + \lambda_{kok}| \leq 1$ and

$|\lambda_{okk} - \lambda_{kok}| \leq 1$.

Since $k \in \{1,\dots,m\}$ was arbitrary, the assertion is proved.□

For dim $\mathfrak{A} = 2$ the necessary condition of theorem 5.23 together with the existence of an idempotent is sufficient for a Gonshor genetic algebra to have a genetic realization.

Theorem 5.24. Let $\mathfrak{A}$ be a 2-dimensional Gonshor genetic algebra over $\mathbb{R}$ with train roots $\lambda_{ooo} = 1$, λ_{o11}, and λ_{1o1}.
Then the following assertions are equivalent

(1) $\mathfrak{A}$ has a genetic realization

(2) $\mathfrak{A}$ has an idempotent and $|\lambda_{o11} + \lambda_{1o1}|, |\lambda_{o11} - \lambda_{1o1}| \leq 1$.

Proof. I. Condition (2) follows from (1) with the help of theorems 5.21 and 5.23.

II. Assume (2). Then there is a basis c_o, c_1 of $\mathfrak{A}$ with the multiplication table

$$c_o^2 = c_o, \quad c_oc_1 = \lambda_{o11}c_1, \quad c_1c_o = \lambda_{1o1}c_1, \quad c_1^2 = 0.$$

Let $a_o := c_o + \alpha c_1$, $a_1 := c_o + \beta c_1$ with $\alpha \neq \beta$ be another basis of $\mathfrak{A}$.
Then the multiplication table with respect to this basis reads

$$a_o^2 = \frac{1}{\alpha-\beta}\left[(\alpha(\lambda_{o11}+\lambda_{1o1})-\beta)a_o + \alpha(1-(\lambda_{o11}+\lambda_{1o1}))a_1\right]$$
$$a_oa_1 = \frac{1}{\alpha-\beta}\left[(\beta\lambda_{o11}+\alpha\lambda_{1o1}-\beta)a_o + (\alpha-(\beta\lambda_{o11}+\alpha\lambda_{1o1}))a_1\right]$$
$$a_1a_o = \frac{1}{\alpha-\beta}\left[(\alpha\lambda_{o11}+\beta\lambda_{1o1}-\beta)a_o + (\alpha-(\alpha\lambda_{o11}+\beta\lambda_{1o1}))a_1\right]$$
$$a_1^2 = \frac{1}{\alpha-\beta}\left[\beta((\lambda_{o11}+\lambda_{1o1})-1)a_o + (\alpha-\beta(\lambda_{o11}+\lambda_{1o1}))a_1\right]$$

Thus $\mathfrak{A}$ has a genetic realization if and only if there are $\alpha,\beta \in \mathbb{R}$, $\alpha > \beta$, such that the following inequalities are satisfied:

$$\beta \leq \beta(\lambda_{o11} + \lambda_{1o1}) \leq \alpha$$

$$\beta \leq \alpha(\lambda_{o11} + \lambda_{1o1}) \leq \alpha$$

$$\beta \leq \beta\lambda_{o11} + \alpha\lambda_{1o1} \leq \alpha$$

$$\beta \leq \alpha\lambda_{o11} + \beta\lambda_{1o1} \leq \alpha \quad .$$

In view of $|\lambda_{o11} + \lambda_{1o1}| \leq 1$ and $|\lambda_{o11} - \lambda_{1o1}| \leq 1$ these inequalities can be satisfied with $\alpha = 1$, $\beta = -1$. □

Finally theorems 4.18 and 4.19 on the convergence of plenary powers can be carried over to the non commutative case, whereby the train roots have to be replaced by the arithmetic means of corresponding left and right train roots.
We quote these theorems for completeness.

Theorem 5.25. Let $\mathfrak{A}$ be an (m+1)-dimensional Gonshor genetic algebra over $\mathbb{R}$ or $\mathbb{C}$.
If the train roots λ_{oii}, λ_{ioi} satisfy $|\lambda_{oii} + \lambda_{ioi}| < 1$, $i=1,\dots,m$, then the sequence of plenary powers of every element $x \in \mathfrak{A}$ of weight 1 converges to the uniquely determined idempotent of $\mathfrak{A}$.

Theorem 5.26. Let $\mathfrak{A}$ be an (m+1)-dimensional Gonshor genetic algebra with a canonical basis $c_o,\dots,c_m$ over $\mathbb{R}$ or $\mathbb{C}$.
Assume that the train roots $\lambda_{ooo} = 1$, λ_{oii}, λ_{ioi}, $i=1,..,m$, satisfy $\lambda_{oii} + \lambda_{ioi} = 1$ exactly for $i = t_1,\dots,t_r$, $t_1 < t_2 < \dots < t_r$, and that the other train roots except λ_{ooo} satisfy $|\lambda_{ojj} + \lambda_{joj}| < 1$. If the product $c_i c_j$ does not have a component in the direction of c_{t_ϱ} , $i,j=0,\dots,t_\varrho - 1$, $\varrho = 1,\dots,r$, then for each element $x = \sum_{i=0}^{m} \xi_i c_i$ of weight 1 the sequence of plenary powers converges towards that idempotent of $\mathfrak{A}$ ($\mathfrak{A}_{\mathbb{C}}$) determined by $\xi_{t_1}, \dots, \xi_{t_r}$.

Remark. If $\mathfrak{A}$ is a non commutative algebra over $\mathbb{K}$, char $\mathbb{K} \neq 2$, which is Gonshor genetic, then the commutative algebra $\mathfrak{A}^+$ with multiplication defined by (2.15) is genetic. Since idempotents and plenary powers are in $\mathfrak{A}^+$ the assertions 5.21, 5.22, 5.25 and 5.26 would have followed directly from section 4. However this is not true for theorems 5.23 and 5.24. For this reason we have chosen the presentation above.

6. CONSTRUCTION OF NEW ALGEBRAS

Suppose we have a "complex" genetic situation in which each particular feature can be described by an algebra.
Then we can try to construct an algebra which models all the features under consideration. Such an approach is successful in various biological situations. We shall encounter the following examples:
A population of polyploid individuals with chromosome and chromatid segregation, cf. section 7C; a population of diploid individuals in which we consider several partially coupled loci where the segregation rates with respect to each locus are known, cf. section 7B; to construct a model for the zygotic population if a model for the gametic population is given, cf. section 6C; to introduce mutation in a population of polyploid individuals, cf. section 7C.

A. Linear combinations of algebras

In the following let $\mathcal{V}$ be an n-dimensional vector space over $\mathbb{K}$. If $\mathfrak{A}$ is an algebra over $\mathcal{V}$ which need not be associative nor commutative then we denote the multiplication in $\mathfrak{A}$ by

$$(x,y) \longmapsto (xy)_{\mathfrak{A}} \quad , \qquad x,y \in \mathcal{V} . \tag{6.1}$$

As indicated by Heuch [59] for the commutative case the linear construction of algebras has the following properties.

Theorem 6.1. Let $\mathcal{A}(\mathcal{V})$ be the family of all algebras over $\mathcal{V}$. For $\mathfrak{A}$, $\mathfrak{B} \in \mathcal{A}(\mathcal{V})$ and $\alpha \in \mathbb{K}$ the sum $\mathfrak{A} + \mathfrak{B}$ and the scalar multiple $\alpha\mathfrak{A}$ are defined as algebras over $\mathcal{V}$ with multiplications given by

$$(x,y) \longmapsto (xy)_{\mathfrak{A}+\mathfrak{B}} := (xy)_{\mathfrak{A}} + (xy)_{\mathfrak{B}} \quad , \qquad x,y \in \mathcal{V} ,$$
$$(x,y) \longmapsto (xy)_{\alpha\mathfrak{A}} := \alpha(xy)_{\mathfrak{A}} \quad , \qquad x,y \in \mathcal{V} .$$

(1) Then $\mathcal{A}(\mathcal{V})$ is a vector space over $\mathcal{V}$ which is called the vector space of all algebras over $\mathcal{V}$. The zero element of this vector space is the zero algebra $\mathcal{O}$ over $\mathcal{V}$.
(2) The family $\mathcal{A}_c(\mathcal{V})$ of all commutative algebras over $\mathcal{V}$ forms a subspace of $\mathcal{A}(\mathcal{V})$.

After the sum and scalar multiples of algebras have been defined in theorem 6.1 it is natural to call

$$\sum_{\varrho=1}^{r} \alpha_\varrho \mathfrak{A}_\varrho = \alpha_1 \mathfrak{A}_1 + \dots + \alpha_r \mathfrak{A}_r, \quad \alpha_1,\dots,\alpha_r \in \mathbb{K}, \tag{6.2}$$

a linear combination of the algebras $\mathfrak{A}_1,\dots,\mathfrak{A}_r \in \mathcal{A}(\mathfrak{V})$. The multiplication in $\sum_{\varrho=1}^{r} \alpha_\varrho \mathfrak{A}_\varrho$ is given by

$$(x,y) \longmapsto (xy)_{\sum_{\varrho=1}^{r} \alpha_\varrho \mathfrak{A}_\varrho} = \sum_{\varrho=1}^{r} \alpha_\varrho (xy)_{\mathfrak{A}_\varrho} \quad . \tag{6.3}$$

<u>Remark.</u> Similarly Rees [97] has introduced the notion of a system of algebras spanned by $\mathfrak{A}_1,\dots,\mathfrak{A}_r$. In the case where $\mathbb{K}$ is $\mathbb{R}$ or $\mathbb{C}$ and the scalars $\alpha_1,\dots,\alpha_r$ are nonnegative and sum up to 1 Holgate [61] calls (6.2) a mixture of algebras. Heuch [59] uses this notion also in the case where the scalars may be negative and he uses the term "proper mixture" if the coefficients are nonnegative. We prefer as in [114] to call such a linear combination a convex combination of algebras.

<u>Definition 6.2.</u> Algebras $\mathfrak{A}_1,\dots,\mathfrak{A}_r \in \mathcal{A}(\mathfrak{V})$ are called <u>linearly independent</u> if only the trivial linear combination of these algebras is the zero algebra.

Let $a_1,\dots,a_n$ be a basis of $\mathfrak{V}$. Then the algebras $\mathfrak{A}_1,\dots,\mathfrak{A}_r \in \mathcal{A}(\mathfrak{V})$ are linearly independent if and only if

$$(a_i a_j)_{\sum_{\varrho=1}^{r} \alpha_\varrho \mathfrak{A}_\varrho} = \sum_{\varrho=1}^{r} \alpha_\varrho (a_i a_j)_{\mathfrak{A}_\varrho} = 0 \quad \text{for all } i,j=1,\dots,n,$$

implies that $\alpha_1 = \alpha_2 = \dots = \alpha_r = 0$.

With respect to a fixed basis $a_1,\dots,a_n$ of $\mathfrak{V}$ any algebra $\mathfrak{A} \in \mathcal{A}(\mathfrak{V})$ is uniquely determined by the array $(\gamma_{ijk}^{(\mathfrak{A})}) \in \mathbb{K}^{n^3}$ of its multiplication constants with respect to this basis. Hence the algebras $\mathfrak{A}_1,\dots,\mathfrak{A}_r \in \mathcal{A}(\mathfrak{V})$ are linearly independent if the arrays $(\gamma_{ijk}^{(\mathfrak{A}_1)}),\dots,(\gamma_{ijk}^{(\mathfrak{A}_r)})$ of their multiplication constants are linearly independent elements of the vector space $\mathbb{K}^{n^3}$.

If we want to determine the dimension of the smallest linear manifold of $\mathcal{A}(\mathfrak{V})$ which contains a given set of algebras then we can either try to find out which multiplication constants of these algebras can be arbitrarily chosen or we can, as Heuch [59] did, find linearly independent algebras which span the corresponding linear subspace of this linear manifold. In proving theorem 6.6 we shall use the first method which does not require further definitions.
Before approaching theorem 6.6 we collect some assertions about linear combinations of baric algebras, of algebras with a genetic realization and of Gonshor genetic algebras. These assertions become immediately obvious if one recalls how these algebras are characterized by properties of the multiplication constants with respect to a given basis, cf. Lemma 1.11 and definitions 1.5 and 5.9 and for the commutative case Heuch [59] .

Theorem 6.3. Let $\mathfrak{A}_1,\dots,\mathfrak{A}_r$ be baric algebras over $\mathfrak{V}$ with the same weight homomorphism ω.
Then every linear combination $\sum_{\varrho=1}^{r}\alpha_\varrho\mathfrak{A}_\varrho$ with $\sum_{\varrho=1}^{r}\alpha_\varrho = 1$ is a baric algebra with weight homomorphism ω.

Theorem 6.4. Let $\mathfrak{V}$ be a vector space over $\mathbb{R}$ or $\mathbb{C}$. Let $\mathfrak{A}_1,\dots,\mathfrak{A}_r$ be algebras over $\mathfrak{V}$ which have a genetic realization with respect to the same basis $a_1,\dots,a_n$ of $\mathfrak{V}$. Then every convex combination of these algebras has a genetic realization with respect to $a_1,\dots,a_n$.

Theorem 6.5. Let $\mathfrak{A}_1,\dots,\mathfrak{A}_r$ be (weakly left,left) Gonshor genetic algebras with respect to a common basis of $\mathfrak{A}_{\mathbb{L}}$, where $\mathbb{L}$ is a suitable extension field of $\mathbb{K}$. Then every linear combination $\sum_{\varrho=1}^{r}\alpha_\varrho\mathfrak{A}_\varrho$ with $\sum_{\varrho=1}^{r}\alpha_\varrho = 1$ is (weakly left, left) Gonshor genetic with respect to this basis and the left (and right) train roots are the corresponding linear combinations of the left (right) train roots of $\mathfrak{A}_1,\dots,\mathfrak{A}_r$.

These theorems enable us to determine the dimension of the smallest linear manifold which contains all algebras with a given property.
In the following theorem 6.6 the results (1), (3) and (6) have been found by Heuch [59] for the commutative case.

Theorem 6.6. Let $\mathfrak{V}$ be an n-dimensional vector space over a field $\mathbb{K}$, $n\geq 2$. Let $\mathcal{A}_{\mathcal{E}}(\mathfrak{V})$ be the family of algebras $\mathfrak{A}$ over $\mathfrak{V}$ with property $\mathcal{E}$. Let $\mathcal{L}_{\mathcal{E}}$ be the smallest linear manifold of $\mathcal{A}(\mathfrak{V})$ which contains $\mathcal{A}_{\mathcal{E}}(\mathfrak{V})$.

Then the dimension of $\mathcal{L}_\ell$ is given by the right column of the following table. In case (2) one has to understand $\mathbb{K} = \mathbb{R}$.

	ℓ	$\dim \mathcal{L}_\ell$
(1)	-	n^3
	commutative	$\frac{1}{2} n^2 (n+1)$
(2)	with genetic realization and the same natural basis	$(n-1)n^2$
	and commutative	$\frac{1}{2}(n-1)n(n+1)$
(3)	baric with the same weight homomorphism	$(n-1)n^2$
	and commutative	$\frac{1}{2}(n-1)n(n+1)$
(4)	weakly left Gonshor genetic with the same canonical basis	$\frac{1}{2}(n-1)(n^2+2)$
(5)	left Gonshor genetic with the same basis	$\frac{1}{3}(n-1)(n^2 + 4n - 3)$
(6)	Gonshor genetic with the same basis	$\frac{1}{6}(n-1)(2n^2+5n)$
	and commutative	$\frac{1}{6}(n-1)(n^2+n+6)$

<u>Proof.</u> I. The assertions (1) are trivial.

II. In view of theorems 6.3 - 6.5 it is sufficient to determine how many linearly independent arrays of multiplication constants are necessary and sufficient to describe all algebras with property ℓ.

a) In case (2) the multiplication arrays are given by all n-tuples of stochastic matrices over $\mathbb{R}$. If the algebras are commutative these have the additional property that the i-th column of the j-th matrix equals the j-th column of the i-th matrix for $i,j=1,\dots,n$. Thus the dimension of $\mathcal{L}_\ell$ is $(n-1)n^2$, and $(n-1)\frac{1}{2}n(n+1)$ in the commutative case.

b) By assumption all algebras in row (3) - (6) are baric with the same weight homomorphism ω. Thus we choose a basis $c_o, c_1, \dots, c_m$, $m:=n-1$, of $\mathcal{V}$ ($\mathcal{V}_{\mathbb{L}}$) with $\omega(c_o) = 1$, $\omega(c_i) = 0$, $i=1,\dots,m$.

The multiplication constants $\lambda_{ijk}^{(\mathfrak{a})}$, $i,j,k=0,\dots,m$, of any baric algebra $\mathfrak{a}$ with weight homomorphism ω satisfy the conditions

$$\lambda^{(\alpha)}_{ooo} = 1, \quad \lambda^{(\alpha)}_{ioo} = 0, \quad \lambda^{(\alpha)}_{ojo} = 0, \quad \lambda^{(\alpha)}_{ijo} = 0, \quad i,j=1,\dots,m,$$

and are arbitrary otherwise. In the commutative case they satisfy the additional condition $\lambda^{(\alpha)}_{ijk} = \lambda^{(\alpha)}_{jik}$, $i,j,k=0,\dots,m$.

Thus the dimension of $\mathcal{L}_\ell$ is $n^3 - n^2 = (n-1)n^2$, and $\frac{1}{2}(n-1)n(n+1)$ in the commutative case.

c) Let the basis $c_o, c_1, \dots, c_m$ be chosen such that all algebras $\mathfrak{A}$ are (weakly left, left) Gonshor genetic with respect to this basis. Then the multiplication constants $\lambda^{(\alpha)}_{ijk}$, $i,j,k=0,\dots,m$, satisfy the conditions of definition 5.9 ((1),(2)), (3).

By simply counting the indeterminate multiplication constants we obtain the assertions (4) - (6) in the non commutative case. If the algebra are also commutative these properties are equivalent with 3.11(i)-(iii). In any case, commutative or not, the dimension of the family of algebras with a genetic realization is larger than the dimension of the family of (weakly left, left) Gonshor genetic algebras.□

Holgate [61] who was the first to consider convex combinations of commutative algebras in connection with genetical problems suggests that a convex combination of special train algebras is a special train algebra. Gonshor [47] gives a counterexample, cf. example 6.7. (In most cases misunderstandings of this kind originate from Etherington's erroneous statement that the existence of a canonical basis ensures that the algebra is a special train algebra.)

Example 6.7. A convex combination of two commutative special train algebras with the same weight homomorphism need not be a special train algebra.

Let $\mathcal{W}$ be the free $\mathbb{R}$-module generated by c_o, c_1, c_2, c_3. In $\mathcal{W}$ two commutative multiplications are defined by:

1	c_o	c_1	c_2	c_3
c_o	c_o	c_1	c_3	c_3
c_1		c_2	c_3	0
c_2			0	0
c_3				0

2	c_o	c_1	c_2	c_3
c_o	c_o	c_1	c_3	c_3
c_1		c_2	$-c_3$	0
c_2			0	0
c_3				0

whereby we obtain the algebras $\mathfrak{A}_1$ and $\mathfrak{A}_2$. In both cases $\mathfrak{N} := \langle c_1,c_2,c_3\rangle$ is an ideal not containing $\mathfrak{A}_1^2$, $\mathfrak{A}_2^2$; thus $\mathfrak{A}_1$ and $\mathfrak{A}_2$ are baric with the same weight homomorphism. The principal powers of $\mathfrak{N}$ in $\mathfrak{A}_i$ satisfy

$$\mathfrak{N}^2_{\mathfrak{A}_i} = \langle c_2,c_3\rangle \quad , \quad \mathfrak{N}^3_{\mathfrak{A}_i} = \langle 0\rangle \; , \qquad i=1,2,$$

and are such ideals in $\mathfrak{A}_1$ and $\mathfrak{A}_2$, respectively. Thus $\mathfrak{A}_1$ and $\mathfrak{A}_2$ are special train algebras.
Now let $\mathfrak{A} := \frac{1}{2}\mathfrak{A}_1 + \frac{1}{2}\mathfrak{A}_2$. Clearly $\mathfrak{A}$ is a baric algebra and again $\mathfrak{N}$ is the kernel of the weight homomorphism. But $\mathfrak{N}^2_{\mathfrak{A}} = \langle c_2\rangle$ is not an ideal of $\mathfrak{A}$, and $\mathfrak{A}$ is not a special train algebra. □

B. Tensor products

Let $\mathfrak{A}_1,\dots,\mathfrak{A}_r$ be algebras over $\mathbb{K}$ with dimensions $n_1,\dots,n_r$ which need not be associative or commutative. Let $\mathfrak{A}_1 \otimes \dots \otimes \mathfrak{A}_r$ be the tensor product of the vector spaces underlying $\mathfrak{A}_1,\dots,\mathfrak{A}_r$ (cf. Greub [49]). Then $\mathfrak{A}_1 \otimes \dots \otimes \mathfrak{A}_r$ is an $n_1\cdot\ldots\cdot n_r$-dimensional vector space over $\mathbb{K}$. Each element of the tensor product can be represented as $\sum_{i\in I} x_{1i} \otimes \dots \otimes x_{ri}$ with linearly independent elements $x_{\varrho i} \in \mathfrak{A}_\varrho$, $i \in I$, for every $\varrho = 1,\dots,r$. If furthermore $a_{\varrho 1},\dots,a_{\varrho n}$ form a basis of $\mathfrak{A}_\varrho$, $\varrho = 1,\dots,r$, then

$$a_{1i_1} \otimes \dots \otimes a_{ri_r} \quad , \qquad i_\varrho = 1,\dots,n_\varrho \; , \; \varrho = 1,\dots,r,$$

form a basis of the tensor product. The following result is well-known.

__Theorem 6.8.__ The tensor product $\mathfrak{A}_1 \otimes \dots \otimes \mathfrak{A}_r$ with the multiplication defined by

$$\Big(\sum_{i\in I} x_{1i} \otimes \dots \otimes x_{ri}\Big)\Big(\sum_{j\in J} y_{1j} \otimes \dots \otimes y_{rj}\Big) = \sum_{\substack{i\in I\\ j\in J}} x_{1i}y_{1j} \otimes \dots \otimes x_{ri}y_{rj} \qquad (6.4)$$

is an algebra. This algebra is called the __tensor product__ of the __algebras__ $\mathfrak{A}_1,\dots,\mathfrak{A}_r$ and is denoted by $\mathfrak{A}_1 \otimes \dots \otimes \mathfrak{A}_r$.

Let the multiplication tables of the factors $\mathfrak{A}_\varrho$ be given by

$$a_{\varrho i}a_{\varrho j} = \sum_{k=1}^{n_\varrho} \lambda^{(\varrho)}_{ijk}\, a_{\varrho k} \; , \qquad i,j=1,\dots,n,$$

then the multiplication table of the tensor product is given by

$$(a_{1i_1}\otimes\dots\otimes a_{ri_r})(a_{1j_1}\otimes\dots\otimes a_{rj_r})$$

$$= \sum_{k_1,\dots,k_r} \lambda^{(1)}_{i_1j_1k_1}\cdot\dots\cdot\lambda^{(r)}_{i_rj_rk_r}\; a_{1k_1}\otimes\dots\otimes a_{rk_r}\,, \qquad (6.5)$$

$$i_\varrho,\ j_\varrho = 1,\dots,n_\varrho\ ,\quad \varrho = 1,\dots,r.$$

Remark. The tensor product of the algebras $\mathfrak{A}_1,\dots,\mathfrak{A}_r$ is isomorphic with the direct product of these algebras in the sense of Albert [5] : With the basis elements $a_{\varrho 1},\dots,a_{\varrho n}$ of $\mathfrak{A}_\varrho$, $\varrho = 1,\dots,r$, form the complexes $a_{1i_1}\cdot a_{2i_2}\cdot\dots\cdot a_{ri_r}$, $i_\varrho = 1,\dots,n_\varrho$, $\varrho = 1,\dots,r$.
Let $\mathfrak{A}_1\cdot\dots\cdot\mathfrak{A}_r$ be the free module over $\mathbb{K}$ generated by these complexes. In $\mathfrak{A}_1\cdot\dots\cdot\mathfrak{A}_r$ define a multiplication by

$$(a_{1i_1}\cdot\dots\cdot a_{ri_r})\ (a_{1j_1}\cdot\dots\cdot a_{1j_r})$$

$$= \sum_{k_1,\dots,k_r} \lambda^{(1)}_{i_1j_1k_1}\cdot\dots\cdot\lambda^{(r)}_{i_rj_rk_r}\; a_{1k_1}\cdot\dots\cdot a_{rk_r}\ .$$

Etherington [27] , [29] and Holgate [64] use the direct product for the description of populations whose individuals may differ in several uncoupled loci. We prefer to use the tensor product which allows us to apply the general theory of tensor products, cf. Greub [49] .

It is well known that the r-fold tensor product of a field $\mathbb{K}$ is isomorphic with $\mathbb{K}$,

$$\mathbb{K}\otimes\dots\otimes\mathbb{K}\cong\mathbb{K}$$

by way of the identification $x_1\otimes\dots\otimes x_r = x_1\dots x_r$.

We introduce the tensor product of algebra homomorphisms : Let $\mathfrak{B}_1,\dots,\mathfrak{B}_r$ be algebras over $\mathbb{K}$ and let $\varphi_\varrho : \mathfrak{A}_\varrho \longrightarrow \mathfrak{B}_\varrho$, $\varrho = 1,\dots,r$, be the algebra homomorphisms. Then by

$$\varphi_1\otimes\dots\otimes\varphi_r\ :\ \mathfrak{A}_1\otimes\dots\otimes\mathfrak{A}_r \longrightarrow \mathfrak{B}_1\otimes\dots\otimes\mathfrak{B}_r$$

with

$$(\varphi_1\otimes\dots\otimes\varphi_r)(x_1\otimes\dots\otimes x_r) = \varphi_1(x_1)\otimes\dots\otimes\varphi_r(x_r)$$

an algebra homomorphism is defined and

$$\mathrm{Im}(\varphi_1 \otimes \ldots \otimes \varphi_r) = \mathrm{Im}\,\varphi_1 \otimes \ldots \otimes \mathrm{Im}\,\varphi_r \,.$$

Now we are prepared to state the following theorem part of which has been proved by Holgate [64] for direct products.

<u>Theorem 6.9.</u> Let $\mathfrak{A}_\varrho$ be a baric algebra with weight homomorphism and corresponding kernel $\mathfrak{N}_\varrho := \ker\omega_\varrho$ for $\varrho = 1,\ldots,r$.
Then the tensor product $\mathfrak{A}_1 \otimes \ldots \otimes \mathfrak{A}_r$ is baric with weight homomorphism

$$\omega := \omega_1 \otimes \ldots \otimes \omega_r$$

and the kernel $\mathfrak{N} := \ker\omega$ can be represented as

$$\mathfrak{N} = \mathfrak{N}_1 \otimes \mathfrak{A}_2 \otimes \ldots \otimes \mathfrak{A}_r + \mathfrak{A}_1 \otimes \mathfrak{N}_2 \otimes \ldots \otimes \mathfrak{A}_r + \mathfrak{A}_1 \otimes \ldots \otimes \mathfrak{A}_{r-1} \otimes \mathfrak{N}_r \,, \qquad (6.6)$$

here "+" denotes the sum of vector spaces.

This representation of $\ker\omega$ leads to the following result. (A similar result has been proved by Holgate [64] for commutative genetic algebras (and not for special train algebras for reasons stated above).

<u>Theorem 6.10.</u> Let $\mathfrak{A}_1,\ldots,\mathfrak{A}_r$ be commutative special train algebras. Then the tensor product $\mathfrak{A} := \mathfrak{A}_1 \otimes \ldots \otimes \mathfrak{A}_r$ is a special train algebra.

<u>Proof.</u> By assumption $\mathfrak{N}_\varrho$ is nilpotent and the plenary powers $\mathfrak{N}_\varrho^{k}$ of $\mathfrak{N}_\varrho$ are ideals of $\mathfrak{A}_\varrho$, $\varrho = 1,\ldots,r$. Since the formation of tensor products is associative we can proceed by induction on r. For r=1 the assertion is trivial. We assume that the assertion has been shown for r-1.
Let $\tilde{\mathfrak{A}}_1 := \mathfrak{A}_1 \otimes \ldots \otimes \mathfrak{A}_{r-1}$ and $\tilde{\omega}_1 := \omega_1 \otimes \ldots \otimes \omega_{r-1}$. Then $\tilde{\mathfrak{N}}_1 := \ker\tilde{\omega}_1$ is nilpotent and the principal powers are ideals of $\tilde{\mathfrak{A}}_1$.
Since the tensor product is associative and distributive we have

$$\mathfrak{A} = \tilde{\mathfrak{A}}_1 \otimes \mathfrak{A}_r \text{ and } \mathfrak{N} = \tilde{\mathfrak{N}}_1 \otimes \mathfrak{A}_r + \tilde{\mathfrak{A}}_1 \otimes \mathfrak{N}_r \,.$$

(In the following calculations we omit $\langle\ \rangle$).
For the second principal power of $\mathfrak{N}$ we have

$$\mathfrak{N}^2 = \tilde{\mathfrak{N}}_1^2 \otimes \mathfrak{A}_r^2 + \tilde{\mathfrak{N}}_1\tilde{\mathfrak{A}}_1 \otimes \mathfrak{N}_r\mathfrak{A}_r + \tilde{\mathfrak{A}}_1^2 \otimes \mathfrak{N}_r^2$$

and by induction it follows that for the k-th principal power

$$\mathfrak{N}^k = \tilde{\mathfrak{N}}_1^k \otimes \mathfrak{A}_r^k + \sum (\ldots(\tilde{\mathfrak{N}}_1 \mathfrak{J}_2)\mathfrak{J}_3 \ldots \mathfrak{J}_k) \otimes (\ldots \mathfrak{N}_r \mathfrak{K}_2)\mathfrak{K}_3 \ldots \mathfrak{K}_k) + \tilde{\mathfrak{A}}_1^k \otimes \mathfrak{N}_r^k, \ k \geq 2,$$

where $\mathfrak{J}_2, \ldots, \mathfrak{J}_k \in \{\tilde{\mathfrak{N}}_1, \tilde{\mathfrak{A}}_1\}$ and $\mathfrak{K}_2, \ldots, \mathfrak{K}_k \in \{\mathfrak{N}_r, \mathfrak{A}_r\}$ are arbitrary and the summation is extended over all possible choices.

Clearly the algebras $\tilde{\mathfrak{N}}_1^k \otimes \mathfrak{A}_r^k$ and $\tilde{\mathfrak{A}}_1^k \otimes \mathfrak{N}_r^k$ are ideals of $\tilde{\mathfrak{A}}_1 \otimes \mathfrak{A}_r$.
For any other term of the above sum

$$[(\ldots(\tilde{\mathfrak{N}}_1 \mathfrak{J}_2)\mathfrak{J}_3 \ldots \mathfrak{J}_k) \otimes (\ldots(\mathfrak{N}_r \mathfrak{K}_2)\mathfrak{K}_3 \ldots \mathfrak{K}_k)]\, \tilde{\mathfrak{A}}_1 \otimes \mathfrak{A}_r$$

$$\subseteq ((\ldots(\tilde{\mathfrak{N}}_1 \mathfrak{J}_3) \ldots \mathfrak{J}_k)\mathfrak{A}_1) \otimes ((\ldots(\mathfrak{N}_r \mathfrak{K}_3) \ldots \mathfrak{K}_k)\mathfrak{A}_r),$$

holds, where the last mentioned set is again one of the summands in the above representation of $\mathfrak{N}^k$. Thus the principal powers of $\mathfrak{N}$ are ideals of $\mathfrak{A}$. Let s and t be the indices of nilpotency of $\tilde{\mathfrak{N}}_1$ and $\mathfrak{N}_r$. Let $p := \min(s,t)$, $q := \max(s,t)$.

Then $\tilde{\mathfrak{N}}_1^{q+1} = \mathfrak{N}_r^{q+1} = \langle 0 \rangle$. Therefore we obtain

$$\mathfrak{N}^{q+1} = (\ldots(\mathfrak{N}_1 \mathfrak{J}_1)\mathfrak{J}_2 \ldots \mathfrak{J}_{q+1}) \otimes (\ldots(\mathfrak{N}_r \mathfrak{K}_2)\mathfrak{K}_3 \ldots \mathfrak{K}_{q+1}) \subseteq \mathfrak{N}_1 \otimes \mathfrak{N}_r.$$

Since $\mathfrak{N}_1 \otimes \mathfrak{N}_r$ is nilpotent of index p we have that $\mathfrak{N}$ is nilpotent of an index not exceeding p+q+1 and thus the assertion is proved. □

We have not been able to derive this implication or a similar implication for (weakly) left special train algebras in the non commutative case, except for the case where the algebras $\mathfrak{A}_1, \ldots, \mathfrak{A}_r$ coincide.
However by an argument similar to the preceding proof one can show that the kernel of $\omega := \omega_1 \otimes \ldots \otimes \omega_r$ is left nilpotent if the kernels $\mathfrak{N}_1, \ldots, \mathfrak{N}_r$ of the weight homomorphisms $\omega_1, \ldots, \omega_r$ have this property.

<u>Theorem 6.11.</u> Let $\mathfrak{A}_1, \ldots, \mathfrak{A}_r$ be (weakly left, left) Gonshor genetic algebras. Then the tensor product $\mathfrak{A}_1 \otimes \ldots \otimes \mathfrak{A}_r$ is (weakly left, left) Gonshor genetic and the left (and right) train roots are all products of r factors where the r factors run over all left (right, respectively) train roots of the algebras $\mathfrak{A}_1, \ldots, \mathfrak{A}_r$.

<u>Proof.</u> Let $c_{\rho 0}, \ldots, c_{\rho m}$, $n_\rho = m_\rho + 1$, be a canonical basis of $\mathfrak{A}_\rho$ with respect to which the algebra has the multiplication constants $\lambda_{ijk}^{(\rho)}$, $i,j,k = 0,\ldots,m_\rho$, $\rho = 1,\ldots,r$.

We show that $c_{1i_1} \otimes \ldots \otimes c_{ri_r}$, $i_\rho = 0,\ldots,m_\rho$, $\rho = 1,\ldots,r$, is a canonical basis of $\mathfrak{A}_1 \otimes \ldots \otimes \mathfrak{A}_r$ with respect to the lexicographic order of the index tuples as given by $(i_1,\ldots,i_r) \leq (k_1,\ldots,k_r)$ if there is an $s \in \{1,\ldots,r\}$ such that $i_1 = k_1,\ldots,i_{s-1} = k_{s-1}$ and $i_s \leq k_s$. The multiplication table of $\mathfrak{A}_1 \otimes \ldots \otimes \mathfrak{A}_r$ is

$$(c_{1i_1} \otimes \ldots \otimes c_{ri_r})(c_{1j_1} \otimes \ldots \otimes c_{rj_r})$$

$$= \sum_{k_1,\ldots,k_r} \lambda^{(1)}_{i_1j_1k_1} \cdots \lambda^{(r)}_{i_rj_rk_r} \; c_{1k_1} \otimes \ldots \otimes c_{rk_r} \; ,$$

$$i_\rho,\; j_\rho = 0,\ldots,m \; , \; \rho = 1,\ldots,r.$$

I. Assume all algebras $\mathfrak{A}_1,\ldots,\mathfrak{A}_r$ are weakly left Gonshor genetic. We have $\lambda^{(1)}_{000} \cdot \ldots \cdot \lambda^{(r)}_{000} = 1 \cdot \ldots \cdot 1 = 1$.

Let $(k_1,\ldots,k_r) < (j_1,\ldots,j_r)$. Then there is an index s with $k_s < j_s$. Since $\mathfrak{A}_s$ is weakly left genetic we have $\lambda^{(s)}_{oj_sk_s} = 0$, thus $\lambda^{(1)}_{oj_1k_1} \cdots \lambda^{(s)}_{oj_sk_s} \cdots \lambda^{(r)}_{ij_rk_r} = 0$.

Let $(i_1,\ldots,i_r) > (0,\ldots,0)$, i.e. there is an index t with $i_t > 0$. Furthermore let $(k_1,\ldots,k_r) \leq (j_1,\ldots,j_r)$. Then either $(k_1,\ldots,k_r) = (j_1,\ldots,j_r)$ or $(k_1,\ldots,k_r) < (j_1,\ldots,j_r)$.
In the latter case there is an s with $k_s < j_s$. In the first case we have $\lambda^{(t)}_{i_tj_tk_j} = 0$ and in the second case $\lambda^{(s)}_{i_sj_sk_s} = 0$.

In any case we have $\lambda^{(1)}_{i_1j_1k_1} \cdots \lambda^{(r)}_{i_rj_rk_r} = 0$ and thus $\mathfrak{A}_1 \otimes \ldots \otimes \mathfrak{A}_r$ is weakly left Gonshor genetic. The left train roots of $\mathfrak{A}_1 \otimes \ldots \otimes \mathfrak{A}_r$ are all r-fold products $\lambda^{(1)}_{oj_1j_1} \cdots \lambda^{(r)}_{oj_rj_r}$, $j_\rho = 0,\ldots,m_\rho$, $\rho = 1,\ldots,r$, of the left train root of $\mathfrak{A}_1,\ldots,\mathfrak{A}_r$.

II. Let $\mathfrak{A}_1,\ldots,\mathfrak{A}_r$ be left Gonshor genetic. In addition to part I of the proof we have to show that for all index tuples $(i_1,\ldots,i_r)$, $(j_1,\ldots,j_r) > 0$ the product

$$(c_{1i_1} \otimes \ldots \otimes c_{1i_r})(c_{1j_1} \otimes \ldots \otimes c_{rj_r})$$

has no components in the direction of $c_{1k_1} \otimes \ldots \otimes c_{rk_r}$ for all $(k_1,\ldots,k_r) \leq \max\{(i_1,\ldots,i_r),\; (j_1,\ldots,j_r)\}$.
If $(j_1,\ldots,j_r)$ is the maximum of both index vectors, then the assertion follows from part I of this proof. Now let $(i_1,\ldots,i_r)$ be the maximal index vector. Either $(k_1,\ldots,k_r) = (i_1,\ldots,i_r)$ or $(k_1,\ldots,k_r) < (i_1,\ldots,i_r)$.

In the first case there is a t with $j_t > 0$, in the second case there is an s with $k_s < i_s$. Since all algebras are left Gonshor genetic we have $\lambda^{(t)}_{i_t j_t k_t} = 0$ or $\lambda^{(s)}_{i_s j_s k_s} = 0$, respectively, thus in any case

$$\lambda^{(1)}_{i_1 j_1 k_1} \cdots \lambda^{(r)}_{i_r j_r k_r} = 0 \quad .$$

III. If all algebras $\mathfrak{A}_1, \ldots, \mathfrak{A}_r$ are Gonshor genetic then they are left and right Gonshor genetic with respect to bases $c_{\varrho o}, \ldots, c_{\varrho m}$ of $\mathfrak{A}_\varrho$, $\varrho = 1, \ldots, r$.
As in part I of the proof it also follows that $\lambda^{(1)}_{i_1 o k_1} \cdots \lambda^{(r)}_{i_r o k_r} = 0$ for all index vectors $(k_1, \ldots, k_r) < (i_1, \ldots, i_r)$.
From this result together with part II of the proof it follows that $\mathfrak{A}_1 \otimes \ldots \otimes \mathfrak{A}_r$ is Gonshor genetic. Similarly one obtains the assertion about the right train roots. □

With respect to the following section we add a simple remark:
If $e_\varrho \in \mathfrak{A}_\varrho$, $= 1, \ldots, r$, are idempotents, then $e_1 \otimes \ldots \otimes e_r$ is an idempotent of the tensor product $\mathfrak{A}_1 \otimes \ldots \otimes \mathfrak{A}_r$.

In section 7B we shall discuss among other topics the linked products, introduced by Holgate [64] .

C. Duplication

Already in section 1 we have seen that the zygotic algebra of a population of diploid individuals with known gametic algebra can be constructed by duplication of the latter. In this section we shall discuss the principle of duplication in greater detail. It was introduced by Etherington [23], [26] and has been taken up by almost all authors in the field. For the commutative case Gonshor [42] has given a basis free definition of duplication.
We shall extend this idea to the non commutative case.

In the following let $\mathfrak{A}$ be an n-dimensional algebra over $\mathbb{K}$ which is not necessarily commutative or associative.

Theorem 6.12. Let $\mathfrak{B}$ be an r-dimensional ideal of $\mathfrak{A}$. Let $\mathfrak{A} \otimes \mathfrak{B}$ be the tensor product of the vector spaces underlying $\mathfrak{A}$ and $\mathfrak{B}$.

Then

$$(\sum_{i\in I} a_i \otimes b_i)(\sum_{k\in K} c_k \otimes d_k) := \sum_{(i,k)\in I\times K} a_i b_i \otimes c_k d_k \qquad (6.7)$$

defines a multiplication in $\mathfrak{A} \otimes \mathfrak{b}$ with respect to which the vector space $\mathfrak{A} \otimes \mathfrak{b}$ becomes an nr-dimensional algebra $\mathfrak{A} \otimes_d \mathfrak{b}$.

Proof. Obviously $\mathfrak{A} \otimes \mathfrak{b}$ is an nr-dimensional vector space over $\mathbb{K}$. Furthermore for all $x = \sum_{i\in I} a_i \otimes b_i$, $y = \sum_{k\in K} c_k \otimes d_k \in \mathfrak{A} \otimes \mathfrak{b}$ the product xy defined by (6.7) is in $\mathfrak{A} \otimes \mathfrak{b}$ since $\mathfrak{b}$ is by assumption an ideal of $\mathfrak{A}$. It remains to show that the product is defined independently of the representation of the elements x and y. It is well known that the tensor product $\mathfrak{A} \otimes \mathfrak{b}$ can be interpreted as the factor space of the free module $\mathbb{K}(\mathfrak{A} \times \mathfrak{b})$ of all pairs $(a,b) \in \mathfrak{A} \times \mathfrak{b}$ over $\mathbb{K}$ modulo the subspace

$$\mathfrak{h} := \langle (\lambda_1 a_1 + \lambda_2 a_2, b) - \lambda_1 (a_1, b) - \lambda_2 (a_2, b),\ (a, \mu_1 + \mu_2 b_2) - \mu_1 (a, b_1) - \mu_2 (a, b_2) \mid$$

$$a, a_1, a_2 \in \mathfrak{A},\ b, b_1, b_2 \in \mathfrak{b},\ \lambda_1, \lambda_2, \mu_1, \mu_2 \in \mathbb{K} \rangle.$$

In $\mathbb{K}(\mathfrak{A} \times \mathfrak{b})$ we define a multiplication by

$$(a,b)(c,d) \longmapsto (ab, cd)$$

and the bilinear extension. The subspace $\mathfrak{h}$ of $\mathbb{K}(\mathfrak{A} \times \mathfrak{b})$ is an ideal of $\mathbb{K}(\mathfrak{A} \times \mathfrak{b})$ with respect to this multiplication because

$$[(\lambda_1 a_1 + \lambda_2 a_2, b) - \lambda_1 (a_1, b) - \lambda_2 (a_2, b)]\,(c,d)$$

$$= (\lambda_1 a_1 + \lambda_2 a_2, b)(c,d) - \lambda_1 (a_1, b)(c,d) - \lambda_2 (a_2, b)(c,d)$$

$$= (\lambda_1 a_1 b + \lambda_2 a_2 b, cd) - \lambda_1 (a_1 b, cd) - \lambda_2 (a_2 b, cd) \in \mathfrak{h}$$

also

$$(c,d)\,[(\lambda_1 a_1 + \lambda_2 a_2, b) - \lambda_1 (a_1, b) - \lambda_2 (a_2, b)] \in \mathfrak{h},$$

and similarly for the remaining generators of $\mathfrak{h}$.
Thus (6.7) defines in fact a multiplication in $\mathfrak{A} \otimes \mathfrak{b}$. This multiplication is distributive:

$$\Big[\sum_{i\in I} a_i \otimes b_i + \sum_{j\in J} a_j \otimes b_j\Big]\big(\sum_{k\in K} c_k \otimes d_k\big)$$

$$= \Big(\sum_{l\in I\cup J} a_l \otimes b_l + \sum_{l\in I\cap J} a_l \otimes b_l\Big)\big(\sum_{k\in K} c_k \otimes d_k\big)$$

$$= \sum_{(l,k)\in(I\cup J)\times K} a_l b_l \otimes c_k d_k + \sum_{(l,k)\in(I\cap J)\times K} a_l b_l \otimes c_k d_k$$

$$= \sum_{(i,k)\in I\times K} a_i b_i \otimes c_k d_k + \sum_{(j,k)\in J\times K} a_j b_j \otimes c_k d_k .$$

The other distribution law follows in the same way.□

Let $\mathfrak{b}$ be an r-dimensional ideal of $\mathfrak{A}$. We choose a basis $a_1,\dots,a_n$ of $\mathfrak{A}$ such that $a_{n-r},\dots,a_n$ form a basis of $\mathfrak{b}$. Then $a_i \otimes a_j$, $i=1,\dots,n$, $j=n-r,\dots,n$, form a basis of $\mathfrak{A}\otimes\mathfrak{b}$.
Let γ_{ijk}, $i,j,k=1,\dots,n$ be the multiplication constants of $\mathfrak{A}$ with respect to the chosen basis. Then the multiplication table of $\mathfrak{A}\otimes_d\mathfrak{b}$ is

$$(a_i\otimes a_j)(a_p\otimes a_q) = \sum_{\substack{k=1,\dots,n\\ s=n-r,\dots,n}} \gamma_{ijk}\gamma_{pqs} a_k \otimes a_s, \qquad (6.8)$$

$$i,p=1,\dots,n,\ j,q=n-r,\dots,n.$$

From here it follows that $\mathfrak{A}\otimes_d\mathfrak{b}$ is isomorphic with the free K-module generated by $a_i a_j$, $i=1,\dots,n$, $j=n-r,\dots,n$, with the multiplication

$$(a_i a_j)(a_p a_q) \longmapsto \sum_{\substack{k=1,\dots,n\\ s=n-r,\dots,n}} \gamma_{ijk}\gamma_{pqs}\, a_k a_s,$$

$$i,p=1,\dots,n,\ j,q=n-r,\dots,n.$$

and bilinear extension.

Definition 6.13. The algebra $\mathfrak{A}\otimes_d\mathfrak{A}$ is called the duplicate of $\mathfrak{A}$ (or the non commutative duplicate of $\mathfrak{A}$ if this algebra has to be distinguished from the commutative duplicate of a commutative algebra).

The duplicate of a commutative algebra is not commutative in general.

The following theorem goes back to Etherington [30] .

Theorem 6.14. The mapping $\varphi : \mathfrak{A} \otimes_d \mathfrak{A} \longrightarrow \mathfrak{A}$ with

$\varphi(\sum_{i\in I} a_i \otimes b_i) := \sum_{i\in I} a_i b_i$ defines an algebra homomorphism and

$$\operatorname{Im}\varphi = \mathfrak{A}^2, \ (\mathfrak{A} \otimes_d \mathfrak{A})\ker\varphi = \ker\varphi\,(\mathfrak{A} \otimes_d \mathfrak{A}) = \langle 0 \rangle .$$

Proof. First we show that φ is well defined. We use the same notation as in the proof of theorem 6.12.
Define $\tilde{\varphi} : \mathbb{K}(\mathfrak{A} \times \mathfrak{A}) \longrightarrow \mathfrak{A}$ by $\tilde{\varphi}(a,b) := ab$ and the linear extension. Then $\mathfrak{h}_\gamma$ is contained in $\ker\tilde{\varphi}$ because of

$$\tilde{\varphi}\left[(\lambda_1 a_1 + \lambda_2 a_2, b) - \lambda_1 (a_1, b) - \lambda_2 (a_2, b)\right]$$

$$= (\lambda_1 a_1 + \lambda_2 a_2) b - \lambda_1 a_1 b - \lambda_2 a_2 b = 0 ,$$

and similarly $\tilde{\varphi}\left[(a, \mu_1 b_1 + \mu_2 b_2) - \mu_1 (a, b_1) - \mu_2 (a, b_2)\right] = 0$.
Thus φ is well defined and φ is linear by definition.

Let $x = \sum_{i\in I} a_i \otimes b_i$, $y = \sum_{k\in K} c_k \otimes d_k \in \mathfrak{A} \otimes_d \mathfrak{A}$ then

$$\varphi(xy) = \varphi\left[(\sum_{i\in I} a_i \otimes b_i)(\sum_{k\in K} c_k \otimes d_k)\right] = \varphi(\sum_{(i,k)\in I\times K} a_i b_i \otimes c_k d_k)$$

$$= \sum_{(i,k)\in I\times K} a_i b_i c_k d_k = (\sum_{i\in I} a_i b_i)(\sum_{k\in K} c_k d_k) = \varphi(x)\varphi(y) .$$

Clearly $\operatorname{Im}\varphi = \mathfrak{A}^2$.
In view of

$$xy = \sum_{(i,k)\in I\times K} a_i b_i \otimes c_k d_k = (\sum_{i\in I} a_i b_i) \otimes (\sum_{k\in K} c_k d_k) = \varphi(x) \otimes \varphi(y)$$

the last assertion follows immediately. □

Theorem 6.15. Let $\mathfrak{A}$ be a baric algebra with weight homomorphism ω, $\mathfrak{N} := \ker\omega$.
Then the duplicate $\mathfrak{A} \otimes_d \mathfrak{A}$ of $\mathfrak{A}$ is baric with weight homomorphism $\omega_d := \omega \circ \varphi$ and $\ker\omega_d = \mathfrak{A} \otimes_d \mathfrak{N} + \mathfrak{N} \otimes_d \mathfrak{A}$.

Proof. Since $\varphi : \mathfrak{A} \otimes_d \mathfrak{A} \longrightarrow \mathfrak{A}$ is an algebra homomorphism, by $\omega_d = \omega \circ \varphi$ an algebra homomorphism of $\mathfrak{A} \otimes_d \mathfrak{A}$ into $\mathbb{K}$ is defined. Since ω is nontrivial we have $\mathfrak{A}^2 \not\subseteq \mathfrak{N}$. Further ω_d is nontrivial because of $\operatorname{Im}\varphi = \mathfrak{A}^2$.

For every element

$\sum_{i\in I} a_i \otimes n_i \in \mathfrak{A} \otimes_d \mathfrak{N}$ we have $\varphi(\sum_{i\in I} a_i \otimes n_i) = \sum_{i\in I} a_i n_i \in \mathfrak{N}$ and thus

$\mathfrak{A} \otimes_d \mathfrak{N} \subseteq \ker \omega_d$. Similarly it follows that $\mathfrak{N} \otimes_d \mathfrak{A} \subseteq \ker \omega_d$. Now the asserted identity follows by counting dimensions. We have
$\dim \ker\omega_d = n^2-1$ and

$$\dim(\mathfrak{A} \otimes_d \mathfrak{N} + \mathfrak{N} \otimes_d \mathfrak{A}) = n(n-1)+(n-1)n-(n-1)^2 = n^2-1. \square$$

Using this representation of $\ker\omega_d$ we get the following lemma by induction

Lemma 6.16. Let $\mathfrak{A}$ be baric with weight homomorphism ω. If $\ker\omega$ is (left) nilpotent of index r then $\ker\omega_d$ is (left) nilpotent of an index not greater than r+1.

The next theorem can be traced back to Etherington [30].

Theorem 6.17. Let $\mathfrak{A}$ be a left train algebra with left rank r and left principal train roots $\lambda_o = 1, \lambda_1, \ldots, \lambda_{r-2}$. Then the duplicate $\mathfrak{A} \otimes_d \mathfrak{A}$ is also a left train algebra of rank at most r+1 and the left principal train roots are contained in the set $\{1, 0, \lambda_1, \ldots, \lambda_{r-2}\}$.

Proof. Let ω be the weight homomorphism of $\mathfrak{A}$ and let

$$^{r}y + \beta_1\, \omega(y)\, (^{r-1}y) + \ldots + \beta_{r-1}\, \omega^{r-1}(y) y = 0$$

be the left rank equation of $\mathfrak{A}$. Let x be a generic element of $\mathfrak{A} \otimes_d \mathfrak{A}$. Then $y := \varphi(x)$ satisfies the left rank equation. Since φ is an algebra homomorphism we have

$$^{r}(\varphi(x)) + \beta_1\, \omega\circ\varphi(x)\, (^{r-1}\varphi(x)) + \ldots + \beta_{r-1} (\omega\circ\varphi(x))^{r-1} \varphi(x)$$

$$= \varphi\left[{}^{r}x + \beta_1\, \omega_d(x)\, (^{r-1}x) + \ldots + \beta_{r-1}\, \omega_d^{r-1}(x) x\right] = 0 ,$$

from which it follows that the element z(x) in the square bracket is contained in $\ker\varphi$. From theorem 6.14 one obtains $xz(x) = 0$. Hence the left rank equation of $\mathfrak{A} \otimes_d \mathfrak{A}$ is a divisor of

$$^{r+1}x + \beta_1\, \omega_d(^{r}x) + \ldots + \beta_{r-1}\, \omega_d^{r-1}(x)\, (^{2}x) \quad . \qquad (6.9)$$

Therefore the left rank of $\mathfrak{A} \otimes_d \mathfrak{A}$ is at most r+1. When factorized the polynomial (6.9) assumes the form

$$xx(x-\omega_d(x))(x-\omega_d(x)\lambda_1) \dots (x-\omega_d(x)\lambda_{r-2}) \ .$$

Thus the assertion about the left principal roots is shown. □

In the proof of the following theorem we follow Gonshor's proof for the fact that the commutative duplicate (cf. 6.21) of a commutative genetic algebra is genetic, cf. [47] .

Theorem 6.18. Let $\mathfrak{A}$ be an (m+1)-dimensional weakly left Gonshor genetic algebra with left train roots λ_{oii}, i=0,...,m. Then the duplicate $\mathfrak{A} \otimes_d \mathfrak{A}$ is weakly left Gonshor genetic and the left train roots are λ_{oii}, i=0,...,m and (m^2+m)-times 0.

Proof. Let $c_o,\dots,c_m$ be a canonical basis of $\mathfrak{A}$ with respect to which $\mathfrak{A}$ is displayed to be weakly left Gonshor genetic. Let λ_{ijk}, i,j,k=0,...,m, be the multiplication constants. We show that $c_i \otimes c_j$, i,j=0,...,m, form a canonical basis of $\mathfrak{A} \otimes_d \mathfrak{A}$ with respect to the following order of index pairs: $(k,s) \leq (p,q)$ if $s < q$ or $s = q$ and $k \leq p$. The multiplication table of $\mathfrak{A} \otimes_d \mathfrak{A}$ is

$$(c_i \otimes c_j)(c_p \otimes c_q) = \sum_{k,s=1}^{m} \lambda_{ijk}\lambda_{pqs} c_k \otimes c_s , \qquad i,j,p,q=0,1,\dots,m.$$

We have $\lambda_{ooo}\lambda_{ooo} = 1 \cdot 1 = 1$.
Furthermore we have to show that

$$\lambda_{ook}\lambda_{pqs} = 0 \qquad \text{for all } (k,s) < (p,q) \tag{6.10}$$

$$\lambda_{ijk}\lambda_{pqs} = 0 \qquad \text{for all } (k,s) \leq (p,q) \text{ if } (i,j) > (0,0). \tag{6.11}$$

Let $(k,s)<(p,q)$, i.e. either $s<q$ or $s = q$ and $k<p$. In the first case we have $\lambda_{pqs} = 0$ and in the second case $\lambda_{pqq} = 0$ since $p>0$ from $k<p$.
Hence (6.10) is always true.
Let $(i,j)>0$, i.e. $i>0$ or $j>0$. Assume $(k,s) \leq (p,q)$, i.e. $s<q$ or $s = q$ and $k \leq p$. In the first case again $\lambda_{pqs} = 0$. In the second case $\lambda_{pqq} = 0$ as long as $p>0$.
If $p = 0$ then also $k = 0$. Since either $j>0$ or $i>0$ it follows for this case that $\lambda_{ijo} = 0$. In any case we have (6.11). Thus we have shown that $\mathfrak{A} \otimes_d \mathfrak{A}$ is weakly left Gonshor genetic.
The left train roots of $\mathfrak{A} \otimes_d \mathfrak{A}$ are the scalar factors of $c_p \otimes c_q$ in the

product $(c_o \otimes c_o)(c_p \otimes c_q)$, i.e. $\lambda_{oop}\lambda_{pqq}$, $p,q = 0,\dots,m$.
For $p = 0$ we have exactly the left train roots of $\mathfrak{A}$ since

$$\lambda_{ooo}\lambda_{oqq} = 1\cdot\lambda_{oqq} = \lambda_{oqq}, \qquad q=0,\dots,m.$$

For $p>0$ we have $\lambda_{pqq} = 0$ and thus the remaining $(m+1)^2 - (m+1) = m^2+m$ left train roots vanish. □

We do not know whether the duplicate of a left Gonshor genetic algebra also has this property. We can only show that the duplicate is weakly left Gonshor genetic and that the kernel of the weight homomorphism is nilpotent.

Theorem 6.19. Let $\mathfrak{A}$ be Gonshor genetic. Then the duplicate is also Gonshor genetic and the left and right train roots of $\mathfrak{A}$ together with the appropriate number of zeros are the left and right train roots of the duplicate.

Proof. Let $c_o,\dots,c_m$ be a canonical basis of $\mathfrak{A}$, and as usual let λ_{ijk}, $i,j,k=0,\dots,m$, be the multiplication constants with respect to this basis.
We define subspaces of the duplicate $\mathfrak{A}\otimes_d\mathfrak{A}$ of $\mathfrak{A}$ by

$$\mathfrak{N}_r := \langle c_i \otimes c_j \mid i \geq r \text{ or } j \geq r \text{ but } (i,j)\neq(r,0),(0,r)\rangle, r=0,1,\dots,m,$$

$$\mathfrak{N}_{m+1} := \langle 0 \rangle .$$

The basis elements of $\mathfrak{N}_o, \mathfrak{N}_1, \dots, \mathfrak{N}_r, \dots, \mathfrak{N}_m$ can be read from the following scheme:

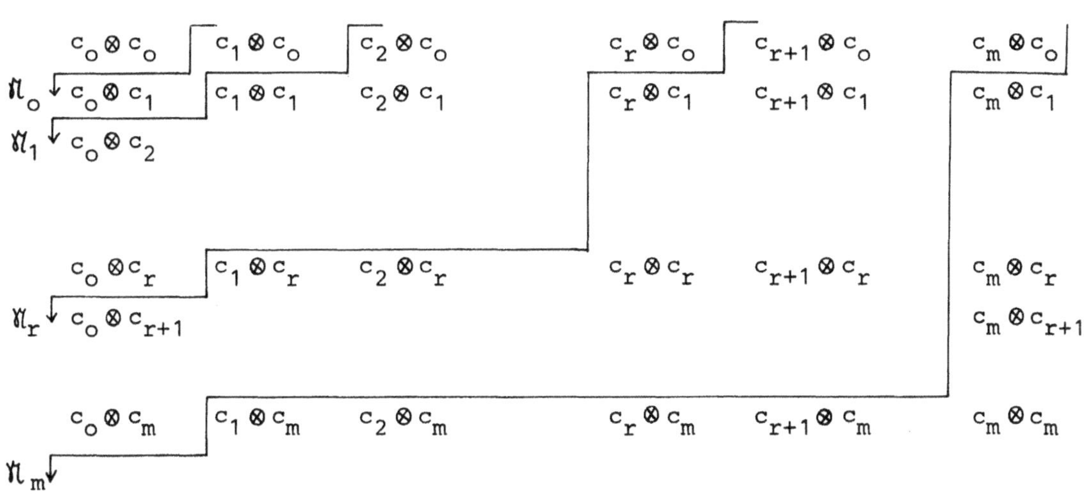

Trivially,

$$\mathfrak{N}_o > \mathfrak{N}_1 > \dots > \mathfrak{N}_{m-1} > \mathfrak{N}_m > \mathfrak{N}_{m+1} = \langle o \rangle .$$

$\mathfrak{N}_o$ is the kernel of the weight homomorphism ω_d which is defined by

$$\omega_d(c_o \otimes c_o) := 1,\ \omega_d(c_i \otimes c_j) := 0 \quad \text{for all } (i,j) \neq (0,0),$$

and linear extension onto $\mathfrak{A} \otimes_d \mathfrak{A}$. We show the relations

$$\mathfrak{N}_o \mathfrak{N}_r,\ \mathfrak{N}_r \mathfrak{N}_o \subseteq \mathfrak{N}_{r+1},\quad r=0,\dots,m, \tag{6.12}$$

and

$$(\mathfrak{A} \otimes_d \mathfrak{A})\mathfrak{N}_r,\ \mathfrak{N}_r(\mathfrak{A} \otimes_d \mathfrak{A}) \subseteq \mathfrak{N}_r,\ r=0,\dots,m. \tag{6.13}$$

Let $c_i \otimes c_j \in \mathfrak{N}_o$, i.e. $(i,j) \neq (0,0)$. Furthermore let $c_p \otimes c_q \in \mathfrak{N}_r$, i.e. $p \geq r$ or $q \geq r$ but $(p,q) \neq (r,0)$, $(0,r)$. Then we have $\lambda_{pqs} = 0$ for all $s \leq r$ and also $\lambda_{ijk} = 0$ for $k = 0$.
Hence the products

$$(c_i \otimes c_j)(c_p \otimes c_q) = \sum_{\substack{k \geq 1 \\ s \geq r+1}} \lambda_{ijk}\lambda_{pqs}\ c_k \otimes c_s$$

and

$$(c_p \otimes c_q)(c_i \otimes c_j) = \sum_{\substack{k \geq r+1 \\ s \geq 1}} \lambda_{pqk}\lambda_{ijs}\ c_k \otimes c_s$$

are contained in $\mathfrak{N}_{r+1}$.
In order to prove (6.13) it is sufficient to show that $c_o \otimes c_o$ when multiplied by any basis element of $\mathfrak{N}_r$ is again in $\mathfrak{N}_r$. Let $c_p \otimes c_q \in \mathfrak{N}$. Then we distinguish four cases

(i) $p \geq r$ and $q > 0$
(ii) $p > 0$ and $q = r$
(iii) $p \geq r+1$ and $q = 0$
(iv) $p = 0$ and $q \geq r+1$.

In view of the special properties of the multiplication constants we have

$$(c_o \otimes c_o)(c_p \otimes c_q) = \sum_{\substack{k \geq 0 \\ s \geq r+1}} \lambda_{ook}\lambda_{pqs}\ c_k \otimes c_s \in \mathfrak{N}_r$$

and the corresponding result for the product $(c_p \otimes c_q)(c_o \otimes c_o)$. Furthermore we find that these products, if they are not already contained in $\mathfrak{N}_{r+1}$, may have components only with respect to the basis vectors $c_r \otimes c_o$

and $c_o \otimes c_r$ of $\mathfrak{N}_r$. For this consider cases (i)-(iv) with the inequality replaced by an equality. Then

$$\begin{aligned}
(c_o \otimes c_o)(c_r \otimes c_q) &= \lambda_{rq,r+1}\, c_o \otimes c_{r+1} + x_1 \\
(c_o \otimes c_o)(c_p \otimes c_r) &= \lambda_{pr,r+1}\, c_o \otimes c_{r+1} + x_2 \\
(c_o \otimes c_o)(c_{r+1} \otimes c_o) &= \lambda_{r+1,o,r+1}\, c_o \otimes c_{r+1} + x_3 \\
(c_o \otimes c_o)(c_o \otimes c_{r+1}) &= \lambda_{o,r+1,r+1}\, c_o \otimes c_{r+1} + x_4
\end{aligned} \qquad x_i \in \mathfrak{N}_{r+1}, i=1,\dots,4.$$

and similarly for the products with factors interchanged. Now we choose a complete ordering of the basis $c_k \otimes c_s$, $k,s=0,\dots,m$ of $\mathfrak{A} \otimes_d \mathfrak{A}$, thereby taking account of the inclusions

$$\mathfrak{A} \otimes_d \mathfrak{A} > \mathfrak{N}_o > \mathfrak{N}_1 > \cdots > \mathfrak{N}_m$$

of ideals such that among the basis vectors of $\mathfrak{N}_r$ which are not contained in $\mathfrak{N}_{r+1}$ the vectors $c_{r+1} \otimes c_o$ and $c_o \otimes c_{r+1}$ appear last, $r=1,\dots,m-1$.

Then to the left and right transformation $\mathcal{L}_{c_i \otimes c_j}$ and $\mathcal{R}_{c_i \otimes c_j}$, $i,j=0,\dots,m$, there correspond lower block triangular matrices with a block partition corresponding to the chain of ideals of $\mathfrak{A} \otimes_d \mathfrak{A}$. Furthermore to all $\mathcal{L}_{c_i \otimes c_j}$ and $\mathcal{R}_{c_i \otimes c_j}$, $(i,j) \neq (0,0)$, there correspond block triangular matrices with only zero diagonal blocks. Finally the diagonal blocks of the matrix representations of $\mathcal{L}_{c_o \otimes c_o}$ and $\mathcal{R}_{c_o \otimes c_o}$ have the form (1),

$$\begin{pmatrix} & & 0 & & \\ & & & & \\ 0 & \cdots & 0 & 0 & 0 \\ * & \cdots & * & \lambda_{ror} & \lambda_{orr} \end{pmatrix} \quad \text{and} \quad \begin{pmatrix} & & 0 & & \\ & & & & \\ * & \cdots & * & \lambda_{ror} & \lambda_{orr} \\ 0 & \cdots & 0 & 0 & 0 \end{pmatrix}$$

respectively, $r=1,\dots,m$. The chain of ideals $\mathfrak{A} \otimes_d \mathfrak{A}$ is invariant under the basis transformation given by

$$c_r \otimes c_o \mapsto \begin{cases} c_r \otimes c_o & \text{if } \lambda_{ror}, \lambda_{orr} = 0 \\ \lambda_{ror} c_r \otimes c_o + \lambda_{orr} c_o \otimes c_r & \text{otherwise} \end{cases}$$

$$c_o \otimes c_r \longmapsto \begin{cases} c_o \otimes c_r & \text{if } \lambda_{ror}, \lambda_{orr} = 0 \\ \lambda_{orr} c_r \otimes c_o - \lambda_{ror} c_o \otimes c_r & \text{otherwise} \end{cases}$$

and the condition that $c_p \otimes c_q$ remains fixed otherwise. One can easily see that with respect to this new basis the transformations $\mathcal{L}_{c_o \otimes c_o}$ and $\mathcal{R}_{c_o \otimes c_o}$ are represented by lower triangular block matrices with triangular blocks (1) and

$$\left(\begin{array}{c|cc} & 0 & \\ \hline * & \lambda_{ror} & 0 \\ & \lambda_{orr} & 0 \end{array}\right), \quad \left(\begin{array}{c|cc} & 0 & \\ \hline * & \lambda_{orr} & 0 \\ & -\lambda_{ror} & 0 \end{array}\right), r=1,..,m.$$

Thus it has been shown that the new basis of $\mathfrak{A} \otimes_d \mathfrak{A}$ with the chosen order is a canonical basis in the sense of definition 5.9(3), i.e. $\mathfrak{A} \otimes_d \mathfrak{A}$ is Gonshor genetic. In addition we have shown the assertion about the left and right train roots. □

From Schafer's [101] example of a genetic algebra which is not a special train algebra one can find that in general the duplicate of a special train algebra need not to be a special train algebra, cf. example 6.26.

The following result goes back to Etherington [30].

<u>Theorem 6.20.</u> For two algebras $\mathfrak{A}_1$ and $\mathfrak{A}_2$ over $\mathbb{K}$ the duplicate of the tensor product is isomorphic with the tensor product of the duplicate, i.e.

$$(\mathfrak{A}_1 \otimes \mathfrak{A}_2) \otimes_d (\mathfrak{A}_1 \otimes \mathfrak{A}_2) \cong (\mathfrak{A}_1 \otimes_d \mathfrak{A}_1) \otimes (\mathfrak{A}_2 \otimes_d \mathfrak{A}_2).$$

<u>Proof.</u> Let $\dim \mathfrak{A}_i = n_i$ for $i=1,2$. Let $a_1,\dots,a_{n_1}$ and $b_1,\dots,b_{n_2}$ be bases of $\mathfrak{A}_1$ and $\mathfrak{A}_2$, respectively. Then the elements $(a_i \otimes b_p) \otimes (a_j \otimes b_q)$, $i,j=1,\dots,n_1$, $p,q=1,\dots,n_2$ form a basis of

$\mathcal{B} := (\mathfrak{A}_1 \otimes \mathfrak{A}_2) \otimes_d (\mathfrak{A}_1 \otimes \mathfrak{A}_2)$ and the elements $(a_i \otimes a_j) \otimes (b_p \otimes b_q)$, $i,j=1,\dots,n_1$, $p,q=1,\dots,n_2$, form a basis of $\mathcal{C} := (\mathfrak{A}_1 \otimes_d \mathfrak{A}_1) \otimes (\mathfrak{A}_2 \otimes_d \mathfrak{A}_2)$.

The mapping $\Phi : \mathfrak{B} \longrightarrow \mathfrak{C}$ defined by

$$\Phi : \quad (a_i \otimes b_p) \otimes (a_j \otimes b_q) \longmapsto (a_i \otimes a_j) \otimes (b_p \otimes b_q), \quad \begin{array}{l} i,j=1,\ldots,n_1 \\ p,q=1,\ldots,n_2, \end{array}$$

and the linear extension is a linear bijective mapping of $\mathfrak{B}$ onto $\mathfrak{C}$. This mapping is compatible with the multiplications in $\mathfrak{B}$ and $\mathfrak{C}$ since we have in $\mathfrak{B}$

$$[(a_{i_1} \otimes b_{p_1}) \otimes (a_{j_1} \otimes b_{q_1})][(a_{i_2} \otimes b_{p_2}) \otimes (a_{j_2} \otimes b_{q_2})]$$

$$= (a_{i_1} a_{j_1} \otimes b_{p_1} b_{q_1}) \otimes (a_{i_2} a_{j_2} \otimes b_{p_2} b_{q_2})$$

and on the other hand in $\mathfrak{C}$

$$[(a_{1_i} \otimes a_{j_1}) \otimes (b_{p_1} \otimes b_{q_1})][(a_{i_2} \otimes a_{j_2}) \otimes (b_{p_2} \otimes b_{q_2})]$$

$$= (a_{i_1} a_{j_1} \otimes a_{i_2} a_{j_2}) \otimes (b_{p_1} b_{q_1} \otimes b_{p_2} b_{q_2}),$$

$$i_k, j_k = 1,\ldots,n_1, \; p_k, q_k = 1,\ldots,n_2, \; k=1,2.$$

We now consider the commutative duplication of a commutative algebra introduced by Etherington [27], [30]. We start with the representation of this algebra given by Gonshor [47] and then we show that Etherington's original construction gives the same result.

In the following let $\mathfrak{A}$ be a commutative, non associative algebra over $\mathbb{K}$ of dimension n.
Let $\mathfrak{B}$ be an r-dimensional ideal of $\mathfrak{A}$. Then

$$\mathfrak{J} := \langle b \otimes d - d \otimes b \mid b,d \in \mathfrak{B} \rangle$$

is an ideal of $\mathfrak{A} \otimes_d \mathfrak{B}$ since

$$(\sum_{i\in I} a_i \otimes b_i)(b \otimes d - d \otimes b) = (\sum_{i\in I} a_i \otimes b_i)(b \otimes d) - (\sum_{i\in I} a_i \otimes b_i)(d \otimes b)$$

$$= \sum_{i\in I} a_i b_i \otimes bd - \sum_{i\in I} a_i b_i \otimes db = 0, \text{ since } bd = db.$$

Thus $\mathfrak{A} \otimes_d \mathfrak{B} / \mathfrak{J} =: \mathfrak{A} \circledast \mathfrak{B}$ is an algebra over $\mathbb{K}$. A simple calculation shows that this algebra has the dimension $nr-r(r-1)/2$.
We denote the representatives of the residue classes of $\mathfrak{A} \otimes_d \mathfrak{B}$ modulo $\mathfrak{J}$ by

$$a \circledast b := a \otimes b + \mathfrak{J}.$$

By the first isomorphism theorem for algebras $\mathfrak{A} \circledast \mathfrak{B}$ is isomorphic with the free $\mathbb{K}$-module $\mathbb{K}(\mathfrak{A} \times \mathfrak{B})$ with the multiplication $(a,b)(c,d) \longmapsto (ab,cd)$ modulo the subspace $\mathfrak{K}$ generated by the elements

$$(\lambda_1 a_1 + \lambda_2 a_2, b) - \lambda_1(a_1,b) - \lambda_2(a_2,b), \quad (c,d)-(d,c),$$

$$a_1, a_2 \in \mathfrak{A}, \; b,c,d \in \mathfrak{B}, \; \lambda_1, \lambda_2 \in \mathbb{K}.$$

This subspace is in fact an ideal of $\mathbb{K}(\mathfrak{A} \times \mathfrak{B})$, cf. also Gonshor [47]. Let $a_1, \ldots, a_n$ be a basis of $\mathfrak{A}$ such that $a_{n-r}, \ldots, a_n$ form a basis of $\mathfrak{B}$. Then

$$a_i \circledast a_j, \quad i=1,\ldots,n, \quad j=n-r,\ldots,n, \quad i \leq j,$$

is a basis of $\mathfrak{A} \circledast \mathfrak{B}$.
Let γ_{ijk}, $i,j,k=1,\ldots,n$, be the multiplication constants of $\mathfrak{A}$ with respect to the said basis then the multiplication table of $\mathfrak{A} \circledast \mathfrak{B}$ reads

$$\begin{aligned}(a_i \circledast a_j)(a_p \circledast a_q) = \sum_{k=n-r}^{n} \gamma_{ijk}\gamma_{pqk}\, a_k \circledast a_k \\ + \sum_{\substack{k=1,\ldots,n \\ s=n-r,\ldots,n \\ k<s}} (\gamma_{ijk}\gamma_{pqs} + \gamma_{ijs}\gamma_{pqk})\, a_k \circledast a_s, \\ i,p=1,\ldots,n \\ j,q=n-r,\ldots,n.\end{aligned} \tag{6.14}$$

Thus it is immediately clear that $\mathfrak{A} \circledast \mathfrak{B}$ is isomorphic with the free $\mathbb{K}$-module over the complexes $a_i a_j$, $i=1,\ldots,n$, $j=n-r,\ldots,n$, $i \leq j$, with a multiplication defined by analogy with (6.14). The latter construction with $\mathfrak{B} = \mathfrak{A}$ was used by Etherington to define the commutative duplicate. Thus we can define

Definition 6.21. The algebra $\mathfrak{A} \circledast \mathfrak{A}$ is the commutative duplicate of the algebra $\mathfrak{A}$.

Remark. For a commutative algebra $\mathfrak{A}$ in the commutative duplicate the elements $a \otimes b$ and $b \otimes a$ are identified but not in the non commutative duplicate.

From the above considerations it follows immediately that a commutative duplicate is an algebra of dimension $n(n+1)/2$. For the commutative duplicate we have results paralleling theorems 6.14, 6.15, 6.17 and 6.19.

Theorem 6.22. The mapping $\psi:\mathfrak{A}\circledast\mathfrak{A}\longrightarrow\mathfrak{A}$ with $\psi(\sum_{i\in I}a_i\circledast b_i) := \sum_{i\in I}a_ib_i$ defines an algebra homomorphism and

$$\operatorname{Im}\psi = \mathfrak{A}^2,\ (\mathfrak{A}\circledast\mathfrak{A})\ker\psi = \langle 0\rangle .$$

Proof. We use the same notation as in theorem 6.14. In view of $\varphi(\mathfrak{J}) = 0$ the mapping φ induces the mapping ψ defined above from $\mathfrak{A}\circledast\mathfrak{A}$ into $\mathbb{K}$ and consequently the latter is a homomorhpism. The remaining assertions follow directly from theorem 6.14. □

Theorem 6.23. Let $\mathfrak{A}$ be a baric algebra with weight homomorphism ω. Let $\mathfrak{N} := \ker\omega$. Then the commutative duplicate $\mathfrak{A}\circledast\mathfrak{A}$ is a baric algebra with weight homomorphism $\omega^{\circledast} := \omega\circ\psi$ and $\ker\omega^{\circledast} = \mathfrak{A}\circledast\mathfrak{N}$.

Theorem 6.24. The commutative duplicate of a train algebra $\mathfrak{A}$ of rank r is a train algebra with rank at most r+1.

Theorem 6.25. Let $\mathfrak{A}$ be an (m+1)-dimensional commutative genetic algebra with train roots $\lambda_o = 1, \lambda_1,\dots,\lambda_m$. Then the commutative duplicate of $\mathfrak{A}$ is again a genetic algebra with train roots $\lambda_o,\dots,\lambda_m$ and (m(m+1)/2)-times 0.

At this point a remark on the history of theorem 6.25 is appropriate. Schafer [101] gave the first proof using the transformation algebra of the commutative duplicate. Later Gonshor [47] showed that for a canonical basis $c_o,\dots,c_m$ of $\mathfrak{A}$ the elements $c_i\circledast c_j$, $i,j=0,\dots,m$, $i\leq j$, form a basis of the commutative duplicate $\mathfrak{A}\circledast\mathfrak{A}$ which is canonical with respect to the order given in the proof of theorem 6.18.
Holgate [67] proved this theorem using the characterization of a genetic algebra by properties of the Lie algebra corresponding to the transformation algebra, cf. theorem 3.19.
Finally Gonshor [48] gave another proof using theorem 3.22.

The following example of Schafer [101] shows that the commutative duplicate of a special train algebra need not have this property.

Example 6.26. (1) The algebra given in the proof of theorem 3.27 is the two-fold commutative duplicate of the gametic algebra $\mathfrak{G}$ for simple Mendelian segregation with two alleles at an autosomal locus, cf. example 1.1, given by

$$\mathfrak{G} := \{\alpha_o a_o + \alpha_1 a_1 \mid \alpha_o, \alpha_1 \in \mathbb{R}\}$$

with

$$a_i a_j = \frac{1}{2}(a_i + a_j), \quad i,j=0,1.$$

With the new basis $d_o := a_o$, $d_1 := a_o - a_1$, the multiplication table reads

$$d_o^2 = d_o, \quad d_o d_1 = \frac{1}{2} d_1, \quad d_1^2 = 0 \quad .$$

Let $\mathfrak{A} := \mathfrak{G} \circledast \mathfrak{G}$ be the commutative duplicate of $\mathfrak{G}$. Then

$$e_o := d_o \circledast d_o, \quad e_1 := d_o \circledast d_1, \quad e_2 := d_1 \circledast d_1$$

form a basis of $\mathfrak{A}$ with multiplication table:

	e_o	e_1	e_2
e_o	e_o	$\frac{1}{2}\, e_1$	0
e_1		$\frac{1}{4}\, e_2$	0
e_2			0

It is immediately clear that this algebra is a special train algebra. The commutative duplicate $\mathfrak{A} \circledast \mathfrak{A}$ of $\mathfrak{A}$ is genetic with

$$c_o := e_o \circledast e_o, \quad c_1 := e_o \circledast e_1, \quad c_2 := e_1 \circledast e_1.$$

$$c_3 := e_o \circledast e_2, \quad c_4 := e_1 \circledast e_2, \quad c_5 := e_2 \circledast e_2 \;,$$

as a canonical basis. The multiplication table of $\mathfrak{A} \circledast \mathfrak{A}$ is given in the proof of 3.27. In that proof we have seen that $\mathfrak{A} \circledast \mathfrak{A}$ is not a special train algebra.

(2) Consider instead of the commutative duplicate the non commutative duplicate $\mathfrak{A} \otimes_d \mathfrak{A}$ of $\mathfrak{A}$ which is spanned by the vectors

$$c_o := e_o \otimes e_o, \quad c_1 := \tfrac{1}{2} e_1 \otimes e_o + \tfrac{1}{2} e_o \otimes e_1, \quad c_2 := \tfrac{1}{2} e_1 \otimes e_o - \tfrac{1}{2} e_o \otimes e_1,$$

$$c_3 := e_1 \otimes e_1, \quad c_4 := e_2 \otimes e_o, \quad c_5 := e_o \otimes e_2, \; c_6 := e_2 \otimes e_1,$$

$$c_7 := e_2 \otimes e_2, \quad c_8 := e_1 \otimes e_2.$$

(Following the proof of theorem 6.19 these elements in the given order form a canonical basis of the duplicate $\mathfrak{A} \otimes_d \mathfrak{A}$).

The ideal $\mathfrak{N} := \langle c_1, \ldots, c_8 \rangle$ of $\mathfrak{A} \otimes_d \mathfrak{A}$ is in fact nilpotent but

$\mathfrak{N}^{(2)} = \langle c_3, c_6, c_7, c_8 \rangle$ is not an ideal of $\mathfrak{A} \otimes_d \mathfrak{A}$ since $c_o c_3 = \frac{1}{4} c_5 \notin \mathfrak{N}^{(2)}$ as can be seen from the following multiplication table of $\mathfrak{A} \otimes_d \mathfrak{A}$. Thus $\mathfrak{A} \otimes_d \mathfrak{A}$ is not a special train algebra.

	c_o	c_1	c_2	c_3	c_4	c_5	c_6	c_7	c_8
c_o	c_o	$\frac{1}{2}c_1 - \frac{1}{2}c_2$	0	$\frac{1}{4}c_5$	0	0	0	0	0
c_1	$\frac{1}{2}c_1 + \frac{1}{2}c_2$	$\frac{1}{4}c_3$	0	$\frac{1}{8}c_8$	0	0	0	0	0
c_2	0	0	0	0	0	0	0	0	0
c_3	$\frac{1}{4}c_4$	$\frac{1}{8}c_6$	0	$\frac{1}{16}c_7$	0	0	0	0	0
c_4	0	0	0	0	0	0	0	0	0
c_5	0	0	0	0	0	0	0	0	0
c_6	0	0	0	0	0	0	0	0	0
c_7	0	0	0	0	0	0	0	0	0
c_8	0	0	0	0	0	0	0	0	0

□

Finally we remark that if $e \in \mathfrak{A}$ is an idempotent then $e \otimes e$ is an idempotent in the duplicate and in the commutative duplicate; of course in the latter case $\mathfrak{A}$ has to be commutative.

Etherington [30] claimed that the assertion of theorem 6.20 is valid also for the commutative duplicate of a commutative algebra. However Heuch [57] found out that this is not true as can be seen by comparing dimensions:
Let $\mathfrak{A}_1$ and $\mathfrak{A}_2$ be commutative algebras of dimension n_1 and n_2. Then on the one hand for the commutative duplicate of the tensor product

$$\dim\left[(\mathfrak{A}_1 \otimes \mathfrak{A}_2) \circledast (\mathfrak{A}_1 \otimes \mathfrak{A}_2)\right] = n_1 n_2 (n_1 n_2 + 1)/2$$

and on the other hand for the tensor product of the commutative duplicate

$$\dim\left[(\mathfrak{A}_1 \circledast \mathfrak{A}_1) \otimes (\mathfrak{A}_1 \circledast \mathfrak{A}_2)\right] = n_1 n_2 (n_1 + 1)(n_2 + 1)/4,$$

and these dimensions are always distinct for $n_1, n_2 > 1$; in fact we have the inequality

$$n_1 n_2 (n_1 + 1)(n_2 + 1)/4 < n_1 n_2 (n_1 n_2 + 1)/2 \quad \text{for } n_1, n_2 > 1.$$

Within this presentation we shall not discuss the partially commutative duplicate of a commutative algebra proposed by Etherington [30] because we do not see an immediate relation to applications.

In section 8 we shall discuss the sex-linked duplication introduced by Gonshor [48] and a generalization proposed in [113].

D. Construction of new algebras by linear mappings

Consider the following example:
Given populations of autopolyploid individuals in which mutations occur. Then it is reasonable (cf. Gonshor [47]) to compute the segregation rates "stepwise", i.e. to compute first the segregation rates without mutation and then to incorporate the action of mutation.
From an algebraic viewpoint this approach takes the form of constructing a new algebra from the gametic algebra for polyploid individuals by way of application of a linear mapping describing mutation.
Similarly in models for sex-linked inheritance it appears reasonable (cf. [113]) to introduce a new multiplication by applying a linear mapping to one factor of the given product.

In the following let $\mathcal{V}$ be an (m+1)-dimensional vector space over a field $\mathbb{K}$ and let $\mathcal{A}_1, \mathcal{A}_2, \mathcal{B} : \mathcal{V} \longrightarrow \mathcal{V}$ be linear mappings.
Furthermore let $\mathfrak{A}$ be a not necessarily associative or commutative algebra over $\mathcal{V}$ with multiplication

$$(x,y) \longmapsto xy \quad .$$

Then by

$$(x,y) \longmapsto x \circ y := \mathcal{B}[(\mathcal{A}_1 x)(\mathcal{A}_2 y)] \tag{6.15}$$

a new multiplication is introduced in $\mathcal{V}$ such that $\mathcal{V}$ becomes an algebra $\mathfrak{A}^{\circ}$. The multiplication is of course distributive since the mappings $\mathcal{A}_1, \mathcal{A}_2$, and $\mathcal{B}$ are linear.

In the example of polyploidy with mutation quoted above the linear mappings $\mathcal{A}_1$ and $\mathcal{A}_2$ are equal to the identity $\mathcal{J}$ and in the example of sex-linked inheritance $\mathcal{A}_1$ and $\mathcal{B}$ are equal to the identity.

<u>Remark.</u> If the linear mappings $\mathcal{A}_1, \mathcal{A}_2$ and $\mathcal{B}$ are one-to-one, then the algebras $\mathfrak{A}$ and $\mathfrak{A}^{\circ}$ are <u>isotopic</u>, cf. Albert [7]. Isotopy is an equivalence relation in the family of all algebras over a vector space $\mathcal{V}$.

Let $a_o,\dots,a_m$ be a basis of $\mathcal{W}$ with respect to which the algebra $\mathfrak{A}$ has the multiplication table

$$a_i a_j = \sum_{k=0}^{m} \gamma_{ijk} a_k, \qquad i,j=0,\dots,m.$$

Assume that the linear mappings $\mathcal{A}_1, \mathcal{A}_2, \mathcal{B} : \mathcal{W} \longrightarrow \mathcal{W}$ are represented by the matrices

$$A_1 = (\alpha_{ij}^{(1)}), \qquad A_2 = (\alpha_{ij}^{(2)}), \qquad \text{and} \quad B = (\beta_{ij}).$$

Then the multiplication constants γ^o_{ijk} of $\mathfrak{A}^o$ are

$$\gamma^o_{ijk} = \sum_{p,q,s=0}^{m} \alpha_{pi}^{(1)} \alpha_{qj}^{(2)} \gamma_{pqs} \beta_{ks}, \quad i,j,k=0,\dots,m. \tag{6.16}$$

Now we investigate which are the properties of the linear mappings $\mathcal{A}_1$, $\mathcal{A}_2$ and $\mathcal{B}$ such that properties of the algebra $\mathfrak{A}$, as for instance baric or genetic, are inherited by $\mathfrak{A}^o$.
The following theorem which has been stated by Gonshor [47] for the special case $\mathcal{A}_1 = \mathcal{A}_2 = \mathfrak{J}$ is obvious.

<u>Theorem 6.27.</u> Let $\mathcal{W}$ be a vector space over $\mathbb{K}$ and $\mathfrak{A}$ a baric algebra over $\mathcal{W}$ with weight homomorphism ω. If the linear mappings $\mathcal{A}_1, \mathcal{A}_2$, $\mathcal{B} : \mathcal{W} \longrightarrow \mathcal{W}$ preserve weight ($\omega(\mathcal{A}_1(x)) = \omega(x)$ etc.) then the algebra $\mathfrak{A}^o$ with multiplication (6.15) is baric with weight homomorphism ω.

If the basis elements $a_o,\dots,a_m$ of $\mathfrak{A}$ have all weight 1 then the linear mappings $\mathcal{A}_1, \mathcal{A}_2, \mathcal{B} : \mathcal{W} \longrightarrow \mathcal{W}$ are weight preserving if and only if they are represented with respect to this basis by matrices A_1, A_2 and B with column sums 1.

If $\mathbb{K} = \mathbb{R}$ and $\mathfrak{A}$ has a genetic realization with respect to a natural basis $a_o,\dots,a_m$, then $\mathfrak{A}^o$ also has a genetic realization with respect to this basis provided that the linear mappings are represented by column stochastic matrices with respect to this basis.
This amounts to the following

<u>Theorem 6.28.</u> Let $\mathfrak{A}$ be an algebra with a genetic realization over $\mathbb{R}$ with a natural basis $a_o,\dots,a_m$. If the linear mappings $\mathcal{A}_1$, $\mathcal{A}_2$, $\mathcal{B} : \mathcal{W} \longrightarrow \mathcal{W}$ leave the simplex

$$\mathcal{S}_m := \left\{ \sum_{i=0}^{m} \alpha_i a_i \;\middle|\; 0 \le \alpha_i \le 1, \quad i=0,\dots,m, \; \sum_{i=0}^{m} \alpha_i = 1 \right\}$$

invariant then $\mathfrak{A}^{o}$ also has a genetic realization.

The next statement given by Gonshor [67] for $\mathcal{A}_1 = \mathcal{A}_2 = \mathfrak{J}$ is also immediately clear.

Theorem 6.29. Let $\mathfrak{A}$ be an algebra over $\mathfrak{V}$ which is Gonshor genetic with a canonical basis $c_o,\dots,c_m$ over $\mathbb{L} \geq \mathbb{K}$. If the linear mappings $\mathcal{A}_1, \mathcal{A}_2, \mathcal{B} : \mathfrak{V} \longrightarrow \mathfrak{V}$ are weight preserving and leave the subspaces $\langle c_k,\dots,c_m \rangle$, k=1,...,m, of $\mathfrak{A}_{\mathbb{L}}$ invariant, then $\mathfrak{A}^{o}$ is also Gonshor genetic.

In terms of matrices this condition says that the linear mappings $\mathcal{A}_1, \mathcal{A}_2$, and $\mathcal{B}$ with respect to a canonical basis are represented by lower triangular matrices

$$\tilde{A}_1 = (\tilde{\alpha}_{ij}^{(1)}), \qquad \tilde{A}_2 = (\tilde{\alpha}_{ij}^{(2)}), \text{ and } \tilde{B} = (\tilde{\beta}_{ij})$$

with the first diagonal element equal to 1.
From this observation it follows that the left and right train roots of $\mathfrak{A}^{o}$ are given by

$$\lambda^{o}_{ooo} = 1, \lambda^{o}_{oii} = \tilde{\alpha}_{ii}^{(2)} \lambda_{oii} \tilde{\beta}_{ii}, \lambda^{o}_{ioi} = \tilde{\alpha}_{ii}^{(1)} \lambda_{ioi} \tilde{\beta}_{ii}, \quad i=1,\dots,m, \quad (6.17)$$

if $\lambda_{ooo} = 1$, λ_{oii}, λ_{ioi}, i=1,...,m, are the left and right train roots of $\mathfrak{A}$.

7. APPLICATIONS I

In this and the following section we apply "the theory of algebras in genetics" to concrete genetic situations.
To begin with we summarize the kinds of populations that have been described by the use of algebras up to the present. We always require that the underlying populations are infinitely large. With the exception of two papers which we shall mention at the end of this section the spatial structure of populations has not been considered.
In general discrete generations are assumed, and random mating between all individuals or all individuals of different sex. The individuals of populations may be diploid or autopolyploid, they can differ in one or several loci which are autosomal or sex linked. In the case of several loci these can be coupled, uncoupled or partially coupled and arbitrary recombination rates can be considered.
Usually investigations are either restricted to the case of two alleles or allow for arbitrarily, but finitely many alleles. In polyploid populations chromosome or chromatid segregation or an arbitrary distribution of double reduction has been investigated.
Up to now for autosomal loci the segregation rates have been assumed to be sex independent. With this hypothesis and the additional assumption that each mating produces both sexes with equal rates, the frequencies of gametes (zygotes) of the different sexes which otherwise carry the same genetic traits are equal from the second generation onward, i.e. from the second generation onward panmixia between individuals of different sexes is equivalent to panmixia of all individuals. On the other hand for a sex-linked locus random mating can be assumed only between individuals of different sex.
Until now segregation rates of diploid or autopolyploid individuals have been assumed symmetric, i.e. no distinction has been made whether a trait of a zygote is of paternal or maternal origin. Segregation rates are called additive if they depend in an additive way on the gametes forming the zygote. Here the special case of simple Mendelian segregation is subsumed. Generally mutation can be included. However, up to now, it has not been possible to model selection in the frame-work of algebras in genetics.

In his fundamental papers Etherington [27], [29], [30] covered many simple cases in which he constructed algebras describing the genetics and he showed that these algebras are train algebras or even special train algebras. In these papers he is mainly interested in train

Table 7.1. Classification of a population

subdivided into sub-populations	generations	individuals	mating	loci	number of loci
no	*separated*	*diploid*	*random*	*autosomal*	*1*
					arbitrarily but finitely many
			random between different sexes		
				sex-linked	
yes				*autosomal and sex-linked*	
		polyploid			
	overlapping				
	continuously overlapping				
			partially random and partially several types of inbreeding		

coupling of loci	number of alleles	segregation	segregation rates	mutation	selection
	2		*sex independent*	*no*	*no*
uncoupled	*arbitrarily but finitely many*		*symmetric additive Mendelian*	*yes*	
coupled with recombination			*asymmetric additive*		
			sex dependent symmetric in the female sex		
			additive in the female sex		
		chromosome segregation	*Mendelian in the male sex*		
		chromatid segregation			
		arbitrary distribution of double reduction			

equations for principal and plenary powers.

After Gonshor [45] had proved the theorems on the existence of idempotents and convergence of plenary powers, described in section 4, another procedure has become appropriate. One constructs an algebra describing the inheritance and then one shows that this algebra or a certain subalgebra is genetic by showing that an explicitly given basis is canonical. Subsequently one computes the train roots and applies Gonshor's convergence theorem or generalizations thereof.
If inheritance can be described by a gametic algebra which, for symmetric segregation rates is commutative, then one obtains the zygotic algebra as the commutative duplicate. If the gametic algebra is genetic then by theorem 6.25 the zygotic algebra is also genetic and then the train roots can be computed in a simple manner from those of the gametic algebra. Again the theorems of section 4 can be applied. Repeated commutative duplication generates the copular algebra.

In the following we shall mainly follow this procedure. Only in a few cases shall we also derive the train equations for principal and plenary powers.
Before entering the discussion of the special cases mentioned above we indicate that Holgate [65], [59] has also treated populations divided into finitely many subpopulations among which individuals migrate with given transition probabilities.
Gonshor [48] and Heuch [52],[55] have described populations with overlapping generations or continuous time, cf. section 7D.
Holgate [70] was the first to describe a non random mating population with the assistance of algebras. He constructed an algebra for a population where a given portion undergoes self-fertilization while the remaining individuals mate at random. Recently Holgate [72] generalized this concept to populations where fixed portions practice several types of inbreeding (selfing, brother-sister-mating, mating between double first cousins etc.) and the remaining part undergoes random mating.

Applying the theorems of sections 4 and 5 we have been able in many cases to derive results on the convergence of sequences of plenary powers or elements of weight 1. We have also introduced mutation in autopolyploid populations with an arbitary distribution of double reduction, cf. section 7C. Finally we can treat diploid populations with asymmetric segregation rates.

In table 7.1 we have collected genetic situations which can be described by algebras. Of course not all combinations are meaningful. On the other hand not all possible combinations have as yet been investigated.

For the traditional matheamtical treatment of some of these genetic situations we refer overall to the books of Crow and Kimura [20], Edwards [25], Karlin [79], Kempthorne [81], and Li [83], and the references given therein.

In section 7 we consider genetic situations which can be described by commutative algebras.

A. Additive segregation rates

Starting from example 1.1 we consider the gametic algebra $\mathfrak{G}$ for simple Mendelian inheritance (cf. Etherington [27],[29], Gonshor [46] for an autosomal locus with alleles $a_o,\ldots,a_m$. Then

$$\mathfrak{G} := \left\{ \sum_{i=0}^{m} \alpha_i a_i \,\middle|\, \alpha_i \in \mathbb{R}, \quad i=0,\ldots,m \right\}$$

is the algebra with the multiplication given by

$$a_i a_j := \tfrac{1}{2}(a_i + a_j), \quad i,j=0,\ldots,m.$$

(Holgate [64] and Heuch [56] call $\mathfrak{G}$ an elementary algebra.) $\mathfrak{G}$ is a baric algebra with weight homomorphism ω defined by

$$\omega(a_i) := 1, \qquad i=0,\ldots,m.$$

Using ω the multiplication in $\mathfrak{G}$ can written as

$$xy = \tfrac{1}{2}(\omega(x)y + \omega(y)x), \qquad x,y \in \mathfrak{G}, \tag{7.1}$$

since for $\quad x = \sum_{i=0}^{m} \xi_i a_i, \quad y = \sum_{j=0}^{m} \eta_j a_j \quad$ one has

$$xy = \sum_{i,j=0}^{m} \xi_i \eta_j a_i a_j = \tfrac{1}{2}\left(\sum_{i=0}^{m} \xi_i\right) \sum_{j=0}^{m} \eta_j a_j + \tfrac{1}{2} \sum_{i=0}^{m} \xi_i a_i \left(\sum_{j=0}^{m} \eta_j\right).$$

<u>Theorem 7.2.</u> The gametic algebra $\mathfrak{G}$ for simple Mendelian inheritance is a genetic algebra. The train roots are $\lambda_o = 1$, $\lambda_i = 1/2$, $i=1,\ldots,m$. The train equation for principal powers and plenary powers of an

element $x \in \mathfrak{G}$ is

$$x^2 - \omega(x)x = 0 \ . \tag{7.2}$$

<u>Proof.</u> The elements

$$c_o := a_o, \quad c_i := a_o - a_i, \quad i=1,\dots,m,$$

form a canonical basis of $\mathfrak{G}$, since $c_o^2 = c_o$, $c_o c_i = \frac{1}{2}c_i$, $c_i c_j = 0$, $i,j=1,\dots,m$. From these equations the train roots are easily obtained. (7.1) implies (7.2).□

From the multiplication table of the canonical basis mentioned above it can be immediately seen that $\mathfrak{G}$ is a special train algebra.

The train equation (7.2) implies that all elements $x \in \mathfrak{G}$ of weight 1 are idempotents. In biological terms this says that every state of the population is an equilibrium state, cf. Holgate [63].

The zygotic algebra $\mathfrak{Z}$ is the commutative duplicate $\mathfrak{G} \circledast \mathfrak{G}$ of the gametic algebra,

$$\mathfrak{Z} = \left\{ \sum_{\substack{i,j=0 \\ i \le j}}^{m} \alpha_{ij} a_i \circledast a_j \;\middle|\; \alpha_{ij} \in \mathbb{R} \right\} \ .$$

With the abbreviations

$$m_{ij} := \min(i,j), \quad M_{ij} := \max(i,j), \quad i,j=0,\dots,m,$$

the multiplication table of $\mathfrak{Z}$ can be written as

$$(a_i \circledast a_j)(a_p \circledast a_q) = \tfrac{1}{4}(a_{m_{ip}} \circledast a_{M_{ip}} + a_{m_{iq}} \circledast a_{M_{iq}} + a_{m_{jp}} \circledast a_{M_{jp}} + a_{m_{jq}} \circledast a_{M_{jq}}),$$

$$i \le j, \ p \le q, \ i,j,p,q=0,1,\dots,m.$$

By theorem 6.23 the algebra $\mathfrak{Z}$ is baric with weight homomorphism given by

$$\omega^{\circledast}(a_i \circledast a_j) := 1, \qquad i \le j, \ i,j=0,\dots,m.$$

The following theorem, essentially going back to Etherington [27], follows directly from section 6C with the exception of (7.4) which follows from (7.3) as a special case of the implication (4.12) $\Longrightarrow$ (4.13).

Theorem 7.3. The zygotic algebra $\mathfrak{Z}$ for simple Mendelian inheritance is a genetic algebra with train roots 1, 1/2 (m-times) and 0 (m(m+1)/2-times).
The rank equation is

$$x^3 + \omega^{\circledast}(x)x^2 = 0 \quad . \tag{7.3}$$

The plenary powers satisfy the train equation

$$x^{[3]} - \omega^{\circledast 2}(x)x^{[2]} = 0 \quad . \tag{7.4}$$

The biological interpretation of (7.4) is the Hardy-Weinberg law (cf. Edwards [25]): the frequencies of zygotes with simple Mendelian segregation at an autosomal locus stabilize after one generation, cf. also Holgate [63].

As we have seen in connection with theorem 6.25 that the elements $c_i \circledast c_j$, $i \leq j$, $i,j=0,\dots,m$, form a canonical basis with respect to the ordering $(i,j) \leq (p,q)$ if $j < q$ or $j = q$ and $i \leq p$.
The multiplication table is (cf. Gonshor [46]):

$$(c_o \circledast c_o)^2 = c_o \circledast c_o,$$

$$(c_o \circledast c_o)(c_o \circledast c_q) = \tfrac{1}{2}c_o \circledast c_q,$$

$$(c_o \circledast c_j)(c_o \circledast c_q) = \tfrac{1}{4}c_{m_{jq}} \circledast c_{M_{jq}} \quad , \quad j,q=1,\dots,m,$$

$$(c_i \circledast c_j)(c_p \circledast c_q) = 0, \; i \leq j, \; p \leq q, \; i,j=0,\dots,m, \; p,q=1,\dots,m.$$

From this it is immediately clear that $\mathfrak{Z}$ is also a special train algebra.

All these results are valid for any field with characteristic different from 2.

Although the following characterizations of the gametic and the zygotic algebra for simple Mendelian inheritance are not of immediate biological interest we include these results here.

Etherington [28] shows that in any commutative train algebra of rank 2 with train equation given by (7.2) the multiplication is given by (7.1). Hence it follows that every commutative train algebra of rank 2 can be interpreted as a gametic algebra for simple Mendelian inheritance.

Schafer [101] and Holgate [63] show that the gametic and also the zygotic algebra for simple Mendelian inheritance are Jordan algebras. The algebras $\mathfrak{G}$ and $\mathfrak{Z}$ are commutative and the Jordan identity (2.13) is easily verified.
Furthermore Schafer [101] and Holgate [63] show that the gametic algebra $\mathfrak{G}$ is a special Jordan algebra, namely that $\mathfrak{G}$ is isomorphic with the Jordan algebra $\mathfrak{A}^+$ corresponding to the associative algebra $\mathfrak{A} := \langle a_o,\dots,a_m\rangle$ with multiplication given by $a_i a_j := a_i$, $i,j=0,\dots,m$.
The zygotic algebra for simple Mendelian inheritance with 2 alleles is also a special Jordan algebra, cf. Schafer [101].

The following characterization of the gametic algebra $\mathfrak{G}$ has been given by Gonshor [46] .

Theorem 7.4. Let $\mathfrak{A}$ be a commutative baric algebra over $\mathbb{K}$, char $\mathbb{K} \neq 2$, with weight homomorphism ω. If every linear, weight preserving mapping $\varphi : \mathfrak{A} \longrightarrow \mathfrak{A}$ is an algebra homomorphism, then $\mathfrak{A}$ is a gametic algebra for simple Mendelian inheritance.

Proof. Let dim $\mathfrak{A}$ = m+1 and $a_o,\dots,a_m$ a basis of $\mathfrak{A}$ such that $\omega(a_i) = 1$, $i=0,\dots,m$.
The multiplication constants γ_{ijk}, $i,j,k=0,\dots,m$, with respect to this basis satisfy the equations $\sum_{k=0}^{m}\gamma_{ijk} = 1$, $i,j=0,\dots,m$.
Let l_1, $l_2 \in \{0,1,\dots,m\}$. By

$$\varphi_{l_1 l_2} : a_i \longmapsto \frac{1}{2}(a_{l_1} + a_{l_2}), \quad i=0,\dots,m,$$

we define a linear mapping of $\mathfrak{A}$ into itself.
This mapping is compatible with the weight homomorphism ω, thus it is an algebra homomorphism. Then on the one hand

$$a_i a_j = \sum_{k=0}^{m}\gamma_{ijk}a_k \overset{\varphi_{l_1 l_2}}{\longmapsto} \frac{1}{2}\sum_{k=0}^{m}\gamma_{ijk}(a_{l_1} + a_{l_2}) = \frac{1}{2}(a_{l_1} + a_{l_2}),$$

and on the other hand

$$\varphi_{l_1 l_2}(a_i)\varphi_{l_1 l_2}(a_j) = \left[\frac{1}{2}(a_{l_1} + a_{l_2})\right]^2 = \frac{1}{4}(a_{l_1}^2 + 2a_{l_1}a_{l_2} + a_{l_2}^2)$$

$$= \frac{1}{4}\sum_{k=0}^{m}(\gamma_{l_1 l_1 k} + 2\gamma_{l_1 l_2 k} + \gamma_{l_2 l_2 k})a_k.$$

On comparing coefficients for $l_1 = l_2 =: l$,

$$\gamma_{llk} = \begin{cases} 1 & \text{for } k = l \\ 0 & \text{otherwise} \end{cases}$$

and then comparing coefficients for $l_1 = l_2$ it follows

$$\gamma_{l_1 l_2 k} = \begin{cases} \frac{1}{2} & \text{for } k=l_1 \text{ or } k=l_2 \\ 0 & \text{otherwise} \end{cases}$$

Thus the assertion is proved. □

Now consider several autosomal loci in a population of diploid individuals. Let $a_o,\dots,a_m$ denote the different types of gametes.
The segregation rates γ_{ijk}, $k=0,\dots,m$, are called <u>additive</u> if they depend in a linear manner on the gametes a_i and a_j forming the zygote $a_i a_j$, $i,j=0,\dots,m$, i.e. if there are real numbers $0 \le \alpha_{ik}, \beta_{jk} \le 1$, such that

$$\gamma_{ijk} = \alpha_{ik} + \beta_{jk}, \qquad i,j,k=0,\dots,m.$$

In the absence of selection one has

$$\sum_{k=0}^{m} \gamma_{ijk} = 1, \qquad i,j=0,\dots,m.$$

If the segregation of the zygotes is symmetric, i.e. independent of the origin of the gametes from the parential or maternal side then

$$\gamma_{ijk} = \gamma_{jik}, \qquad i,j,k=0,\dots,m.$$

In this case α_{ik} and β_{ik} differ at most by an additive term ε_k independent of i, i.e.

$$\alpha_{ik} = \beta_{ik} + \varepsilon_k, \quad i,k=0,\dots,m.$$

Thus, introducing

$$\sigma_{ik} := 2\beta_{ik} + \varepsilon_k, \quad i,k=0,\dots,m,$$

we can write the segregation rates as

$$\gamma_{ijk} = \frac{1}{2}(\sigma_{ik} + \sigma_{jk}), \qquad i,j,k=0,\dots,m, \tag{7.5}$$

where $S := (\sigma_{ik})$ is a row stochastic matrix. Therefore the gametic algebra $\mathfrak{G} := \langle a_o,\ldots,a_m\rangle_{\mathbb{R}}$ for autosomal inheritance with additive, symmetric segregation rates has the form of a gametic algebra with mutation, cf. example 1.2.

Theorem 7.5. The gametic algebra $\mathfrak{G}$ with additive symmetric segregation rates (7.5) is a special train algebra. Hence $\mathfrak{G}$ is a genetic algebra and the train roots of $\mathfrak{G}$ are $\lambda_o = 1$, $\lambda_i = \frac{1}{2}\mu_i$, $i=1,\ldots,m$, where $1,\mu_1,\ldots,\mu_m$ are the eigenvalues of the stochastic matrix $S = (\sigma_{ik})$.

Proof. $\mathfrak{G}$ is a baric algebra with weight homomorphism given by $\omega(a_i) := 1$, $i=0,\ldots,m$.
The kernel of ω is $\mathfrak{N} := \langle a_o - a_i \mid i=1,\ldots,m\rangle$ and $\mathfrak{N}^2 = \langle 0\rangle$.
Hence $\mathfrak{G}$ is a special train algebra and a genetic algebra too. The train roots of $\mathfrak{G}$ are the eigenvalues of any left transformation $\mathcal{L}_x$, with $x = \sum_{i=0}^{m}\xi_i a_i \in \mathfrak{G}$ and $\sum_{i=0}^{m}\xi_i = 1$.

To this left transformation corresponds the matrix

$$L_x = \frac{1}{2}\,(S + e(\xi_o,\ldots,\xi_m)S)^T$$

with $e := (1,\ldots,1)^T \in \mathbb{R}^{m+1}$. The eigenvalues of this matrix have the stipulated form since the matrix $(e(\xi_o,\ldots,\xi_m)S)^T$ has rank 1 with eigenvalue 1 and eigenvector e.□

Applying theorems 4.4 and 4.18 we have the following result

Theorem 7.6. Let $\mathfrak{G}$ be a gametic algebra with additive, symmetric segregation rates (7.5).
Let the stochastic matrix $S = (\sigma_{ik})$ be regular, i.e. the eigenvalue 1 has multiplicity 1 and all other eigenvalues are less than 1 in modulus. Then the sequence of plenary powers of any element of weight 1 converges towards the idempotent of $\mathfrak{G}$.

As far as elements representing populations are concerned the convergence result can be shown without using genetic algebras.
Let $x = \sum_{i=0}^{m}\xi_i a_i \in \mathfrak{G}$ be an element in the simplex $\mathcal{T}_m$ spanned by $a_o,\ldots,a_m$.
Then to the sequence $\{x^{[t]}\}$ of plenary powers of x corresponds a Markov chain with transition matrix S and initial state $(\xi_o,\xi_1,\ldots,\xi_m)$.

From the theory of Markov chains, for a regular transition matrix, it follows that the sequence $\{x^{[t]}\}$ converges towards the uniquely determined equilibrium distribution in $\mathcal{T}_m$.

B. Partially coupled loci

Before entering the general situation we consider the gametic algebra $\mathfrak{G}$ for a population of diploid individuals with two coupled autosomal loci each with two alleles A,a and B,b, respectively, as described in example 1.3.

Let Θ be the recombination rate. With respect to the basis

$$a_1 := AB, \quad a_2 := Ab, \quad a_3 := aB, \quad a_4 := ab$$

the algebra $\mathfrak{G} = \langle a_1, a_2, a_3, a_4 \rangle_{\mathbb{R}}$ has the multiplication table given in example 1.3.

Etherington [27] and Bertran ([16], troisième partie, Chapitre I) have investigated the corresponding gametic algebra for an arbitrary finite number of alleles at each of the two loci. For the case of two alleles in both loci the result is as follows.

Theorem 7.7. The gametic algebra $\mathfrak{G}$ for two coupled loci, each with two alleles, and recombination rate Θ, $0 \leq \Theta \leq 1/2$, is a special train algebra with train roots $\lambda_0 = 1$, $\lambda_1 = \lambda_2 = 1/2$, $\lambda_3 = \frac{1}{2}(1-\Theta)$.
For $\Theta \neq 0$ the rank equation is

$$x(x-\omega(x))(x - \tfrac{1}{2}(1-\Theta)\,\omega(x)) = 0$$

and the plenary powers satisfy the train equation

$$x^{[3]} - (2-\Theta)\,\omega^2(x)x^{[2]} + (1-\Theta)\,\omega^3(x)x^{[1]} = 0 \quad .$$

Proof. The first two statements of this theorem follow immediately after transition to the basis

$$c_0 := a_1, \quad c_1 := a_1 - a_2, \quad c_2 := a_1 - a_3, \quad c_3 := a_1 - a_2 + a_4$$

with multiplication table

	c_o	c_1	c_2	c_3
c_o	c_o	$\frac{1}{2}c_1$	$\frac{1}{2}c_2$	$\frac{1}{2}(1-\Theta)c_3$
c_1		0	$\frac{1}{2}\Theta c_3$	0
c_2			0	0
c_3				0

The rank equation for principal powers is easily checked and the train equation for the plenary powers follows by the implication (4.12)$\Rightarrow$(4.13). □

From the proof it is obvious that c_o, c_1, c_2, c_3 form a canonical basis of $\mathfrak{G}$. If the recombination rate $\Theta \neq 0$ then exactly two train roots have the value 1/2 and they follow directly after the train root $\lambda_o = 1$. The gametic algebra $\mathfrak{G}$ satisfies all assumptions of theorems 4.8 and 4.19, hence we have:
For any element $x = c_o + \xi_1 c_1 + \xi_2 c_2 + \xi_3 c_3 \in \mathfrak{G}$ the sequence $\{x^{[t]}\}$ of plenary powers converges towards the idempotent
$e = c_o + \xi_1 c_1 + \xi_2 c_2 + \xi_1 \xi_2 c_3$ determined by ξ_1 and ξ_2.
This statement can be expressed in terms of the natural basis $a_1, \ldots, a_4$ as follows: The sequence of plenary powers of

$$x = (1+\xi_1 + \xi_2 + \xi_3)a_1 - (\xi_1 + \xi_3)a_2 - (\xi_2 + \xi_3)a_3 + \xi_3 a_4$$

converges towards the idempotent

$$e = (1+\xi_1)(1+\xi_2)a_1 - \xi_1(1+\xi_1)a_2 - \xi_2(1+\xi_1)a_3 + \xi_1\xi_2 a_4 \;.$$

We apply this result to elements $x \in \mathfrak{G}$ representing actual populations, i.e. elements with non negative coordinates: With non vanishing recombination rate the gamete frequencies converge towards the products of allele frequencies of the initial population. This fact has been mentioned by Jennings [78] and has been proved by Robbins [100].

It is not difficult to show that similar statements are true for the case of $n_1 = m_1 + 1$ and $n_2 = m_2 + 1$ alleles in the two coupled loci, cf. Etherington [27] . Then the train roots are 1, 1/2 ((m_1+m_2)-times) and $(1-\Theta)/2$ ($m_1 m_2$-times). The gametic frequences converge towards the products of allele frequencies in the initial population, cf. example 7.9.

Furthermore Etherington [27] and [31] has given the gametic algebra for three coupled loci with recombination. He has shown that this algebra is a special train algebra, and he has given the train equations for principal and plenary powers. Applying theorems 4.6, 4.19 one obtains the results of Geiringer [40] and Bennett [12] for the case considered.
Etherington [27] has represented the multiplication in the gametic algebra for r = 3 coupled loci using an elementary abelian p-group (p = 2) of rank r - 1 = 2. Shikita [106] also expresses the recombination among r loci by an elementary abelian p-group (p = 2) of rank r-1. But from these representations one does not get much information about the convergence behavior of sequences of plenary powers.

Now we turn to the general case for r partially coupled autosomal loci $L_1,\dots,L_r$ each with an arbitrary but finite number of alleles. We assume that the zygotes are formed by random union of gametes and that selection does not occur. Then the inheritance can be described by a gametic algebra $\mathfrak{G}$. Various representations of this algebra have been given by the following authors in chronological order:
in the case of uncoupled loci by Etherington [27] and in the general case by Reiersøl [98], Holgate [64], and Heuch [54], [56].

Suppose $a_{\varrho o},\dots,a_{\varrho m_\varrho}$ are the possible alleles at the locus L_ϱ, $\varrho = 1,\dots,r$. Let $\mathfrak{V}_\varrho$ be the $\mathbb{R}$-vector space spanned by the elements $a_{\varrho o},\dots,a_{\varrho m_\varrho}$, $\varrho = 1,\dots,r$.
Then a gamete which carries the allele $a_{\varrho i_\varrho}$ at the ϱ-th locus, $\varrho = 1,\dots,r$, is represented by the basis element

$$a_{1i_1} \otimes \dots \otimes a_{ri_r}$$

of the tensor product

$$\mathfrak{V} := \mathfrak{V}_1 \otimes \dots \otimes \mathfrak{V}_r$$

of the vector spaces $\mathfrak{V}_\varrho$, $\varrho = 1,\dots,r$, and to each gametic population corresponds an element $x \in \mathfrak{V}$. Hence the gametic algebra $\mathfrak{G}$ will be an algebra over $\mathfrak{V}$.

First we consider the case where all r loci are uncoupled:
If the segregation behavior in each locus is known and is described by a commutative algebra $\mathfrak{G}_\varrho$ over $\mathfrak{V}_\varrho$, $\varrho = 1,\dots,r$, then the genetics

of the population is described by the tensor product $\mathcal{G}_1 \otimes \ldots \otimes \mathcal{G}_r$ of the algebras $\mathcal{G}_1,\ldots,\mathcal{G}_r$, cf. Etherington [27] and Holgate [64]. As we have seen in theorems 6.10 and 6.11 the tensor product is a special train algebra or a genetic algebra if the given algebras $\mathcal{G}_1,\ldots,\mathcal{G}_r$ have these properties and the train roots are the r-fold products of the train roots of $\mathcal{G}_1,\ldots,\mathcal{G}_r$. Furthermore we have: If $c_{\varrho o},\ldots,c_{\varrho m_\varrho}$ is a canonical basis of $\mathcal{G}_\varrho$, $\varrho = 1,\ldots,r$, then

$$c_{1i_1} \otimes \ldots \otimes c_{ri_r}, \quad i_\varrho = 0,\ldots,m_\varrho, \varrho = 1,\ldots,r, \tag{7.6}$$

with respect to lexicographic ordering form a canonical basis of the tensor product.

In particular, if we have Mendelian segregation at each single locus then each algebra $\mathcal{G}_\varrho$ is a gametic algebra for simple Mendelian inheritance. Thus it is a special train algebra, and the elements

$$c_{\varrho o} := a_{\varrho o}, \quad c_{\varrho i_\varrho} := a_{\varrho o} - a_{\varrho i_\varrho}, \qquad i_\varrho = 1,\ldots,m_\varrho, \tag{7.7}$$

form a canonical basis of $\mathcal{G}_\varrho$, $\varrho = 1,\ldots,r$. The tensor products of these basis elements form a canonical basis of $\mathcal{G}$ and the multiplication is given by

$$(c_{1i_1} \otimes \ldots \otimes c_{ri_r})(c_{1j_1} \otimes \ldots \otimes c_{rj_r}) \tag{7.8}$$

$$= \begin{cases} (\frac{1}{2})^{\alpha(i,j)} \quad c_{1k_1} \otimes \ldots \otimes c_{rk_r} & \text{if for every } \varrho \in \{1,\ldots,r\} \\ & \text{either } i_\varrho = 0 \text{ or } j_\varrho = 0 \\ 0 & \text{otherwise} \end{cases}$$

where

$$c_{\varrho k_\varrho} := \begin{cases} c_{\varrho i_\varrho} & \text{if } j_\varrho = 0 \\ c_{\varrho j_\varrho} & \text{if } i_\varrho = 0 \end{cases}, \quad \varrho = 1,\ldots,r.$$

Here $\alpha(i,j)$ is the number of non vanishing coefficients of $(i_1,\ldots,i_r, j_1,\ldots,j_r)$, $i_\varrho, j_\varrho = 0,\ldots,m_\varrho$, $\varrho = 1,\ldots,r$.

The distinct train roots of $\mathcal{G}$ are $1, \frac{1}{2}, (\frac{1}{2})^2, \ldots, (\frac{1}{2})^r$, their multiplicities can be computed from $m_1,\ldots,m_r$. In particular the train root $\frac{1}{2}$ has multiplicity $\sum_{\varrho=1}^{r} m_\varrho$ and the corresponding basis elements are

$$d_{ri_r} := c_{10} \otimes \dots \otimes c_{r-1,0} \otimes c_{ri_r} \quad , \quad i_r = 1,\dots,m_r$$

$$d_{r-1,i_{r-1}} := c_{10} \otimes \dots \otimes c_{r-1,i_{r-1}} \otimes c_{ro} \quad , \quad i_{r-1} = 1,\dots,m_{r-1} \tag{7.9}$$

$$d_{1i_1} := c_{1i_1} \otimes c_{20} \otimes \dots \otimes c_{ro} \quad , \quad i_1 = 1,\dots,m_1 \quad .$$

Since the product of two basis elements which, with respect to the lexicographic order, precedes some $d_{\varrho i_\varrho}$ has no component in the direction of $d_{\varrho i_\varrho}$, $i_\varrho = 1,\dots,m_\varrho$, $\varrho = r,\dots,1$, the assumptions of theorems 4.6 and 4.19 are satisfied. Thus $\mathfrak{G}$ has a $\sum_{\varrho=1}^{m} m_\varrho$ - parametric family of idempotents

$$e = \sum \eta_{1i_1,\dots,ri_r} \, c_{1i_1} \otimes \dots \otimes c_{ri_r} \quad .$$

The parameters, namely the coordinates $\gamma_{\varrho i_\varrho}$ of $d_{\varrho i_\varrho}$, can be arbitrarily chosen, then the remaining coordinates have to be computed from

$$\eta_{1i_1,\dots,ri_r} := \gamma_{1i_1} \cdot \dots \cdot \gamma_{ri_r} \; , \; i_\varrho = 0,\dots,m_\varrho , \; \varrho = 1,\dots,r,$$

with $\gamma_{\varrho o} := 1$, $\varrho = 1,\dots,r$.
For every element $x = \sum \xi_{1i_1,\dots,ri_r} \, c_{1i_1} \otimes \dots \otimes c_{ri_r}$

of weight 1 the sequence of plenary powers converges towards the idempotent of $\mathfrak{G}$ determined by the coordinates of $d_{\varrho i_\varrho}$, $i_\varrho = 1,\dots,m_\varrho$, $\varrho = 1,\dots,r$.

Consider the other extreme case that all loci $L_1,\dots,L_r$ are strictly coupled. In the absence of recombination the gametic algebra $\mathfrak{G} = \mathfrak{G}^*$ over $\mathfrak{W}$ describing the genetic is isomorphic to the gametic algebra of a random mating population of diploid individuals with a single locus with $\prod_{\varrho=1}^{r} (m_\varrho+1)$ alleles and the same segregation rates.
In particular for the case of simple Mendelian inheritance the multiplication in $\mathfrak{G}^*$ is

$$(a_{1i_1} \otimes \dots \otimes a_{ri_r})(a_{1j_1} \otimes \dots \otimes a_{rj_r}) \tag{7.10}$$
$$= \frac{1}{2}(a_{1i_1} \otimes \dots \otimes a_{ri_r} + a_{1j_1} \otimes \dots \otimes a_{rj_r}),$$
$$i_\varrho , j_\varrho = 0,\dots,m_\varrho , \; \varrho = 1,\dots,r.$$

A basis free representation of the multiplication in $\mathcal{G}^*$ is obtained from

$$(x,y) \longmapsto xy := \frac{1}{2}(\omega(x)y + \omega(y)x), \quad x,y \in \mathfrak{V} , \tag{7.11}$$

where $\omega := \omega_1 \otimes \ldots \otimes \omega_r$ is the tensor product of the linear mappings $\omega_\varrho : \mathfrak{V}_\varrho \longrightarrow \mathbb{R}$ given by

$$\omega_\varrho(a_{\varrho i_\varrho}) := 1, \quad i_\varrho = 0,1,\ldots,m_\varrho , \quad \varrho = 1,\ldots,r. \tag{7.12}$$

Holgate [64] calls the algebra $\mathcal{G}^*$ the linked product of the algebras $\mathcal{G}_1,\ldots, \mathcal{G}_r$ describing simple Mendelian inheritance.

The linked product $\mathcal{G}^*$ is a special train algebra with train roots 1 and $(\prod_{\varrho=1}^{r}(m_\varrho+1)-1)$-times 1/2. A canonical basis of $\mathcal{G}^*$ is given by the vectors (7.6) with $c_{\varrho i_\varrho}$, $i_\varrho = 0,\ldots,m$, $\varrho = 1,\ldots,r$, defined in (7.7).

To describe the general situation of r coupled loci with arbitrary recombination and Mendelian segregation rates we can proceed along two different paths: Either we can follow Holgate [64] and represent $\mathcal{G}$ as a linear combination of tensor products of linked algebras or we can follow Reiersøl [98] and Heuch [54] who constructed the gametic algebra $\mathcal{G}$ as a convex combination of recombination algebras weighted by the linkage distribution. These two methods are based on the two traditional representations of recombination namely by crossover distribution or linkage distribution, for an extensive discussion we refer to Geiringer [40]. (A further description in terms of a semigroup generated by single recombination values has been given by Strobeck [107].)
Let us first indicate Holgate's method, cf. [64] :
Assume that the r loci $L_1,\ldots,L_r$ fall into p groups of k_i loci, $i=1,\ldots,p$, with strict coupling within the groups but no coupling between the groups. Number these groups from 1 to p such that to each coupling group there corresponds a set of successive indices

$$V_i = \{k_1 + \ldots + k_{i-1} + 1,\ldots, k_1 + \ldots + k_i\} , \quad i=1,\ldots,p.$$

Assume Mendelian segregation within each group of coupled loci. Then the gametic algebra $\mathcal{G} = \mathcal{G}_V$, $V = V_1 \mid V_2 \mid \ldots \mid V_p$, the given p-partition of $I = \{1,\ldots,r\}$, is the tensor product of p linked products of $\mathcal{G}_\mu$, $\mu \in V_i$, $i=1,\ldots,p$. Holgate proved that $\mathcal{G}_V$ is a special train algebra with a canonical basis given by (7.6) and (7.7). The distinct train roots are $1, \frac{1}{2},\ldots,(\frac{1}{2})^p$, their multiplicities can be computed from $k_1,\ldots,k_p$

and $m_1,\ldots,m_r$.
Let $Z(I)$ be the set of all p-partitions $V = V_1 \mid V_2 \mid \ldots \mid V_p$ of I, respecting the natural secession of the numbers $1,\ldots,r$, $p=1,\ldots,r$.
Then $Z(I)$ contains 2^{r-1} elements. Choose $\alpha(V) \in \mathbb{R}$, $V \in Z(I)$, with

$$\sum_{V\in Z(i)} \alpha(V) = 1 \quad .$$

Since all algebras $\mathcal{G}_V$, $V \in Z(I)$, are genetic with respect to the same basis it follows by theorem 6.5 that the algebra

$$\sum_{V\in Z(I)} \alpha(V)\, \mathcal{G}_V \tag{7.13}$$

is genetic with a canonical basis given by (7.6) and (7.7). Now assume that the r loci $L_1,\ldots,L_r$ are linearly ordered on a chromosome according to their numbering. Let $V = V_1 \mid V_2 \mid \ldots \mid V_p \in Z(I)$ be a p-partition of I.
We assume that during meiosis recombination takes place with probability $p(V)$ between the k_1-th and (k_1+1)-th loci,..., and between the $(k_1 + \ldots + k_{p-1})$-th and $(k_1 + \ldots + k_{p-1})$-th loci, and nowhere else.
Holgate proves that for any crossover distribution $p(V)$, $V \in Z(I)$, the gametic algebra $\mathcal{G}$ is a linear combination of algebras $\mathcal{G}_V$ of the above kind and he gives an expression for the $\alpha(V)$ in terms of the $p(V)$, $V \in Z(I)$.
Hence the gametic algebra for r linked loci with arbitrary recombination is a genetic algebra.

Now we define in more details a second representation of the gametic algebra $\mathcal{G}$ where the probabilities of crossovers enter directly.
Every recombination can be represented by an unordered 2-partition $U = U_1 \mid U_2$ of I and the 1-partition written as $I\mid\emptyset$, whereby $U_1 \mid U_2$ says that during meiosis recombination occurs between loci L_i, $i\in U_1$, and loci L_j, $j \in U_2$, and nowhere else, and $I\mid\emptyset$ says that there is no recombination at all.
Let $W(I)$ be the set of all unordered 1- and 2-partitions. We assume that in each partition $U \in W(I)$, $U = U_1 \mid U_2$, the component U_1 is that containing $1 \in I$. There is a one-to-one correspondence between the elements $V = V_1 \mid V_2 \mid \ldots \mid V_p$ of $Z(I)$ and the elements of $W(I)$ given by $V \longmapsto U_1 \mid U_2$ with

$$U_1 := \bigcup_{\substack{i=1\\ i \text{ odd}}}^{p} V_i \qquad \text{and} \qquad U_2 := \bigcup_{\substack{j=2\\ j \text{ even}}}^{p} V_j$$

The crossover distribution represented with respect to all 1-and 2-partitions, i.e. represented by

$$p(U), \quad U \in W(I),$$

is called linkage distribution, cf. Geiringer [40].
Now we are going to define an algebra $\mathfrak{g}(U)$ over $\mathcal{W}$ for every $U \in W(I)$. In view of this we associate with every $U = U_1 | U_2 \in W(I)$, $|U_1| = k$, a permutation $\pi_U \in \mathfrak{S}_r$, the symmetric group on I, such that the elements of $\{1,\dots,k\}$ are mapped onto the elements of U_1 and the elements of $\{k+1,\dots,r\}$ are mapped onto U_2, always with preservation of their natural order. For instance to the partition $U = U_1|U_2$, $U_1 = \{1,4,5\}$, $U_2 = \{2,3\}$ corresponds the permutation $\pi_U = (24)(35)$.

To a permutation $\pi \in \mathfrak{S}_r$ corresponds a bijective linear mapping

$$\Phi_\pi : \mathcal{W}_1 \otimes \dots \otimes \mathcal{W}_r \longrightarrow \mathcal{W}_{\pi(1)} \otimes \dots \otimes \mathcal{W}_{\pi(r)} \tag{7.14}$$

defined by

$$x_1 \otimes \dots \otimes x_r \longmapsto x_{\pi(1)} \otimes \dots \otimes x_{\pi(r)} .$$

Trivially we have the following

<u>Lemma 7.8.</u> Let $\mathcal{W}$ and $\mathcal{M}$ be vector spaces over $\mathbb{K}$ with a bijective linear mapping $\Phi : \mathcal{W} \longrightarrow \mathcal{M}$. Let $\mathcal{B}$ be an algebra over $\mathcal{M}$ with multiplication $(w,z) \longmapsto wz$. Then by

$$(x,y) \longmapsto xy := \Phi^{-1}(\Phi x)(\Phi y), \qquad x,y \in \mathcal{W}$$

a multiplication in $\mathcal{W}$ is defined and $\mathcal{W}$ becomes an algebra isomorphic with $\mathcal{B}$.

Let $U = U_1 | U_2 \in W(I)$ be a partition with $|U_1| = k$. We set

$$\omega_{U_1} := \omega_{\pi_U^{-1}(1)} \otimes \dots \otimes \omega_{\pi_U^{-1}(k)} , \quad \omega_{U_2} := \omega_{\pi_U^{-1}(k+1)} \otimes \dots \otimes \omega_{\pi_U^{-1}(r)} ,$$

where ω_{U_2} is not defined if $U_2 = \emptyset$. In the tensor product

$$\mathcal{M}_1 \otimes \dots \otimes \mathcal{M}_r := \Phi(\mathcal{W}_1 \otimes \dots \otimes \mathcal{W}_r) = \mathcal{W}_{\pi(1)} \otimes \dots \otimes \mathcal{W}_{\pi(r)}$$

we define a multiplication by

$$(w_1 \otimes \dots \otimes w_r)(z_1 \otimes \dots \otimes z_r) \tag{7.15}$$

$$:= \tfrac{1}{2}\omega_{U_1}(w_1 \otimes .. \otimes w_k)\, \omega_{U_2}(z_{k+1} \otimes \dots \otimes z_r) z_1 \otimes \dots \otimes z_k \otimes w_{k+1} \otimes \dots \otimes w_r$$

$$+ \tfrac{1}{2}\omega_{U_1}(z_1 \otimes .. \otimes z_k)\, \omega_{U_2}(w_{k+1} \otimes \dots \otimes w_r) w_1 \otimes \dots \otimes w_k \otimes z_{k+1} \otimes \dots \otimes z_r .$$

Since Φ_π is a linear bijective mapping, in view of lemma 7.8, in $\mathcal{W} = \mathcal{W}_1 \otimes \ldots \otimes \mathcal{W}_r$ a multiplication is defined by

$$(x,y) \longmapsto xy = \Phi_\pi^{-1}(\Phi_\pi x)(\Phi_\pi y), \qquad x,y \in \mathcal{W}, \tag{7.16}$$

and $\mathcal{W}$ becomes an algebra $\mathfrak{g}(U)$, a <u>recombination algebra</u> in the sense of Reiersøl [98] and Heuch [54] .
(Reiersøl, who calls the multiplication in $\mathfrak{g}(U)$ a recombination operator, introduces the multiplication with differential operators. Heuch introduces the multiplication without explicitely using permutations, but he uses permutations implicitely since he introduces the multiplication with respect to a given basis whose components have two indices and they always have to be called in their natural order).

For a given linkage distribution $p(U), U \in W(I)$, the genetics of the population is described by the gametic algebra $\mathfrak{g}$ which is the convex combination

$$\mathfrak{g} = \sum_{U \in W(I)} p(U)\, \mathfrak{g}(U) \tag{7.17}$$

of the recombintation algebras $\mathfrak{g}(U)$, $U \in W(I)$.

Before investigating the algebra $\mathfrak{g}$ we extend example 7.7.

<u>Example 7.9.</u> Suppose in $r = 2$ coupled loci there are alleles $a_{10},\ldots,a_{1m_1}$ and $a_{20},\ldots,a_{2m_2}$, respectively. The index set $I = \{1,2\}$ has partitions

$$\{1,2\}|\emptyset \quad =: \quad (1,2), \qquad \{1\}\big|\{2\} \quad =: \quad (1|2) \ .$$

Let Θ be the recombination probability, then the linkage distribution is given by

$$p(1,2) = 1-\Theta, \qquad p(1|2) = \Theta \quad .$$

In view of (7.17) the gametic algebra is the convex combination

$$\mathfrak{g} = (1-\Theta)\ \mathfrak{g}(1,2) + \Theta\ \mathfrak{g}(1|2) \quad .$$

With respect to the basis (7.6) for $r = 2$ with $c_{\rho i_\rho}$ given by (7.7) the multiplication tables of these algebras read

$\mathcal{G}(1,2)$	$c_{1o}\otimes c_{2o}$	$c_{1o}\otimes c_{2j}$	$c_{1i}\otimes c_{2o}$	$c_{1i}\otimes c_{2j}$	
$c_{1o}\otimes c_{2o}$	$c_{1o}\otimes c_{2o}$	$\frac{1}{2}c_{1o}\otimes c_{2j}$	$\frac{1}{2}c_{1i}\otimes c_{2o}$	$\frac{1}{2}c_{1i}\otimes c_{2j}$	
$c_{1o}\otimes c_{2j}$		0	0	0	$i,j>0$
$c_{1i}\otimes c_{2o}$			0	0	
$c_{1i}\otimes c_{2j}$				0	

$\mathcal{G}(1\|2)$	$c_{1o}\otimes c_{2o}$	$c_{1o}\otimes c_{2j}$	$c_{1i}\otimes c_{2o}$	$c_{1i}\otimes c_{2j}$	
$c_{1o}\otimes c_{2o}$	$c_{1o}\otimes c_{2o}$	$\frac{1}{2}c_{1o}\otimes c_{2j}$	$\frac{1}{2}c_{1i}\otimes c_{2o}$	0	
$c_{1o}\otimes c_{2j}$		0	$\frac{1}{2}c_{1i}\otimes c_{2j}$	0	$i,j>0$
$c_{1i}\otimes c_{2o}$			0	0	
$c_{1i}\otimes c_{2j}$				0	

Both algebras are genetic with $c_{1i_1}\otimes c_{2i_2}$, $i_k = 0,\dots,m_k$, $k = 1,2$, as a canonical basis. The train roots of $\mathcal{G}$ are the convex combinations of the train roots of $\mathcal{G}(1,2)$ and $\mathcal{G}(1|2)$, i.e. 1, $(m_1 + m_2)$-times 1/2, and m_1m_2 - times $(1 + \Theta)/2$, in accordance with the results mentioned above. □

The following theorem was given by Heuch [54] .

Theorem 7.10. For an arbitrary linkage distribution $p(U)$, $U\in W(I)$, the gametic algebra

$$\mathcal{G} = \sum_{U\in W(I)} p(U)\,\mathcal{G}(U)$$

is a genetic algebra with (7.6), (7.7) as a canonical basis.

Proof. Because of theorem 6.5 the assertion is proved if we have shown that all algebras $\mathcal{G}(U)$, $U\in W(I)$, are genetic with (7.6), (7.7) as a canonical basis in some fixed order, for example the lexicographic order. For each $k\in\{1,\dots,r\}$ we consider the partition $U^k := U_1|U_2 \in W(I)$ with

$$U_1 := \{1,\dots,k\}\ ,\quad U_2 := \{k+1,\dots,r\}.$$

To this partition corresponds the identical permutation on I and thus the identity on $\mathcal{W} = \mathcal{W}_1\otimes\dots\otimes\mathcal{W}_r$. Hence the multiplication in

$\mathfrak{g}(U^k)$ is

$$(x_1 \otimes \dots \otimes x_r)(y_1 \otimes \dots \otimes y_r)$$
$$= \frac{1}{2} \prod_{i=1}^{k} \omega_i(x_i) \prod_{j=k+1}^{r} \omega_j(y_j) y_1 \otimes \dots \otimes y_r \otimes x_{r+1} \otimes \dots \otimes x_r$$
$$+ \frac{1}{2} \prod_{i=1}^{k} \omega_i(y_i) \prod_{j=k+1}^{r} \omega_j(x_j) x_1 \otimes \dots \otimes x_r \otimes y_{r+1} \otimes \dots \otimes y_r .$$

For a given permutation $\pi \in \mathfrak{S}_r$ we define the lexicographic π-ordering by $(i_1,\dots,i_r) \leq_\pi (j_1,\dots,j_r)$ if there is an $s \in \{1,\dots,r\}$ such that $i_{\pi(1)} = j_{\pi(1)},\dots,i_{\pi(s-1)} = j_{\pi(s-1)}, i_{\pi(s)} \leq j_{\pi(s)}$.

We show that the basis (7.6), (7.7) is a canonical basis for the algebras $\mathfrak{g}(U^k)$, $k=1,\dots,r$, with respect to all lexicographic π-orderings. For every index vector $i=(i_1,\dots,i_r)$, $i_\varrho = 0,\dots,m_\varrho$, $\varrho = 1,\dots,r$, we define $\alpha(i) \in \{0,1,\dots,r\}$ as the number of indices which are different from 0. One easily checks that for any two index vectors $i=(i_1,\dots,i_r)$ and $j=(j_1,\dots,j_r)$ with $\alpha(i) < \alpha(j)$ we have $i <_\pi j$ for all $\pi \in \mathfrak{S}_r$.

Consider the multiplication table of $\mathfrak{g}(U^k)$ with respct to the basis $d_i := c_{1i_1} \otimes \dots \otimes c_{ri_r}$, $i=(i_1,\dots,i_r)$, with $c_{\varrho i_\varrho}$ given by (7.7),

$i_\varrho = 0,\dots,m_\varrho$, $\varrho = 1,\dots,r$. We define $0 := (0,\dots,0)$.

Then we have

$$d_0 d_i = \begin{cases} d_0 & \text{for } i = 0 \\ \frac{1}{2} d_i & \text{for } i \neq 0 \text{ and } i_1,\dots,i_k = 0 \\ & \quad \text{or } i_{k+1},\dots,i_r = 0 \\ 0 & \text{otherwise} \end{cases}$$

and for $i,j > 0$

$$d_i d_j = \begin{cases} \frac{1}{2} d_l & \text{for (a) } j_1,\dots,j_k,\ i_{k+1},\dots,i_r = 0 \\ & \text{or (b) } i_1,\dots,i_k,\ j_{k+1},\dots,j_r = 0 \\ 0 & \text{otherwise} \end{cases}$$

with

$$d_l = \begin{cases} c_{1i_1} \otimes \dots \otimes c_{ki_k} \otimes c_{k+1,j_{k+1}} \otimes \dots \otimes c_{rj_r} & \text{for (a)} \\ c_{1j_1} \otimes \dots \otimes c_{kj_k} \otimes c_{k+1,i_{k+1}} \otimes \dots \otimes c_{ri_r} & \text{for (b).} \end{cases}$$

Hence the product $d_0 d_i$ has no components in the direction of d_j for all j with $j <_\pi i$, $\pi \in \mathfrak{S}_r$.

If $d_i d_j = \frac{1}{2} d_l$ for i,j >0, then we have $\alpha(l) > \max(\alpha(i), \alpha(j))$, thus $d_i d_j$ has no components in the direction of d_k for all $k \leq_\pi \max(i,j)$, $\pi \in \mathfrak{S}_r$.
Therefore d_i, $i=0,\dots,(m_r,\dots,m_r)$, is a canonical basis of $\mathfrak{G}(U^k)$ with respect to every lexicographic π-ordering, $\pi \in \mathfrak{S}_r$.
Hence in every recombination algebra $\mathfrak{G}(U)$, $U \in W(I)$, this basis is canonical with respect to the lexicographic order. The assertion follows with theorem 6.5. □

Heuch [54] has also shown that the gametic algebra $\mathfrak{G}$ is a special train algebra for every linkage distribution.

From the proof of the theorem 7.10 it follows that $\mathfrak{G}(U^k)$, $k \in \{1,\dots,r\}$, and thus every recombination algebra $\mathfrak{G}(U)$, $U \neq I \mid \emptyset$, has the train roots 1 (once), 1/2 and 0.
In his paper [54] Heuch associates with each index tuple $i=(i_1,\dots,i_r)$ the set of all 1- and 2-partitions $U = U_1|U_2$ of I with the property that all $l \in I$ with $i_l \neq 0$ are either in U_1 or U_2, this set is denoted by $Q_i(I)$.

If the index vector $i=(i_1,\dots,i_r)$ has exactly one coefficient different from 0 then $Q_i(I) = W(I)$ and thus

$$\sum_{U \in Q_i(I)} p(U) = 1 \quad \text{for all } i \text{ with } \alpha(i) = 1 \ . \tag{7.18}$$

It is easy to see that in the gametic algebra $\mathfrak{G}$ (7.17) there belong to the basis elements c_0 and c_i the train roots 1 and

$$\frac{1}{2} \sum_{U \in Q_i(I)} p(U), \tag{7.19}$$

respectively, where i runs over all $(\prod_{\rho=1}^{r}(m_\rho +1)-1)$ index vectors different from the zero vector.
In view of (7.18), in particular, those train roots equal 1/2 which belong to basis elements c_i with $\alpha(i) = 1$, i.e. the basis elements given by (7.9).

If the linkage distribution p(U) satisfies the condition

$$\sum_{U \in Q_i(I)} p(U) < 1 \qquad \text{for all } i \text{ with } \alpha(i) = 2, \tag{7.20}$$

and thus also for i with $\alpha(i) > 2$, then exactly those train roots equal

1/2 which belong to the $\sum_{\varrho=1}^{r} m_\varrho$ basis elements c_i with $\alpha(i) = 1$. Condition (7.20) expresses the condition that between each of the coupled loci, recombination occurs with positive probability.

From the proof of theorem 7.10 it follows that all products of basis elements c_i and c_j which precede c_l, $\alpha(l) = 1$, in lexicographic order, have no component in the direction of c_l, $l=(0,\dots,0,1),\dots,(m_1,0,\dots,0)$. Therefore we can apply theorems 4.6 and 4.19 and obtain the following result of Heuch [57].

Theorem 7.11. If the linkage distribution $p(U)$, $U \in W(\underline{I})$, satisfies (7.20) then the gametic algebra $\mathcal{G}$ given by (7.17) has a $(\sum_{\varrho=1}^{r} m_\varrho)$- parametric family of idempotents

$$\mathcal{E} = \Big\{ e = \sum_i \eta_i c_i \;\Big|\; \eta_0 = 1,\ \eta_i \in \mathbb{R} \quad \text{for } i \text{ with } \alpha(i) = 1 \text{ and}$$
$$\eta_j = \eta_{j_1,\dots,j_r} = \prod_{\substack{i=(0,\dots,0,j_\varrho,0,\dots,0)\\ \varrho=1,\dots,r}} \eta_i \Big\}$$

and for each element $x \in \mathcal{G}$ of weight 1 the sequence of plenary powers converges towards the idempotent of $\mathcal{G}$ given by the coordinates of x with respect to c_i, $\alpha(i) = 1$.

This theorem has the following genetic interpretation: If for any two loci of finitely many coupled loci the recombination probability is positive then under panmixia the population converges towards an equilibrium state, to which the population would converge without coupling.

Heuch [54], [57] shows that theorems 7.10 and 7.11 can be modified to include mutation. We give an idea of the proof. For any locus L_ϱ mutation can be described by a linear mapping $\mathcal{M}_\varrho : \mathcal{W}_\varrho \longrightarrow \mathcal{W}_\varrho$, $\varrho = 1,\dots,r$. Let

$$\mathcal{M} := \mathcal{M}_1 \otimes \dots \otimes \mathcal{M}_r$$

be the tensor product of these mappings.
With the assumption that mutation in the considered loci occurs independently (which is natural to assume for biological reasons) then inheritance is described by the gametic algebra

$$\mathcal{G}_{\mathcal{M}} := \sum_{U \in W(I)} p(U)\, \mathcal{G}_{\mathcal{M}}(U) \qquad ,$$

where the multiplication in $\mathcal{G}_{\mathcal{M}}(U)$ is defined as

$$(x,y) \longmapsto \mathcal{M}(xy),$$

here xy is the ordinary product of x and y in $\mathcal{G}(U)$, $U \in W(I)$, cf. section 6D.
In the algebra $\mathcal{G}_\varrho$ over $\mathcal{W}_\varrho$ for simple Mendelian inheritance (with multiplication $(x,y) \mapsto \frac{1}{2}(\omega_\varrho(x)y + \omega_\varrho(y)x))$ every basis $c_{\varrho 0}, c_{\varrho 1}, \ldots, c_{\varrho m_\varrho}$ with $\omega_\varrho(c_{\varrho 0}) = 1$, $\omega_\varrho(c_{\varrho i_\varrho}) = 0$, $i_\varrho = 1,\ldots,m_\varrho$, is canonical, $\varrho = 1,\ldots,r$. Therefore theorem 7.10 remains valid if the mutation mappings $\mathcal{M}_\varrho$, $\varrho = 1,\ldots,r$, are diagonizable. This is easy to see: in $\mathcal{W}_\varrho$ choose a canonical basis with respect to which $\mathcal{M}_\varrho$ is represented by a diagonal matrix, $\varrho = 1,\ldots,r$. The train roots of $\mathcal{G}_\varrho$ are modified according to the eigenvalues of $\mathcal{M}_\varrho$.
Heuch [57] has shown that these statements are true also for the limiting case of non diagonizable mutation matrices $\mathcal{M}_\varrho$, $\varrho = 1,\ldots,r$.

If instead of one coupling group several coupling groups are present then the gametic algebra is the tensor product of algebras for the single coupling groups and the results carry over with the theorems of section 6B.

Holgate [69] has considered in addition a subdivision of the population into finitely many subpopulations with migration between subpopulations alternating with random mating within each subpopulation. We also refer to Holgate's recently published paper [71] on a canonical multiplication in the genetic algebra for linked loci.

Finally we indicate that Reiersøl [98] and Heuch [53], [54], [56], [57] have considered also recombination for sex-linked or partially sex-linked loci and for chromosomes with an additional incompatibility locus.

C. Autopolyploidy

The theory of genetic algebras is of great advantage in the investigation of populations of autopolyploid individuals. With direct methods, in general, populations of tetra- and hexaploid individuals with chromosome segregation have been investigated with great computation effort, e.g. by the following authors in chronological order: Haldane [60] , Geiringer [41] , Bennett [12] , Li [83] , and Seyffert [104] , [105] ; Geiringer [42] and Seyffert [105] also have covered the case of chromatid segregation and of an arbitrary amount of double reduction, respectively.

For autopolyploid individuals of degree $r > 1$, in short 2r-ploid individuals, we have to distinguish between several modes of segregation, namely chromosome segregation, in which only chromosomes can be independently distributed among the offspring and chromatid segregation, in which all chromosomes have split into two chromatids and all chromatids can be independently distributed. The latter case is also called complete double reduction. Of course all mixed types of segregation between these extreme cases, so called partial double reductions, may occur. In natural populations the mode of segregation depends mainly on the distance of the genetically different loci from the centromere. A mathematical classification of segregation types has been given by Fisher [34] . For a comprehensive discussion of these segregation modes in autotetraploid populations compare Seyffert [104],[105] .

Again Etherington [27] introduced the investigation of such populations through algebras. He considered the gametic algebra for a population of tetraploid individuals with one locus and two alleles in the case of pure chromosome segregation. Gonshor [45],[46],[47] carried these investigations further and he took mutation into account. Furthermore Gonshor extended the theory to the case of an arbitrary but finite number of alleles at one locus. He showed that the gametic and zygotic algebras are genetic and computed the train roots. Heuch [58] gave an explicit formula for the gametic frequencies in the t-th generation of a random mating 2r-ploid population with chromosome segregation. Holgate [61] also constructed the so called segregation algebras $\mathcal{A}(r,s)$, $s \in \mathbb{N}$, for two alleles at one locus describing pure chromosome segregation (s=1) as well as pure chromatid segregation (s=2). He showed that these algebras are genetic and computed their train roots.

Recently Heuch [60] and the author [114] have constructed the gametic algebra of a population of 2r-ploid individuals with an arbitrary amount of double reduction. Heuch considers the case of an arbitrary but finite number of alleles at one locus. He uses a technique introduced by Gonshor [46], [47], avoiding the explicit use of multinomial coefficients. Heuch shows that the gametic algebra is genetic and he computes the train roots. However he does not consider the convergence properties of sequences of plenary powers.

In the following we restrict ourselves to the case of two alleles and follow the presentation in [114]. In addition we allow mutation.

Suppose the r-ploid gametes of the population differ at one locus, with alleles A and a. As usual we denote the gametes by

$$D_i := A^i a^{r-i}, \qquad i = 0,1,\dots,r.$$

Then the zygotes of the population are $A^{i+j}a^{2r-(i+j)}$, $i,j=0,1,\dots,r$. For every integer ϱ, $0 \le \varrho \le 2r$, let $p(\varrho)$ be the probability that ϱ of the 2r chromosomes of a zygote split into two chromatids which can be independently distributed among the offspring. Hence the function p is the distribution of double reduction. The distributions

$$p_0(\varrho) = \begin{cases} 1 & \text{if } \varrho = 0 \\ 0 & \text{otherwise} \end{cases} \qquad \text{and} \qquad p_{2r}(\varrho) = \begin{cases} 1 & \text{if } \varrho = 2r \\ 0 & \text{otherwise} \end{cases}$$

give the two extreme cases of pure chromosome and pure chromatid segregation.

Now let $\varrho \in \{0,1,\dots,2r\}$ be fixed. Consider the zygote $A^{i+j}a^{2r-(i+j)}$. Then the probability that exactly σ chromosomes, $0 \le \sigma \le \min(\varrho, i+j)$, which carry the allele A, split into two chromatids is given by

$$\frac{\binom{i+j}{\sigma}\binom{2r-(i+j)}{\varrho-\sigma}}{\binom{2r+\varrho}{r}}$$

In this case gametes $D_l = A^l a^{r-l}$, $0 \le l \le \min(r, i+j+\sigma)$, are produced with probability

$$\frac{\binom{i+k+\sigma}{l}\binom{2r+\varrho-(i+j+\sigma)}{r-1}}{\binom{2r+\varrho}{r}}$$

For convenience we define $\binom{n}{\sigma} := 0$ if $\sigma < 0$ or $\sigma > n$.
Let $\mathfrak{V}$ be the free $\mathbb{R}$-module generated by $D_0, D_1, \ldots, D_r$. For every $\varrho \in \{0,\ldots,2r\}$ we introduce in $\mathfrak{V}$ a multiplication by $(D_i, D_j) \longmapsto f_\varrho(D_i,D_j) :=$

$$\sum_{l=0}^{r} \sum_{\sigma=0}^{2r} \frac{\binom{i+j}{\sigma}\binom{2r-(i+j)}{\varrho-\sigma}}{\binom{2r-\varrho}{r}} \frac{\binom{i+j+\sigma}{l}\binom{2r+\varrho-(i+j+\sigma)}{r-1}}{\binom{2r+\varrho}{r}} D_l, \quad i,j=0,\ldots,r, \tag{7.21}$$

and its bilinear extension onto $\mathfrak{V} \times \mathfrak{V}$.
With this multiplication $\mathfrak{V}$ becomes a commutative algebra which is denoted by $\mathfrak{G}_\varrho$.

Then the gametic algebra $\mathfrak{G}$ for a random mating population of 2r-ploid individuals and a given distribution $\varrho \longmapsto p(\varrho)$ of double reduction is given by

$$\mathfrak{G} = \sum_{\varrho=0}^{2r} p(\varrho)\mathfrak{G}_\varrho \quad . \tag{7.22}$$

For the special distribution p_0 and p_{2r} of double reduction the gametic algebra $\mathfrak{G}$ coincides with the gametic algebras for pure chromosome segregation and pure chromatid segregation, respectively, which have been constructed by Gonshor [45] and Holgate [61] , respectively.

Let us consider again the general case. In the vector space $\mathfrak{V}$ we choose a new basis

$$c_u := \sum_{i=0}^{u} (-1)^i \binom{u}{i} D_i, \qquad u = 0,\ldots,r. \tag{7.23}$$

The corresponding linear transformation is self-inverse, hence we have

$$D_i = \sum_{u=0}^{i} (-1)^u \binom{i}{u} c_u, \qquad i = 0,\ldots,r. \tag{7.24}$$

<u>Theorem 7.12.</u> Every algebra $\mathfrak{G}_\varrho$, $\varrho \in \{1,\ldots,r\}$, is genetic and possesses $c_0, c_1,\ldots,c_r$ as a canonical basis. The multiplication table is given by

$$f_\varrho(c_u,c_v) = \sum_{t=m}^{2m} \lambda_{mt}^{(\varrho)} c_t, \quad m := u + v, \quad u,v=0,\ldots,m, \tag{7.25}$$

with

$$\lambda_{mt}^{(\varrho)} := (-1)^{m+t}\binom{2r}{\varrho}^{-1}\binom{2r+\varrho}{r}^{-1}\binom{2r+\varrho-t}{r-t}\sum_{\nu=0}^{t}\binom{2r-\nu}{\varrho-\nu}\binom{m}{t-\nu}\binom{t-\nu}{m-\nu} \qquad (7.26)$$

Proof. We calculate the multiplication table of $\mathfrak{g}_\varrho$ with respect to the basis $c_o, c_1,\ldots,c_r$.

$$f_\varrho(c_u,c_v) = \sum_{i=0}^{u}\sum_{j=0}^{v}(-1)^{i+j}\binom{u}{i}\binom{v}{j}f_\varrho(D_i,D_j) \quad .$$

Since the right-hand side of the expression for $f_\varrho(D_i,D_j)$ in (7.21) only depends on k := i+j, with the further substitution m := u + v and the well known formula

$$\sum_{i=0}^{k}\binom{u}{i}\binom{m-u}{k-i} = \binom{m}{k} \quad , \qquad (7.27)$$

we get

$$f_\varrho(c_u,c_v) = \sum_{k=0}^{m}\sum_{i=0}^{k}(-1)^k\binom{u}{i}\binom{m-u}{k-i}f_\varrho(D_i,D_j)$$

$$= \sum_{k=0}^{m}(-1)^k\binom{m}{k}f_\varrho(D_i,D_j) \quad .$$

By the definition (7.21) and relation (7.24) we obtain

$$f_\varrho(c_u,c_v)$$

$$= \binom{2r}{\varrho}^{-1}\binom{2r+\varrho}{r}^{-1}\sum_{l=0}^{r}\sum_{\sigma=0}^{\varrho}\sum_{k=0}^{m}(-1)^k\binom{m}{k}\binom{k}{\sigma}\binom{2r-k}{\varrho-\sigma}\binom{k+\sigma}{l}\binom{2r+\varrho-(k+\sigma)}{r-l}D_l$$

$$= \binom{2r}{\varrho}^{-1}\binom{2r+\varrho}{r}^{-1}\sum_{t=0}^{r}(-1)^t\sum_{l=0}^{r}\sum_{\sigma=0}^{\varrho}\sum_{k=0}^{m}(-1)^k\binom{m}{k}\binom{k}{\sigma}\binom{2r-k}{\varrho-\sigma}\binom{k+\sigma}{l}\binom{2r+\varrho-(k+\sigma)}{r-l}c_t$$

From the formula

$$\binom{p}{l}\binom{l}{t} = \binom{p}{t}\binom{p-t}{l-t} \qquad (7.28)$$

for $\varrho = k+\sigma$ and formula (7.27) we get for

$$f_\varrho\,(c_u,c_v)$$

$$= \binom{2r}{\varrho}^{-1}\binom{2r+\varrho}{r}^{-1}\sum_{t=0}^{r}(-1)^t\sum_{\sigma=0}^{\varrho}\sum_{k=0}^{m}(-1)^k\binom{m}{k}\binom{k+\sigma}{t}\binom{k}{\sigma}\binom{2r-k}{\varrho-\sigma}\sum_{l=0}^{r}\binom{k+\sigma-t}{l-t}\cdot$$

$$\binom{2r+\varrho-(k+\sigma)}{r-l}c_t$$

$$= \binom{2r}{\varrho}^{-1}\binom{2r+\varrho}{r}^{-1}\sum_{t=0}^{r}(-1)^t\binom{2r+\varrho-t}{r-t}s_{mt}^{(\varrho)}c_t$$

with

$$S_{mt}^{(\varrho)} := \sum_{k=0}^{m} (-1)^k \binom{m}{k} \sum_{\sigma=0}^{\varrho} \binom{k+\sigma}{t} \binom{k}{\sigma} \binom{2r-k}{\varrho-\sigma} \quad . \tag{7.29}$$

By the above definition we have $\binom{k+\sigma}{t} = 0$ if $t > k+\sigma$ and $\binom{k}{\sigma} = 0$ if $\sigma > k$. Therefore in every case $S_{mt}^{(\varrho)} = 0$ if $t > 2m$. Since the multiplication constant $\lambda_{mt}^{(\varrho)}$ is a multiple of $S_{mt}^{(\varrho)}$ we know that $\lambda_{mt}^{(\varrho)} = 0$ for $t > 2m$.

We are going to prove that $S_{mt}^{(\varrho)} = 0$ if $t < m$, and we use the following formulas for the product of two binomial coefficients (see for example Riordan [99] , p.15)

$$\binom{m}{p}\binom{n}{q} = \sum_{k=0}^{n} \binom{m-n+q}{p-k+q} \binom{k}{q} \binom{n}{k} \tag{7.30}$$

$$\binom{n}{p}\binom{p}{n-q} = \sum_{k=0}^{n} (-1)^{n+k} \binom{n}{k} \binom{k}{p} \binom{k}{q} \tag{7.31}$$

The summation with respect to σ in (7.29) can be taken from 0 to k. Hence with (7.27), (7.30), and (7.31) we get

$$\begin{aligned} S_{mt}^{(\varrho)} &= \sum_{k=0}^{m} (-1)^k \binom{m}{k} \sum_{\sigma=0}^{k} \binom{k+\sigma}{t} \binom{k}{\sigma} \binom{2r-k}{\varrho-\sigma} \\ &= \sum_{\nu=0}^{t} \sum_{k=0}^{m} (-1)^k \binom{m}{k} \binom{k}{t-\nu} \sum_{\sigma=0}^{k} \binom{\sigma}{\nu} \binom{k}{\sigma} \binom{2r-k}{\varrho-\sigma} \\ &= \sum_{\nu=0}^{t} \sum_{k=0}^{m} (-1)^m \binom{m}{k} \binom{k}{t-\nu} \binom{k}{\nu} \binom{2r-k}{\varrho-\nu} \\ &= (-1)^m \sum_{\nu=0}^{t} \binom{m}{t-\nu} \binom{t-\nu}{m-\nu} \binom{2r-\nu}{\varrho-\nu} \quad . \end{aligned}$$

Since $\binom{t-\nu}{m-\nu} = 0$ if $t < m$, we know that $S_{mt}^{(\varrho)} = 0$ for $t < m$ and therefore $\lambda_{mt}^{(\varrho)} = 0$ for $t < m$. Furthermore the multiplication constants $\lambda_{mt}^{(\varrho)}$ are of the asserted form (7.26).

Since $\lambda_{mt}^{(\varrho)} = 0$ if $t < m$ and $\lambda_{00}^{(\varrho)} = 1$ the algebra $\mathfrak{G}_\varrho$ is genetic with $c_o, c_1, \ldots, c_r$ as a canonical basis. □

Since $\mathfrak{G}$ is a convex combination of the 2r+1 genetic algebras $\mathfrak{G}_\varrho$, which have a common canonical basis, we obtain from theorem 6.5

Theorem 7.13. The gametic algebra $\mathfrak{G}$ for a random mating population of 2r-ploid individuals with an arbitrary distribution of double reduction is a genetic algebra and $c_o,\dots,c_r$ (7.23) form a canonical basis of $\mathfrak{G}$. The multiplication table reads

$$c_u c_v = \sum_{t=m}^{2m} \lambda_{mt} c_t, \qquad m = u + v, \qquad u,v = 0,\dots,r, \tag{7.32}$$

with

$$\lambda_m = \sum_{\varrho=0}^{2r} p(\varrho)\lambda_{mt}^{(\varrho)}, \quad \lambda_{mt}^{(\varrho)} \quad \text{given by (7.26).}$$

For reasons of completeness we mention that the algebras $\mathfrak{G}_o,\dots,\mathfrak{G}_{2r}$ are not linearly independent in the sense of definition 6.2. One easily checks that the algebras $\mathfrak{G}_{2r}$ and $\mathfrak{G}_{2r-1}$ coincide, cf. [114]. For a further investigation of the kind of linear dependence of these algebras we refer to Heuch [60].

Theorem 7.14. (1) The train roots of the genetic algebras $\mathfrak{G}_\varrho$ are

$$\lambda_i^{(\varrho)} = \binom{2r}{\varrho}^{-1}\binom{2r+\varrho}{r}^{-1}\binom{2r+\varrho-i}{r-i}\sum_{\nu=0}^{i}\binom{i}{\nu}\binom{2r-\nu}{\varrho-\nu}, \quad i=0,1,\dots,r. \tag{7.33}$$

In particular

$$\lambda_0^{(\varrho)} = 1, \quad \lambda_1^{(\varrho)} = 1/2 .$$

(2) The train roots satisfy the inequality

$$\lambda_i^{(\varrho)}\lambda_j^{(\varrho)} > \lambda_{i+j}^{(\varrho)} > 0 \text{ if } \quad i,j > 0 \text{ and } i+j \leq r. \tag{7.34}$$

(Part I of this theorem is proved for $\varrho = 0$ by Gonshor [45] and for $\varrho = 2r$ by Holgate [61], and the inequality (2) is mentioned for $\varrho = 0$ by Heuch [58].

Proof. (I) Since the train roots $\lambda_i^{(\varrho)}$ of $\mathfrak{G}_\varrho$ are the mulitplication constants $\lambda_{ii}^{(\varrho)}$ of $\mathfrak{G}_\varrho$ mentioned in (7.26), we get (7.33) directly from (7.26). Especially $\lambda_0^{(\varrho)} := \lambda_{00}^{(\varrho)} = 1$ and

$$\lambda_1^{(\varrho)} := \lambda_{11}^{(\varrho)} = \binom{2r}{\varrho}^{-1}\binom{2r-\varrho}{r}^{-1}\binom{2r+\varrho-1}{r-1}\left[\binom{2r}{\varrho}+\binom{2r-1}{\varrho-1}\right]$$

$$= \binom{2r}{\varrho}^{-1}\binom{2r-\varrho}{r}^{-1}\binom{2r+\varrho}{r}\frac{2r}{2r+\varrho}\left[1+\frac{\varrho}{2r}\right]\binom{2r}{\varrho} = 1/2.$$

II. To prove part (2) of the above theorem we need the following

Lemma 7.15. Let $b,B \in \mathbb{N}$ with $b \leq B$.
For all $i,j=0,1,\dots,B-1$ such that $i+j < B-b+1$ we have the inequality

$$\binom{B-i}{b}\binom{B-j}{b} \geq \binom{B}{b}\binom{B-(i+j)}{b} > 0 \tag{7.35}$$

and equality holds if and only if $i = 0$ or $j = 0$.

Proof. For $i = 0$ or $j = 0$ (7.35) holds trivially with equality. Assume now $i,j \in \{1,\dots,B-1\}$ with $i+j < B$. Then the strong version of (7.35) is equivalent to

$$(B-i)_b(B-j)_b > (B)_b(B-(i+j))_b > 0 .$$

A sufficient condition for the last inequality to be fulfilled is that

$$(B-i-c)(B-j-c) > (B-c)(B-(i+j)-c) > 0 \qquad \text{for every } c = 0,1,\dots,b-1,$$

which is satisfied for all $i,j \in \{1,\dots,B-1\}$ if $i+j < B-b+1$. □

Proof of Theorem 7.14(2). Take $i,j > 0$ with $i+j \leq r$. Then (7.34) is equivalent to

$$\frac{\binom{2r+\varrho-i}{r-i}\sum_{\nu=0}^{i}\binom{i}{\nu}\binom{2r-\nu}{\varrho-\nu}}{\binom{2r+\varrho}{r}\binom{2r}{\varrho}} \cdot \frac{\binom{r+\varrho-j}{r-j}\sum_{\mu=0}^{j}\binom{j}{\mu}\binom{2r-\mu}{\varrho-\mu}}{\binom{2r+\varrho}{r}\binom{2r}{\varrho}}$$

$$> \frac{\binom{2r+\varrho-(i+j)}{r-(i+j)}\sum_{\varkappa=0}^{i+j}\binom{i+j}{\varkappa}\binom{2r-\varkappa}{\varrho-\varkappa}}{\binom{2r+\varrho}{r}\binom{2r}{\varrho}} > 0,$$

or, if we multiply by $\binom{2r+\varrho}{r}^2\binom{2r}{\varrho}^2$ and take account of

$$\binom{B}{b} = \binom{B}{B-b} \qquad \text{for } b,B \geq 0, \tag{7.36}$$

then (7.34) is equivalent to

$$\left[\binom{2r+\varrho-i}{r+\varrho}\binom{2r+\varrho-j}{r+\varrho}\right]\left[\sum_{\nu=0}^{i}\binom{i}{\nu}\binom{2r-\nu}{\varrho-\nu}\sum_{\mu=0}^{j}\binom{j}{\mu}\binom{2r-\mu}{\varrho-\mu}\right]$$

$$> \left[\binom{2r+\varrho}{r+\varrho}\binom{2r+\varrho-(i+j)}{r+\varrho}\right]\left[\binom{2r}{\varrho}\sum_{\varkappa=0}^{i+j}\binom{i+j}{\varkappa}\binom{2r-\varkappa}{\varrho-\varkappa}\right] > 0. \tag{7.37}$$

From lemma 7.15 with $B = 2r + \varrho$ and $b = r + \varrho$ we know that

$$\binom{2r+\varrho-i}{r+\varrho}\binom{2r+\varrho-j}{r+\varrho} > \binom{2r+\varrho}{r+\varrho}\binom{2r+\varrho-(i+j)}{r+\varrho} > 0, \qquad (7.38)$$

since $B-b+1 = 2r+\varrho-(r+\varrho)+1 = r+1 > i+j$.

Now we can compare the remaining factors of (7.37). Since $\binom{2r-\nu}{\varrho-\nu} = 0$ if $\varrho < \nu$, it is sufficient in (7.37) to sum over $\nu = 0,\dots,\min(i,\varrho)$ and similarly over $\mu = 0,\dots,\min(j,\varrho)$ and $\varkappa = 0,\dots,\min(i+j,\varrho)$.

Then we can use (7.36) for $\binom{2r-\nu}{\varrho-\nu}$, $\binom{2r-\mu}{\varrho-\mu}$ and $\binom{2r-\varkappa}{\varrho-\varkappa}$. With the substitution $i + j = m$ and $\nu+\mu=\varkappa$ we obtain

$$\sum_{\nu=0}^{i}\binom{i}{\nu}\binom{2r-\nu}{\varrho-\nu}\sum_{\mu=0}^{j}\binom{j}{\mu}\binom{2r-\mu}{\varrho-\mu}$$

$$= \sum_{\nu=0}^{\min(i,\varrho)}\sum_{\mu=0}^{\min(j,\varrho)}\binom{i}{\nu}\binom{j}{\mu}\binom{2r-\nu}{2r-\varrho}\binom{2r-\mu}{2r-\varrho}$$

$$= \sum_{\varkappa=0}^{\min(i,\varrho)+\min(j,\varrho)}\sum_{\nu=0}^{\varkappa}\binom{i}{\nu}\binom{m-i}{\varkappa-\nu}\binom{2r-\nu}{2r-\varrho}\binom{2r-(\varkappa-\nu)}{2r-\varrho}$$

$$=: \Sigma_{ij}\,.$$

Since

$$\min(i,\varrho) + \min(j,\varrho) \geq \min(m,\varrho) \qquad \text{for } m = i + j$$

and all terms of the above sum are nonnegative we get a lower bound if we sum over $\varkappa = 0,\dots,\min(m,\varrho)$. Then we can apply (7.35) for $B = 2r$, $b = 2r-\varrho$ and ν and $\varkappa-\nu$ instead of i and j, respectively, since $B-b+1 = 2r-(2r-\varrho)+1 = \varrho+1 > \min(m,\varrho) \geq \varkappa$. Therefore

$$\Sigma_{ji} \geq \sum_{\varkappa=0}^{\min(m,\varrho)}\sum_{\nu=0}^{\varkappa}\binom{i}{\nu}\binom{m-i}{\varkappa-\nu}\binom{2r}{2r-\varrho}\binom{2r-\varkappa}{2r-\varrho}$$

$$= \binom{2r}{\varrho}\sum_{\varkappa=0}^{\min(m,\varrho)}\binom{m}{\varkappa}\binom{2r-\varkappa}{2r-\varrho}$$

$$= \binom{2r}{\varrho}\sum_{\varkappa=0}^{m}\binom{m}{\varkappa}\binom{2r-\varkappa}{\varrho-\varkappa}\,.$$

The last sum is positive. Hence (7.37) follows if we multiply (7.38) with

$$\Sigma_{ij} = \binom{2r}{\varrho}\sum_{\varkappa=0}^{i+j}\binom{i+j}{\varkappa}\binom{2r-\varkappa}{\varrho-\varkappa} > 0. \qquad \square$$

In every genetic algebra $\mathcal{G}_\varrho$ the second train root is $\lambda_1^{(\varrho)} = 1/2$. Therefore we get by induction from Theorem 7.14 (2) the following

Corollary 7.16. The train roots $\lambda_i^{(\varrho)}$ of $\mathcal{G}_\varrho$ satisfy

$$\lambda_0^{(\varrho)} = 1, \; \lambda_1^{(\varrho)} = 1/2 \text{ and } 0 < \lambda_i^{(\varrho)} < (1/2)^i \quad \text{for } i = 2,..,r.$$

Since $\mathcal{G}$ is a convex combination of the algebras $\mathcal{G}_\varrho$, which have a common canonical basis, the train roots of $\mathcal{G}$ are the corresponding convex combinations of train roots of $\mathcal{G}_\varrho$. Therefore we have the following

Theorem 7.17. The train roots of $\mathcal{G}$ are

$$\lambda_i = \sum_{\varrho=0}^{2r} p(\varrho)\, \lambda_i^{(\varrho)}, \quad \text{with} \quad \lambda_i^{(\varrho)} \text{ given by (7.33)}, \quad i=0,1,\dots,r.$$

In particular we have

$$\lambda_0 = 1, \quad \lambda_1 = 1/2 \quad \text{and } 0 < \lambda_i < (1/2)^i \quad \text{for } i = 2,\dots,r.$$

Hence the gametic algebra $\mathcal{G}$ has exactly one train root $\lambda_1 = 1/2$ and this train root succeeds directly $\lambda_0 = 1$. Since $c_0^2 = c_0$, we can apply theorem 4.8 and 4.19 and obtain

Theorem 7.18. The gametic algebra $\mathcal{G}$ for a random mating population of 2r-ploid individuals with an arbitrary distribution of double reduction possesses an 1-dimensional manifold of idempotents

$$\mathcal{E} = \left\{ e(\eta_1) = c_0 + \eta_1 c_1 + \sum_{i=2}^{r} f_i(\eta_1) c_i \;\middle|\; \eta_1 \in \mathbb{R} \right\},$$

where f_i is a polynomial in η_1 of degree not greater than i, i=2,...,r. For every $x = \sum_{i=0}^{r} \xi_i c_i \in \mathcal{G}$, $\xi_0 = 1$, the sequence of plenary powers converges towards the idempotent $e(\xi_1)$.

Proof. It remains to show that the degree of the polynomial f is at most i, and not, as in the general case, at most 2^{i-1}, cf. theorem 4.8. This statement follows by induction from the fact that the product $c_i c_j$ may have only components in the direction of c_t for t = m,...,2m, where m := i+j, cf. theorem 7.13. □

In the case of pure chromosome segregation the polynomial f_i reduces to $f_i(\eta_1) = \eta_1^i$, i=2,...,r, cf. Gonshor [45] .

For this case with $r = 2$ and $r = 3$ Geiringer [42] has already computed the equilibrium states of the population.

In view of applications of theorem 7.18 we ask for the genetic meaning of the coefficient ξ_1 of c_1. From (7.24) we obtain

$$x = \sum_{i=0}^{r} \alpha_i D_i = \sum_{l=0}^{r} (-1)^l \left(\sum_{i=0}^{r} \alpha_i \binom{i}{l}\right) c_l ,$$

in particular

$$\xi_1 = (-1) \sum_{i=0}^{r} i \alpha_i = - r \left(\frac{1}{r} \sum_{i=0}^{r} i \alpha_i\right),$$

i.e. ξ_1 is (-1)-times the arithmetic mean of the number of alleles A per genotype or (-r)-times the frequency of the A-allele in the initial generation. (This result is in accordance with theorem 6.3 by Gonshor [45] for the presentation of the 1-dimensional manifold of idempotents of the gametic algebra for pure chromosome segregation).

In the case of autotetraploid individuals, i.e. $r = 2$, this algebraic model yields the known difference equations for the relative frequencies of the gametes D_o, D_1, D_2, D_3, D_4 and their convergence behavior, cf. Seyffert [105].

Now we generalize the above model such that it includes mutation (for the case of pure chromosome segregation cf. Gonshor [45], [47]).

Assume that all the allele A mutates into the allele a with probability σ and that the converse mutation occurs with probability τ,

$$0 \leq \sigma, \tau \leq 1 \quad .$$

Furthermore we assume that mutations on the r chromosomes of a gamete occur independently.
From simple combinatorial calculations we obtain that the mutation can be described by a linear mapping $\mathcal{B} : \mathcal{W} \longrightarrow \mathcal{W}$, whereby $\mathcal{W}$ is the vector space underlying the gametic algebra $\mathcal{G}$, given by

$$D_i = \sum_{\varkappa=0}^{r} \sum_{\mu=0}^{\varkappa} \binom{i}{\mu} \binom{r-i}{\varkappa-\mu} (1-\sigma)^{\mu} \sigma^{i-\mu} \tau^{\varkappa-\mu} (1-\tau)^{r-i-\varkappa+\mu} D_{\varkappa}, \quad i=0,\dots,m. \tag{7.40}$$

With respect to the canonical basis $c_o,\dots,c_m$ (7.23) the linear mapping is represented by

$$\mathcal{B}\, c_k = (1-\sigma-\tau)^k \sum_{l=k}^{r} \binom{r-k}{l-k} \tau^{l-k} c_l, \qquad k=0,\dots,m. \tag{7.41}$$

For the derivation of (7.41) from (7.40) we mention two possibilities: Either one computes this implication directly, cf. Gonshor [45] (notice that Gonshor's canonical basis coincides with (7.23) after interchanging A and a) or one proceeds along the following path indicated by Gonshor [47] : Consider c_i (7.23) as the formal product $c_i = (A-a)^i a^{r-i}$, $i=0,\dots,m$. Then one easily checks that $\mathcal{B}$ acts on c_i as if the linear mapping $\mathcal{f} : \langle A,a\rangle_{\mathbb{R}} \longrightarrow \langle A,a\rangle_{\mathbb{R}}$ with

$$\mathcal{f}A := (1-\sigma)A + \sigma a, \qquad \mathcal{f}a := \tau A + (1-\tau)a$$

acts on every factor of c_i, i.e.

$$\mathcal{B}\, c_i = (\mathcal{f}(A-a))^i (\mathcal{f}a)^{r-i}, \qquad i=0,\dots,m.$$

From (7.41) we obtain immediately that $\mathcal{B}$, with respect to the canonical basis, is represented by a lower triangular matrix with diagonal

$$(1,\ (1-\sigma-\tau),\ (1-\sigma-\tau)^2,\dots,(1-\sigma-\tau)^m)\ .$$

Then the genetics of a population of 2r-ploid individuals with an arbitrary distribution of double reduction and arbitrary mutation probabilities is described by the gametic algebra $\mathcal{G}_{\mathcal{B}}$ whose multiplication is given by

$$(x,y) \longmapsto \mathcal{B}(xy)\ , \qquad x,y \in \mathcal{W}\ .$$

With the assistance of theorems 6.29 and 4.18 we obtain immediately

<u>Theorem 7.19.</u> The gametic algebra $\mathcal{G}_{\mathcal{B}}$ for a random mating population of 2r-ploid individuals with an arbitrary distribution of double reduction and mutation rates σ and τ is a genetic algebra with the train roots, 1, $\lambda_i(1-\sigma-\tau)^i$, λ_i given by (7.39), $i=1,\dots,m$.
If the mutation rates σ and τ are different from 0 and 1, then the sequence of plenary powers of every element $x \in \mathcal{G}_{\mathcal{B}}$ of weight 1 converges towards the idempotent of $\mathcal{G}_{\mathcal{B}}$.

The general algebraic model allows us to investigate autopolyploid populations whose individuals differ in several uncoupled loci,where the degree of polyploidy of the corresponding chromosomes may be equal or different. In this case the gametic algebra is the tensor product of these algebras describing the inheritance with respect to each one of the considered loci.

Finally we mention that Fortini and Barakat [35] have considered an infinitely large population of tetraploid individuals which differ in n linked loci between which recombination may occur. In every locus there may be an arbitrary finite number of alleles and, furthermore, there may be chromosome segregation, chromatid segregation or an arbitrary mixture of these segregation modes. The authors construct an algebra which describes the inheritance of this population and they indicate a proof that this algebra is genetic.
The algebra $\mathfrak{G}$ which is the gametic algebra of the population is a baric algebra with weight homomorphism ω. Let $\mathfrak{N} := \ker \omega$. Consider in $\mathfrak{G}$ the set $\mathfrak{C}$ of all elements which can be written symbolically as arrays of alleles or contrasts, i.e. differences of two alleles. Elements of $\mathfrak{C}$ with exactly one contrast are called unit contrasts. Then the ideal $\mathfrak{N}$ of $\mathfrak{G}$ is spanned by all unit contrasts. The proof that $\mathfrak{G}$ is genetic is essentially based on the following statements: Every product of two elements of $\mathfrak{C}$ which have together s, $0 \leq s \leq 4$, contrasts in one fixed locus is a linear combination of elements of $\mathfrak{C}$ with min(s,2) contrasts in this locus, and, in addition, for $s \in \{3,4\}$ it vanishes.
Every product of $\mathfrak{G}$ with at least 2n+1 (n-number of loci) elements of $\mathfrak{N}$ (which can be written as linear combination of products of (2n+1) unit contrasts) is a linear combination of products each having at least 3 contrasts in one locus.
Hence these products vanish and $\mathfrak{G}$ is genetic by a theorem of Gonshor [48] , cf. theorem 3.25.

D. Overlapping and continuously overlapping generations

As mentioned as the begin of this section Gonshor [48] and Heuch [52] , [55] have investigated populations with overlapping or continuously overlapping generations. We give a short account of their results.

We start with a population with genotypes $a_o,\dots,a_m$. We assume that random mating of genotypes a_i and a_j produces offspring of genotype a_k

with probability γ_{ijk}, where $0 \leq \gamma_{ijk} \leq 1$, $\gamma_{ijk} = \gamma_{jik}$, $k=0,\ldots,m$, $\sum_{k=0}^{m} \gamma_{ijk} = 1$, $i,j=0,\ldots,m$.

Gonshor [48] makes the hypothesis that reproduction occurs at discrete moments such that at each moment a fixed portion $1-\Theta$, $0 \leq \Theta \leq 1$, of the population survives whereas the remaining portion Θ of the population is replaced by offspring. To simplify he assumes that the surviving individuals and the new born individuals are indistinguishable. Thus the model does not take account of the age structure.
With these assumptions the genetics is described by an algebra $\mathfrak{G} = \langle a_0,\ldots,a_m \rangle_{\mathbb{R}}$ with the multiplication defined by

$$a_i a_j := \Theta \sum_{k=0}^{m} \gamma_{ijk} a_k + \tfrac{1}{2}(1-\Theta)(a_i+a_j), \quad i,j=0,\ldots,m, \tag{7.42}$$

and bilinear extension.
This algebra is the convex combination

$$\mathfrak{G} = \Theta \mathfrak{A} + (1-\Theta)\mathfrak{B}, \quad , \tag{7.43}$$

of the algebra $\mathfrak{A}$ with multiplication constants γ_{ijk}, $i,j,k=0,\ldots,m$, and the $(m+1)$-dimensional algebra $\mathfrak{B}$ for simple Mendelian segregation. By construction those algebras $\mathfrak{A}$ and $\mathfrak{B}$ have a genetic realization with $a_0,\ldots,a_m$ as a natural basis, and in particular these algebras are baric with the homomorphism given by $\omega(a_i) = 1$, $i=0,\ldots,m$.
Since the algebra $\mathfrak{B}$ is genetic and any basis a_0, $c_1,\ldots,c_m$ with $\omega(c_i) = 0$, $i=1,\ldots,m$, is canonical, theorem 6.5 implies that $\mathfrak{G}$ is a genetic algebra whenever $\mathfrak{A}$ is a genetic algebra. It is also true, cf. [48], that $\mathfrak{G}$ is a special train algebra whenever $\mathfrak{A}$ has this property.
Let us now assume that $\mathfrak{A}$ is a genetic algebra. Then we can make the following assertions: The train roots of $\mathfrak{G}$ are given by

$$\lambda_0 = 1, \quad \lambda_i = \Theta\mu_i + \tfrac{1}{2}(1-\Theta), \quad i=1,\ldots,m, \tag{7.44}$$

where $1, \mu_1,\ldots,\mu_m$ are the train roots of $\mathfrak{A}$. Since $\mathfrak{G}$ is a commutative genetic algebra with a genetic realization the train roots $\lambda_1,\ldots,\lambda_m$ of $\mathfrak{G}$ do not exceed 1/2 in absolute value by theorem 4.14.
If $\Theta \neq 0$, then 1/2 is a train root of $\mathfrak{G}$ if and only if 1/2 is a train root of $\mathfrak{A}$.
If $0 < \Theta < 1$ and not train root of $\mathfrak{A}$ equals 1/2, then the train roots $\lambda_1,\ldots,\lambda_m$ of $\mathfrak{G}$ are less than 1/2 in absolute value. Applying theorem 4.12 and 4.18 we obtain

Theorem 7.20. Let the segregation rates γ_{ijk}, $i,j,k=0,\dots,m$, be such that the algebra $\mathfrak{A}$ is genetic. Then also the algebra $\mathfrak{G}$ (7.43) for populations with overlapping generations is genetic.
If $0<\Theta<1$ and if all train roots of $\mathfrak{A}$ are different from 1/2, then for every $x \in \mathfrak{G}$ of weight 1 the sequence of plenary powers converges towards the idempotent of $\mathfrak{G}$.

By theorem 4.9 we know that a genetic algebra has exactly one idempotent if and only if 1/2 is not a train root. Thus we can reformulate the theorem:
If the algebra for separated generations has exactly one idempotent then in the algebra for overlapping generations all sequences of plenary powers of elements of weight 1 converge.
Roughly speaking this means that overlapping of generations enhances convergence.
If train roots 1/2 of the algebra $\mathfrak{A}$ occur, then the possible application of theorem 4.19 requires more detailed arguments.

Remark. The effect that in the case of a genetic algebra with genetic realization overlapping of generations enhances convergence is similar to the following result of Stein and Rosenberg (cf. Varga [111]) in matrix analysis.
Consider the linear matrix equation

$$Ax = b$$

where A is non singular and has no zeros in the diagonal. Decompose

$$A = D - E - F$$

into a diagonal and strictly lower and upper matrices D, E, F, respectively, and define

$$L := D^{-1}E \quad \text{and} \quad U := D^{-1}F \ .$$

If the matrix $J := L + U$ is non negative, the total-step iterative method of Jacobi can be compared with the single-step iterative method of Gauss and Seidel.
In particular, if the total-step iterative method converges, then the single-step method converges too, and the latter converges faster than the former.

One year before Gonshor published the above mentioned model Heuch [52] had written a paper on "sequences in genetic algebras for overlapping generations". In this paper Heuch defined a sequence of elements in a genetic algebra depending on the p initial generations. The (p+1)-th

generation is composed in fixed portions of the offspring resulting from random mating between every pair of two populations of the preceding p generations.
This model takes account of the ageing of individuals insofar as every population participates in random mating during p successive generations and then no more. But this model says nothing about the age distribution among the genotypes in each generation, in contrast to Heuch's model with continuous time which we will mention later in this section, cf. (7.56).

In detail the model is the following:
Assume as at the begin of section 7D that there are m+1 genetically different genotypes represented by $a_o,\dots,a_m$ and that random mating of genotypes a_i and a_j produces offspring of type a_k with probability γ_{ijk}, k=0,...,m, j=0,...,m. Let $\mathfrak{A} := \langle a_o,\dots,a_m\rangle_{\mathbb{R}}$ the resulting algebra.
Then $\mathfrak{A}$ is an algebra with a genetic realization, hence it is baric and the weight homomorphism ω is given by $\omega(a_i) = 1$, i=0,...,m.
Let the p initial generations of the population be represented by $x_o,\dots,x_{p-1} \in \mathfrak{A}$, hence $\omega(x_i) = 1$, i=0,...,p-1.
Assume that a portion $\beta_{\nu\mu}$ of the (p+l)-th generation, $l \in \mathbb{N}_o$, represented by $x_{p+j} \in \mathfrak{A}$, is the offspring resulting from random mating between the $(l+\nu)$-th and the $(l+\mu)$-th generation, $\nu,\mu = 0,\dots,p-1$. Then $0 \leq \beta_{\nu\mu} \leq 1$, $\nu,\mu = 0,\dots,p-1$, $\sum_{\nu,\mu=0}^{p-1} \beta_{\nu\mu} = 1$ and assume, in addition,

$$\beta_{\nu\mu} = \beta_{\mu\nu} \text{ for } \nu,\mu = 0,\dots,p-1.$$

Then in terms of the algebra $\mathfrak{A}$ the element $x_{p+1} \in \mathfrak{A}$ can be written as

$$x_{p+l} = \sum_{\nu,\mu=0}^{p-1} \beta_{\nu\mu} x_{l+\nu} x_{l+\mu}, \quad l \in \mathbb{N}_o \quad . \tag{7.45}$$

In view of $\sum_{\nu,\mu=0}^{p-1} \beta_{\nu\mu} = 1$ all elements x_{p+l}, $l \in \mathbb{N}_o$, are of weight 1.

Heuch [52] has shown that, if the algebra $\mathfrak{A}$ is genetic, then the sequence $\{x_i\}$ defined by $x_o,\dots,x_{p-1}$ and (7.45) forms a train, i.e. it satisfies a polynomial identity with constant coefficients. Furthermore Heuch has indicated how to calculate the train roots of this sequence, i.e. the roots of the corresponding polynomial, from the train roots of the algebra and the real numbers

$$\beta_\nu := \sum_{\mu=0}^{p-1} \beta_{\nu\mu}, \qquad \nu = 0,\dots,p-1, \tag{7.46}$$

and he has given their multiplicities.

Subsequently, Heuch studied a more general sequence of generations, depending on the p initial generations, where one fixed portion of the population in the (p+1)-th generation is obtained as above, while the remaining portion is a weighted sum of the genotype distribution in the preceding p generations. This model includes the preceding one as limiting case.
Heuch proved that this sequence, too, forms a train and he made some considerations on its convergence behavior. He gave a sufficient condition for the convergence of these sequences. This condition is partially redundant in view of bounds of the train roots in a genetic algebra with genetic realization. Concerning the former model for overlapping generations the following statement is true.

Theorem 7.21. Assume that the commutative algebra $\mathfrak{A} = \langle a_o,\dots,a_m\rangle_{\mathbb{R}}$ with multiplication constants γ_{ijk}, $0 \leq \gamma_{ijk} \leq 1$, $i,j,k=0,\dots,m$, $\sum_{k=0}^{m}\gamma_{ijk} = 1$, $i,j=0,\dots,m$, is genetic and its train roots are $\lambda_o = 1$, $\lambda_1,\dots,\lambda_m$.
Let $x_o,\dots,x_{p-1}\in\mathfrak{A}$ be elements of weight 1. If $|\lambda_i| \neq 1/2$ for all $i=1,\dots,m$, then the sequence $\{x_i\}$ defined by $x_o,\dots,x_{p-1}$ and (7.45) converges towards the idempotent of $\mathfrak{A}$.

Proof. Assume $|\lambda_i| \neq 1/2$, $i=1,\dots,m$. Then in particular $\lambda_i \neq 1/2$, $i=1,\dots,m$, hence the algebra $\mathfrak{A}$ has exactly one idempotent, cf. theorem 4.4, which we can choose as first element of a canonical basis $c_o,\dots,c_m$ of $\mathfrak{A}$. Let the multiplication table of $\mathfrak{A}$, relative to this basis, be given by λ_{ijk}, $i,j,k=0,\dots,m$. Then the train roots of $\mathfrak{A}$ are $\lambda_i = \lambda_{oii}$, $i=0,\dots,m$. Let the element x_i of the sequence $\{x_i\}$ be represented by

$$x_i = \sum_{i=0}^{m}\xi_{ik}c_k, \qquad i\in\mathbb{N}_o.$$

Since every element of this sequence has weight 1, we have

$$\xi_{io} = 1 \qquad \text{for all } i\in\mathbb{N}_o.$$

From (7.4) and the special properties of the multiplication constants λ_{ijk} of $\mathfrak{A}$ we obtain the following recurrence relations for the last m coordinates of x_{p+1}:

$$\xi_{p+1,k} = \sum_{\nu,\mu=0}^{p-1} \beta_{\nu\mu} \sum_{i,j=0}^{m} \xi_{l+\nu,i} \lambda_{ijk} \xi_{l+\mu,j}$$

$$= 2\lambda_{okk} \sum_{\nu,\mu=0}^{p-1} \beta_{\nu\mu} \xi_{l+\nu,k} + \sum_{i,j=0}^{k-1} \lambda_{ijk} \sum_{\nu,\mu=0}^{p-1} \xi_{l+\nu,i} \beta_{\nu\mu} \xi_{l+\mu,j} ,$$

$$l \in \mathbb{N}_o, \quad k=1,\ldots,m.$$

Hence in view of (7.46)

$$\xi_{p+1,k} = 2\lambda_k \sum_{\nu=0}^{p-1} \beta_\nu \xi_{l+\nu,k} + \sum_{i,j=0}^{k-1} \lambda_{ijk} \sum_{\nu,\mu=0}^{p-1} \xi_{l+\nu,i} \beta_{\nu\mu} \xi_{l+\mu,j}, \tag{7.47}$$

$$l \in \mathbb{N}_o, \quad k=1,\ldots,m.$$

We are going to prove by induction on k that every coordinate sequence $\{\xi_{i,k}\}$ converges to zero, $k=1,\ldots,m$.
For $k=1$ equation (7.47) reduces to

$$\xi_{p+1,1} = 2\lambda_1 \sum_{\nu=0}^{p-1} \beta_\nu \xi_{l+\nu,1} , \quad l \in \mathbb{N}_o, \tag{7.48}$$

i.e. a linear homogeneous difference equation for ξ_{i1}, $i \in \mathbb{N}_o$.
The characteristic polynomial of this equation is

$$q(\xi) = \xi^p - 2\lambda_1 (\beta_{p-1} \xi^{p-1} + \ldots + \beta_1 \xi + \beta_o) . \tag{7.49}$$

Since by construction $\mathfrak{A}$ has a genetic realization, we know from theorem 4.14 that $|\lambda_i| \leq 1/2$, $i=1,\ldots,m$, and with the above assumption follows

$$|\lambda_i| < 1/2, \qquad i=1,\ldots,m.$$

Form a theorem of Gershgorin (cf. Marcus and Minc [89]) applied to the Frobenius matrix of (7.48), we know that the roots $\mu_1,\ldots,\mu_p$ of q satisfy

$$|\mu_\varrho| \leq \max\left\{1,\ 2|\lambda_1| \sum_{\nu=0}^{p-1} \beta_\nu\right\} = \max\{1, 2|\lambda_1|\} , \quad \varrho = 1,\ldots,p,$$

hence

$$|\mu_\varrho| \leq 1, \qquad \varrho = 1,\ldots,p.$$

In view of this inequality we obtain from (7.49)

$$|\mu_\varrho|^p \leq 2|\lambda_1| \sum_{\nu=0}^{p-1} \beta_\nu |\mu_\varrho|^\nu \leq 2|\lambda_1| \sum_{\nu=0}^{p-1} \beta_\nu = 2|\lambda_1| \ , \varrho = 1,\ldots,p.$$

Hence from $|\lambda_1| < 1/2$ it follows that

$$|\mu_\rho| < 1 \qquad \text{for all } \rho = 1,\ldots,p.$$

Consequently every solution $\{\xi_{i,1}\}$ of (7.48) converges to zero. Now assume, that all sequences $\{\xi_{i,1}\}$,..., $\{\xi_{i,k-1}\}$ converge to zero and consider the k-th difference equation of (7.47). It is a linear inhomogeneous difference equation in $\xi_{p+l,k}$, $l \in \mathbb{N}_0$ with the inhomogeneous part

$$\sum_{i,j=0}^{k-1} \lambda_{ijk} \sum_{\nu,\mu=0}^{p-1} \xi_{l+\nu,i}\beta_{\nu\mu}\xi_{l+\mu,j} \ ,$$

which is a quadratic polynomial in $\xi_{l+\nu,i}$, $\nu = 1,\ldots,p-1$, $i=0,\ldots k-1$, with vanishing absolute term in view of $\lambda_{okk} = 0$. Since we know that every one of these sequences converges to zero, it follows that the inhomogeneous term converges to zero, too.
The characteristic polynomial corresponding to the related homogeneous difference equation is given by (7.49) with λ_k instead of λ_1. Since $|\lambda_k| < 1/2$, it follows by the same arguments as above that the roots of this polynomial are less than one in modulus, and with the above result on the inhomogeneous term we obtain that $\lim_{i\to\infty} \xi_{i,k} = 0$.

Hence the sequence $\{x_i\}$ converges to c_o, the uniquely determined idempotent of $\mathfrak{A}$. □

This result can be easily extended to the second of Heuch's models. Beyond that it can be sharpened in the following way. If the algebra $\mathfrak{A}$ is genetic and a positive part of the preceding p generations do not take part in random mating but only rests alive the following generation(s), then the generation sequence converges in every case.

Next we discuss the model for continuously overlapping generations proposed by Heuch [55].
Let $\mathfrak{A}$ be the algebra with multiplication table γ_{ijk}, $i,j,k=0,\ldots,m$, mentioned at the beginning of section 7D, with a natural basis a_o, $a_1,\ldots,a_m$. We assume the the genotype frequencies are time dependent. At time t the population is represented by

$$x(t) = \sum_{i=0}^{m} \alpha_i(t) a_i, \qquad t \in \mathbb{R}_+.$$

(Formally x is an element of the set $\mathfrak{A}^{\mathbb{R}_+}$ of all functions with domain $\mathbb{R}_+$ and range $\mathfrak{A}$).

During the time interval $[t,t+h]$ a portion $h+o(h)$ of the population is replaced by the offspring resulting from random mating.
Then the frequency of the genotype a_k at time $t+h$ is given by

$$\alpha_k(t+h) = (1-k-o(h))\alpha_k(t) + h \sum_{i,j=0}^{m} \alpha_i(t)\gamma_{ijk}\alpha_j(t) + o(h), \; k=0,\dots,m.$$

In the limit $h \longrightarrow 0$ we obtain the following set of differential equations

$$\alpha_k'(t) = -\alpha_k(t) + \sum_{i,j=0}^{m} \alpha_i(t)\gamma_{ijk}\alpha_j(t), \quad k=0,\dots,m. \tag{7.50}$$

Using the algebra multiplication we can write these equations as a Riccati differential equation

$$x'(t) = -x(t) + x^2(t) \tag{7.51}$$

for functions in $\mathfrak{A}^{\mathbb{R}_+}$, cf. Andreoli [10] .
Summing over k, we obtain from (7.50) the differential equation

$$\frac{d}{dt}\left(\sum_{k=0}^{m}\alpha_k(t)\right) = -\sum_{k=0}^{m}\alpha_k(t) + \left(\sum_{k=0}^{m}\alpha_k(t)\right)^2 .$$

From the existence and uniqueness theorem for systems of differential equations it follows that any solution with initial conditions $\alpha_o(0),\dots,\alpha_m(0) \geqq 0$ with $\sum_{k=0}^{m}\alpha_k(0) = 1$ has the property $\sum_{k=0}^{m}\alpha_k(t) = 1$ for all $t > 0$.
For a genetic algebra $\mathfrak{A}$ Heuch [55] describes the general solution of (7.51) and discusses the stability of solutions. Following his arguments and using the fact that the algebra $\mathfrak{A}$ has a genetic realization we obtain the following theorem.

Theorem 7.21. If the algebra $\mathfrak{A}$ is genetic and all its train roots are different from 1/2 then the uniquely determined idempotent of $\mathfrak{A}$ is a globally stable stationary point of the differential equation (7.51).

Proof. With the assumptions of the theorem the algebra $\mathfrak{A}$ has exactly one idempotent. This can be chosen as first element of a canonical basis $c_o,\dots,c_m$. Let λ_{ijk}, $i,j,k=0,\dots,m$, be the multiplication constants with respect to this basis. Then $\lambda_i := \lambda_{oii}$, $i=0,\dots,m$, are the train roots of $\mathfrak{A}$. Let the function $t \longmapsto x(t) = \sum_{i=0}^{m}\alpha_i(t)a_i$ has the representation

$$x(t) = \sum_{i=0}^{m} \xi_i(t) c_i$$

with respect to the chosen canonical basis. Because of the special structure of the multiplication constants the equivalent differential system for the functions $t \longmapsto \xi_i(t)$, $i=0,\dots,m$, assumes the form

$$\xi_k'(t) = -\xi_k(t) + \sum_{i,j=0}^{k} \xi_i(t)\lambda_{ijk}\xi_j(t), \qquad k=0,\dots,m. \tag{7.52}$$

Since $c_o \in \mathfrak{A}$ is an idempotent, the differential equation for ξ_o reads

$$\xi_o'(t) = 0 \quad .$$

For the initial data $\xi_o(0) = 1, \xi_1(0),\dots,\xi_m(0)$ we obtain

$$\xi_o(t) = 1, \qquad t \in \mathbb{R}_+ \; .$$

If we substitute this expression in the remaining m differential equations of (7.52) we have

$$\xi_k'(t) = (2\lambda_k - 1)\xi_k(t) + \sum_{i,j=0}^{k-1} \xi_i(t)\lambda_{ijk}\xi_j(t) \;, \; k=1,\dots,m. \tag{7.53}$$

By induction on k we see that the solution $t \longmapsto \xi_k(t)$ has the form

$$\xi_k(t) = \beta_k e^{(2\lambda_k - 1)t} + \sum_{\substack{i,j=0 \\ \lambda_{ijk} \neq 0}}^{k-1} \beta_{ij}(t) e^{2(\lambda_i + \lambda_j - 1)t} \;, \; k=1,\dots,m. \tag{7.54}$$

where $\beta_{ij}(t)$ are polynomials in t.
Since $\mathfrak{A}$ has a gametic realization the train roots $\lambda_1,\dots,\lambda_m$ are not greater than 1/2 in absolute value. If we assume $\lambda_i \neq 1/2$, $i=1,\dots,m$, then the real parts of the exponents in (7.49) are negative and the functions $t \longmapsto \xi_k(t)$, $k=1,\dots,m$, converge to zero.
Thus $\lim_{t \to \infty} x(t) = c_o$ for all initial data $\xi_o(0) = 1, \xi_1(0), \dots, \xi_m(0)$. □

Heuch [55] applies this model to a population of tetraploid individuals with an arbitrary mixture of chromosome- and chromatid segregation and computes the general solution of the corresponding initial value problem.

He considers two further mating schemes for which we shall give a short description:

I. The portion of dying individuals of the population is replaced by offspring resulting from random mating of the population at time t with the initial population. Then one has a linear differential equation

$$x'(t) = - c(t) + x(t)x(0) \qquad (7.55)$$

in $\mathfrak{A}^{\mathbb{R}_+}$.

II. Let the population be subdivided in two age classes A_1 and A_2. Let the genotype frequencies in the i-th age class be represented by $x_i(t) \in \mathfrak{A}$, i=1,2. Suppose the aged portion of the first age class is replaced by offspring from random mating of the i-th and j-th age class with the portion β_{ij}, $0 \leq \beta_{ij} \leq 1$, $\beta_{ij} = \beta_{ji}$, i,j=1,2, $\sum_{i,j=1}^{2} \beta_{ij} = 1$.

The dying portion of the second age class is replaced by the aged individuals of the first age class. With these assumptions one obtains the following system of differential equations

$$x_1'(t) = - x_1(t) + \sum_{i,j=1}^{2} \beta_{ij} x_i(t) x_j(t) \qquad (7.56)$$

$$x_2'(t) = - x_2(t) + x_1(t) \quad .$$

8. APPLICATIONS II

In this section we consider genetic situations which can be described or investigated with the assistance of not necessarily commutative algebras.

A. Asymmetric segregation rates

As several times before we consider an infinitely large population of diploid or 2r-ploid individuals $a_i a_j$, $i,j=0,\dots,m$, with the same fertility and random mating. However we drop the hypothesis that the zygotes $a_i a_j$ and $a_j a_i$ have the same segregation behavior, i.e. in general the segregation rates γ_{ijk} and γ_{jik}, for $i \neq j$, do not agree, $i,j,k=0,\dots,m$. Then the gametic algebra

$$\mathcal{G} = \left\{ \sum_{i=0}^{m} \xi_i a_i \,\middle|\, \xi_i \in \mathbb{R}, \quad i=0,\dots,m \right\}$$

is, in general, non commutative.
The zygotic algebra $\mathcal{Z}$ is obtained by (non commutative) duplication $\mathcal{G} \otimes_d \mathcal{G}$ of the gametic algebra $\mathcal{G}$.
To such situations we can only apply the theorems of Section 5 and 6. We restrict ourselves to the case of <u>additive segregation rates</u> γ_{ijk}, i.e.

$$\gamma_{ijk} = \alpha_{ik} + \beta_{jk}, \qquad i,j,k=0,\dots,m, \tag{8.1}$$

for real numbers

$$0 \leq \alpha_{ik}, \ \beta_{jk} \leq 1, \ i,j,k=0,\dots,m,$$

with

$$\sum_{k=0}^{m} (\alpha_{ik} + \beta_{jk}) = 1 \quad \text{for all } i,j=0,\dots,m. \tag{8.2}$$

From (8.2) it follows that $\sum_{k=0}^{m} \alpha_{ik}$ and $\sum_{k=0}^{m} \beta_{jk}$ are independent of i and j, respectively, hence we have

$$\alpha := \sum_{k=0}^{m} \alpha_{ik} \quad \text{and} \quad \beta := \sum_{k=0}^{m} \beta_{jk} \quad \text{for all } i,j=0,\dots,m, \tag{8.3}$$

furthermore

$$0 \leq \alpha, \beta \leq 1 \text{ and } \alpha + \beta = 1.$$

Let us define the matrices

$$A := (\alpha_{ik}) \quad \text{and} \quad B := (\beta_{jk}) \ . \tag{8.4}$$

In view of (8.3) these matrices have constant row sums α and β, respectively, and their sum A + B is a row stochastic matrix.

Paralleling theorem 7.5 we have

Theorem 8.1. The gametic algebra $\mathcal{G}$ with additive segregation rates (8.1) is a special train algebra. Hence it is a train algebra and the left and right principle train roots are contained among the complex numbers $1, \nu_1, \dots, \nu_m$ and $1, \mu_1, \dots, \mu_m$, respectively, where $\alpha, \mu_1, \dots, \mu_m$ and β, $\nu_1, \dots, \nu_m$ are the eigenvalues of A and B, respectively.

Proof. $\mathcal{G}$ is a baric algebra with weight homomorphism ω defined by $\omega(a_i) := 1$, $i=0,\dots,m$. One easily checks that the kernel $\mathfrak{N} := \langle a_0 - a_i \mid i=1,\dots,m \rangle$ of ω is nilpotent of index 2, thus $\mathcal{G}$ is a special train algebra. Theorem 5.8 and 5.7 imply that $\mathcal{G}$ is a train algebra, i.e. the left rank as well as the right rank polynomial of a generic element x of $\mathcal{G}$ only depend on $\omega(x)$. From Section 3C we know that these polynomials are divisors of $xg_x(x)$ and $f_x(x)x$, respectively, where g_x and f_x are the characteristic polynomials of $\mathcal{L}_x$ and $\mathcal{R}_x$, respectively. Hence the left and right principle train roots of $\mathcal{G}$, cf. definition 5.1, are among the eigenvalues of $\mathcal{L}_x$ and $\mathcal{R}_x$ for $x = \sum_{i=0}^{m} \xi_i a_i \in \mathcal{G}$ with $\omega(x) = 1$. With respect to the natural basis to these transformations correspond the matrices

$$L_x = (B + e(\xi_0,\dots,\xi_m)A)^T \quad \text{and} \quad R_x = (A + e(\xi_0,\dots,\xi_m)B)^T, \tag{8.5}$$

respectively, where $e := (1,\dots,1)^T \in \mathbb{R}^{m+1}$.
From the properties of the matrices B and A it follows that β and α are eigenvalues with right eigenvector e. Furthermore the matrices $e(\xi_0,\dots,\xi_m)A$ and $e(\xi_0,\dots,\xi_m)B$ are of rank 1 and, in view of $\sum_{i=0}^{m} \xi_i = 1$, their non trivial eigenvalues are α and β, respectively, with right eigenvector e. Hence the eigenvalues of L_x and R_x are $\beta + \alpha = 1$, $\nu_1, \dots, \nu_m$ and $\alpha + \beta = 1, \mu_1, \dots, \mu_m$, respectively, and the final assertion follows immediately. □

From theorem 8.1 we know that the gametic algebra $\mathcal{G}$ is Schafer genetic,

cf. theorem 5.8.
Furthermore the algebra $\mathcal{G}$ is left and right Gonshor genetic, cf. theorem 5.10, but $\mathcal{G}$ in general, is not Gonshor genetic, since, without further restrictions on the matrices A and B, there does not exist a non singular matrix which transforms the matrices

$$L_{a_o} = (B + e(1,0,\dots,0)A)^T \text{ and } R_{a_o} = (B + e(1,0,\dots,0)B)^T$$

simultaneously into lower triangular form. A sufficient condition for this property is that the two matrices commute, cf. Frobenius [36] , i.e. in terms of algebras

$$a_o(xa_o) = (a_o x)a_o \quad \text{for all } x \in \mathcal{G} . \tag{8.6}$$

Later on a necessary and sufficient condition was given by McCoy [90] . He proved that two matrices L and R over an algebraically closed field $\mathbb{K}$ can be simultaneously transformed into lower triangular matrices if and only if all matrices

$$f(L,R)(LR-RL), \qquad f \in \mathbb{K}\{X_1,X_2\}$$

are nilpotent, where f runs over all polynomials in two non commuting inderminates over $\mathbb{K}$.

Although in the present case the gametic algebra $\mathcal{G}$ is not in general Gonshor genetic, we can say something about the convergence behavior of the sequence of plenary powers of an element of weight 1.
We can proceed along two different paths: Either we construct the algebra $\mathcal{G}^+$, as indicated at the end of Section 5B, or we can associate with every sequence of plenary powers of elements of unit weight a linear difference equation, cf. Section 7A.
Let us first consider $\mathcal{G}^+$, whose multiplication is given by

$$(x,y) \longmapsto x\cdot y = \frac{1}{2}(xy + yx) \quad ,$$

cf. Section 2A. Then the multiplication constants of $\mathcal{G}^+$ with respect to the basis $a_o,\dots,a_m$ are

$$\frac{1}{2}\left[(\alpha_{ik} + \beta_{ik}) + (\alpha_{jk} + \beta_{jk})\right] \quad , \; i,j,k=0,\dots,m. \tag{8.7}$$

Hence $\mathcal{G}^+$ is a gametic algebra for additive, symmetric segregation, whose associated stochastic matrix is A + B, cf. section 7A.

In particular $\mathfrak{G}^+$ is a genetic algebra, cf. theorem 7.5.
Since for every $x \in \mathfrak{G}$ the squares x^2 and $x \cdot x$ taken in $\mathfrak{G}$ and $\mathfrak{G}^+$ coincide, we can consider the sequence $x^{[t]}$ of plenary powers of $\mathfrak{G}^+$ instead of $\mathfrak{G}$. From theorem 7.6 we obtain

Theorem 8.2. Let $\mathfrak{G}$ be a gametic algebra for asymmetric inheritance with additive segregation rates (8.1).
If the row stochastic matrix A + B, defined by (8.4) is regular, then the sequence of plenary powers of every element $x \in \mathfrak{G}$ of weight 1 converges towards the idempotent of $\mathfrak{G}$.

The second approach mentioned above is the following:

Take $x = \sum_{i=0}^{m} \xi_i a_i \in \mathfrak{G}$ with $\omega(x) = \sum_{i=0}^{m} \xi_i = 1$, and let

$$x^{[t]} = \sum_{i=0}^{m} \xi_i^{[t]} a_i$$

be the t-th plenary power of $x = x^{[1]}$. Then in view of (8.1) we have

$$x^{[t+1]} = x^{[t]}\, x^{[t]} = \sum_{k=0}^{m} \sum_{i,j=0}^{m} \xi_i^{[t]} (\alpha_{ik} + \beta_{jk}) \xi_j^{[t]} a_k$$

$$= \sum_{k=0}^{m} \left[\sum_{i=0}^{m} \xi_i^{[t]} \alpha_{ik} \sum_{j=0}^{m} \xi_j^{[t]} + \sum_{i=0}^{m} \xi_i^{[t]} \sum_{j=0}^{m} \xi_j^{[t]} \beta_{jk} \right] a_k$$

$$= \sum_{k=0}^{m} \left[\sum_{i=1}^{m} \xi_i^{[t]} (\alpha_{ik} + \beta_{ik}) \right] a_k, \qquad t \in \mathbb{N},$$

which is equivalent to the linear difference equation

$$(\xi_0^{[t+1]}, \ldots, \xi_m^{[t+1]}) = (\xi_0^{[t]}, \ldots, \xi_m^{[t]})(A + B), \quad t \in \mathbb{N},$$

with the row stochastic matrix A + B and the initial state $(\xi_0, \ldots, \xi_m) \in \mathbb{R}^{m+1}$ with $\sum_{i=0}^{m} \xi_i = 1$, whose solution is

$$(\xi_0^{[t+1]}, \ldots, \xi_m^{[t+1]}) = (\xi_0, \ldots, \xi_m)(A + B)^t, \qquad t \in \mathbb{N}.$$

Hence the assertion of theorem 8.2 follows by well known theorems on stochastic matrices, cf. e.g. Gantmacher [38].

B. Sex-linked inheritance

We consider an infinitely large population of diploid individuals which are female or male. We assume that females are homogametic (XX) and males are heterogametic (XY).
Furthermore we assume that the genotypes differ in finitely many loci each with finitely many alleles on the X-chromosome, i.e. the genotypes differ in sex linked loci.
Such a situation has been discussed in general or in special cases with the assistance of algebras by the following authors in chronological order:
Etherington [29] started such investigations and Gonshor [45] considered two alleles at one sex linked locus with and without mutation, cf. also Bertran [16]. Reiersøl [98] considered the generation sequence for finitely many sex linked loci with recombination. Gonshor [46] constructed a model for finitely many alleles at one locus. Holgate [66] provided a new classification of algebras describing sex linked and partially sex linked inheritance. Gonshor [48] introduced the principle of sex linked duplication whereby he assumed simple Mendelian segregation in the male sex. W.-B. [112] considered finitely many alleles at one locus with arbitrary segregation rates in both sexes. Heuch [56] investigated completely and partially sex linked loci with recombination. W.-B. [113] considered finitely many sex linked loci with arbitrary segregation rates in both sexes and generalized the concept of sex-linked duplication.

Before entering the general situation we demonstrate basic ideas at the following example, which has been introduced by Etherington[29] and further dealt with by Gonshor [45], Bertran [16]and Holgate [66].

Example 8.3. Assume the alleles A and a at a sex linked locus on the X-chromosome. Then the female and male genotypes are AA, Aa, aa, A := AY and a := aY. We assume that simple Mendelian segregation occurs in both sexes, i.e. $AA \longrightarrow A$, $Aa \longrightarrow \frac{1}{2}A + \frac{1}{2}a$, $aa \longrightarrow a$, $AY \longrightarrow \frac{1}{2}A + \frac{1}{2}Y$, $aY \longrightarrow \frac{1}{2}a + \frac{1}{2}Y$.
Let $\mathfrak{z}$ be the free $\mathbb{R}$-module spanned by AA, Aa, aa, A,a,

$$\mathfrak{z} := \{ x = \alpha_1 AA + \alpha_2 Aa + \alpha_3 aa + \beta_1 A + \beta_2 a \mid \alpha_i, \beta_i \in \mathbb{R} \} .$$

To each actual population with equal portions of female and male zygotes there corresponds an element $x \in \mathfrak{z}$ with $\sum_{i=1}^{3} \alpha_i = \sum_{i=1}^{2} \beta_i = 1$.

In $\mathfrak{Z}$ we introduce a multiplication such that the product of two basis elements, corresponding to zygotes, represents the offspring of these zygotes. Since only zygotes of different sex mate, all products of basis elements corresponding to zygotes of the same sex equal zero. The remaining products form the segregation rates. Then the multiplication table reads

	AA	Aa	aa	A	a
AA	0	0	0	$\frac{1}{2}AA + \frac{1}{2}A$	$\frac{1}{2}Aa + \frac{1}{2}A$
Aa		0	0	$\frac{1}{4}AA+\frac{1}{4}Aa+\frac{1}{4}A$	$\frac{1}{4}Aa + \frac{1}{4}aa+ \frac{1}{4}A + \frac{1}{4}a$
aa			0	$\frac{1}{2}AA + \frac{1}{2}a$	$\frac{1}{2}aa + \frac{1}{2}a$
A				0	0
a					0

The vector space $\mathfrak{Z}$ with this multiplication is called the <u>zygotic algebra</u> for sex <u>linked inheritance</u> for two alleles with <u>simple Mendelian segregation rates.</u>

Of course $\mathfrak{Z}$ is not a baric algebra since the specified basis elements are nilpotent of index 2.

We switch to another basis in $\mathfrak{Z}$:

$$d := A, \quad d_o := A+AA, \quad d_1 := A-a, \quad d_2 := AA-Aa, \quad d_3 := AA-2Aa+aa$$

Then the multiplication table reads

	d	d_o	d_1	d_2	d_3
d	0	$\frac{1}{2}d_o$	0	$\frac{1}{4}d_1 + \frac{1}{4}d_2$	0
d_o		d_o	$\frac{1}{2}d_2$	$\frac{1}{4}d_1 + \frac{1}{4}d_1$	0
d_1			0	$\frac{1}{4}d_3$	0
d_2				0	0
d_3					0

Define $\mathfrak{B} := \mathfrak{Z}^2 = \langle d_o, d_1, d_2, d_3\rangle$ and $\mathfrak{N} := \langle d_1, d_2, d_3\rangle$. Then $\mathfrak{N}$ is an ideal of $\mathfrak{B}$ (and of $\mathfrak{Z}$) which does not contain $\mathfrak{B}^2$. By Lemma 1.11 $\mathfrak{B}$ is a baric algebra with the weight homomorphism $\omega : \mathfrak{B} \longrightarrow \mathbb{R}$ defined by $\omega(d_o) = 1$, $\omega(d_i) = 0$, $i=1,2,3$. Furthermore $\mathfrak{N}$ is nilpotent and the principal powers of $\mathfrak{N}$, namely $\mathfrak{N}^2 = \langle d_3\rangle$, $\mathfrak{N}^3 = \langle 0\rangle$, are ideals of $\mathfrak{B}$. Thus $\mathfrak{B}$ is a special train algebra. By theorem 3.25 it follows that $\mathfrak{B}$ is also a genetic algebra. The multiplication table of $\mathfrak{B}$ with respect to the basis

$$c_o := d_o, \quad c_1 := d_1 + 2d_2, \quad c_2 := d_1 - d_2, \quad c_3 := d_3$$

reads

	c_o	c_1	c_2	c_3
c_o	c_o	$\frac{1}{2}c_1$	$\frac{1}{4}c_2$	0
c_1		c_3	$\frac{1}{4}c_3$	0
c_2			$-\frac{1}{2}c_3$	0
c_3				0

Hence these elements form a canonical basis of $\mathfrak{b}$ and the train roots of $\mathfrak{b}$ are $\lambda_o = 1$, $\lambda_1 = 1/2$, $\lambda_2 = -1/4$, $\lambda_3 = 0$.

Since the train root $\lambda_1 = 1/2$ is an immediate successor of the train root $\lambda_o = 1$ and $c_o^2 = c_o$ we can apply theorem 4.8 to obtain that $\mathfrak{b}$ has a 1-parametric family of idempotents

$$\mathcal{E} = \{ e(\eta_1) = c_o + \eta_1 c_1 + \eta_1^2 c_3 \mid \eta_1 \in \mathbb{R} \} .$$

Of course the elements of $\mathcal{E}$ are also idempotents of $\mathfrak{Z}$.
With respect to the original basis these idempotents have the form

$$e(\eta_1) = (1+\eta_1)^2 AA - 2\eta_1(1+\eta_1)Aa + \eta_1^2 aa + (1+\eta_1)A - \eta_1 a.$$

Calling in theorem 4.19 we obtain the known result that the sequence of plenary powers of an element

$$x = \alpha_1 AA + \alpha_2 Aa + \alpha_3 aa + \beta_1 A + \beta_2 a \quad \text{with} \sum_{i=1}^{3} \alpha_i = \sum_{j=1}^{2} \beta_j = 1$$

converges towards that idempotent $e(\eta_1)$, which is characterized by the relative frequency $\varrho := \frac{1}{3}(\alpha_2 + 2\alpha_3 + \beta_2)$, $\eta_1 = -\varrho$, of the allele a in the initial generation, cf. Edwards [25] . □

Now we consider the general situation:
Let $a_o, a_1, \dots, a_m$ be the genetically different X-chromosomes; the differences can arise from different alleles at one locus or at several loci.
Then there are $(m+1)^2$ female genotypes $a_{ij} := a_i a_j$, $i,j=0,\dots,m$. In the genotype $a_i a_j$ the chromosome a_i is of maternal origin and a_j is of paternal origin. The male genotypes are represented by $a_i := a_i Y$, $i=0,\dots,m$. Assume the female and male genotypes a_{ij} and a_i produces gametes a_k with segregation rates

$$\gamma_{ijk} \text{ and } \gamma_{ik}, \text{ respectively, } k=0,\dots,m.$$

If selection is absent and the male genotypes produce X- and Y-gametes with the same frequency, then

$$\sum_{k=0}^{m} \gamma_{ijk} = 1, \qquad \sum_{k=0}^{m} \gamma_{ik} = 1/2, \qquad i,j=0,\ldots,m. \tag{8.8}$$

If

$$\gamma_{ijk} = \frac{1}{2}(\delta_{ik} + \delta_{jk}) \text{ or } \gamma_{ik} = \frac{1}{2}\delta_{ik}, \quad i,j,k=0,\ldots,m, \tag{8.9}$$

then we speak of simple Mendelian segregation in the female or male sex, respectively.

If

$$\gamma_{ijk} = \gamma_{jik}, \qquad i,j,k=0,\ldots,m, \tag{8.10}$$

holds, then inheritance in the female sex is symmetric. Thus simple Mendelian segregation in the female sex is symmetric. In the case of symmetric inheritance we can identify zygotes a_{ij} and a_{ji} and the number of female genotypes is reduced to $(m+1)(m+2)/2$.

In the following we do not assume symmetry.

In the vector space

$$\mathfrak{Z} := \left\{ \sum_{i,j=0}^{m} \alpha_{ij} a_{ij} + \sum_{i=0}^{m} \alpha_i a_i \,\middle|\, \alpha_{ij}, \alpha_i \in \mathbb{R},\ i,j=0,\ldots,m \right\} \tag{8.11}$$

of dimension $(m+1)^2 + m + 1 = m^2 + 3m + 2$ we introduce a multiplication by

$$\begin{aligned} a_{ij}a_l = a_l a_{ij} &= \sum_{k,r=0}^{m} \gamma_{ijk}\gamma_{lr} a_{lr} + \frac{1}{2}\sum_{k=0}^{m} \gamma_{ijk} a_k \\ a_{ij}a_{lp} = a_l a_p &= 0, \qquad i,j,l,p=0,\ldots,m. \end{aligned} \tag{8.12}$$

With this multiplication $\mathfrak{Z}$ becomes a commutative, in general non associative algebra. $\mathfrak{Z}$ is the zygotic algebra (for asymmetric) sex-linked inheritance.

$\mathfrak{Z}$ is not a baric algebra since the specified basis elements are nilpotent of index 2. Now we can proceed along two different paths. Either we can follow Gonshor [45], [46], [48] and W.-B. [112], [113] and find a baric ideal of $\mathfrak{Z}$, cf. theorem 8.6, or we can follow Holgate [66] and generalize the notion of a baric algebra. First we shall follow the latter approach.

Definition 8.4. Let $\mathfrak{V} := \langle w,m \rangle_{\mathbb{K}}$ be a 2-dimensional commutative algebra over $\mathbb{K}$, char $\mathbb{K} \neq 2$, with multiplication table

$$w^2 = m^2 = 0, \qquad wm = \frac{1}{2}(w+m) \quad .$$

Then $\mathfrak{V}$ is called the sex differentiation algebra.

One easily checks that $\mathfrak{V}^2 = \langle w + m \rangle_{\mathbb{K}}$ is an ideal of $\mathfrak{V}$ which is isomorphic with $\mathbb{K}$. Thus, trivially, the algebra $\mathfrak{V}^2$ is baric.

Definition 8.5. An algebra $\mathfrak{A}$ over $\mathbb{K}$, char $\mathbb{K} \neq 2$, is called dibaric, if it possesses an algebra homomorphism $\chi: \mathfrak{A} \longrightarrow \mathfrak{V}$ onto $\mathfrak{V}$.

In a dibaric algebra $\mathfrak{A}$ the homomorphism $\chi : \mathfrak{A} \longrightarrow \mathfrak{V}$ induces an algebra homomorphism $\tilde{\chi}: \mathfrak{A}^2 \longrightarrow \mathfrak{V}^2 \cong \mathbb{K}$ and thus $\mathfrak{A}^2$ is a baric algebra.

The zygotic algebra $\mathfrak{Z}$ defined before is dibaric with respect to the mapping $\chi: \mathfrak{Z} \longrightarrow \mathfrak{V}$ defined by

$$\chi(a_{ij}) := w \qquad \chi(a_i) := m, \quad i,j=0,\dots,m.$$

We could now apply our machinery for baric algebras to the algebra $\mathfrak{Z}^2$. However, since the dimension of $\mathfrak{Z}^2$ depends on the multiplication in $\mathfrak{Z}$, we prefer the following ideal $\mathfrak{L}$ of $\mathfrak{Z}$ which has the dimension $m^2 + 3m + 1$ independent of the multiplication in $\mathfrak{Z}$, cf [112].

Theorem 8.6. The set

$$\mathfrak{L} := \left\{ \sum_{i,j=0}^{m} \alpha_{ij} a_{ij} + \sum_{i=0}^{m} \alpha_i a_i \;\middle|\; \sum_{i,j=0}^{m} \alpha_{ij} = \sum_{i=0}^{m} \alpha_i \right\}$$

is an algebra of $\mathfrak{Z}$ which contains $\mathfrak{Z}^2$. Furthermore $\mathfrak{L}$ is a baric algebra with the weight homomorphism $\varphi : \mathfrak{L} \longrightarrow \mathbb{R}$ given by

$$\varphi\left(\sum_{i,j=0}^{m} \alpha_{ij} a_{ij} + \sum_{i=0}^{m} \alpha_i a_i\right) := \sum_{i=0}^{m} \alpha_i .$$

Every actual population is given by an element x of the $(m^2 + 3m)$-dimensional simplex

$$\mathcal{Y} = \left\{ \sum_{i,j=0}^{m} \alpha_{ij} a_{ij} + \sum_{i=0}^{m} \alpha_i a_i \in \mathfrak{Z} \;\middle|\; 0 \le \alpha_{ij}, \alpha_i \le 1, \; \sum_{i,j=0}^{m} \alpha_{ij} = \sum_{i=0}^{m} \alpha_i = 1 \right\}. \quad (8.13)$$

Thus the ideal $\mathfrak{L}$ of $\mathfrak{Z}$ contains all elements corresponding to actual

populations, hence it is sufficient to consider $\mathcal{B}$ for any genetic purpose. On the other hand the algebra $\mathfrak{Z}^2$ contains only those actual populations which result from one step of random mating.
Before discussing sequences of plenary powers in $\mathcal{B}$ we give a basis free representation of $\mathfrak{Z}$, cf. Gonshor [48] and W.-B. [119].

Let $\mathfrak{A}$ be the $\mathbb{R}$-algebra spanned by $a_o,\ldots,a_m$ with multiplication table

$$a_i a_j := \sum_{k=0}^{m} \gamma_{ijk} a_k, \qquad i,j=0,\ldots,m. \tag{8.14}$$

$\mathfrak{A}$ is a baric algebra with weight homomorphism $\omega : \mathfrak{A} \longrightarrow \mathbb{R}$ defined by $\omega(a_i) := 1$, $i=0,\ldots,m$.

Let $\mathfrak{N} := \ker\omega$. Let $\ell : \mathfrak{A} \longrightarrow \mathfrak{A}$ the linear mapping induced by the matrix Γ^T, where $\Gamma := (\gamma_{il})$, i.e.

$$\ell a_i = \sum_{l=0}^{m} \gamma_{il} a_l, \qquad i=0,\ldots,m. \tag{8.15}$$

Then $\omega(\ell a_i) = 1/2$ for all $i=0,\ldots,m$, in view of (8.8). Thus ℓ leaves the ideal $\mathfrak{N}$ invariant., i.e. $\ell\mathfrak{N} \subseteq \mathfrak{N}$.

We introduce the vector space $\tilde{\mathfrak{Z}} := \mathfrak{A}\otimes\mathfrak{A}\oplus\mathfrak{A}$, where $\oplus$ denotes the direct sum and $\otimes$ denotes the tensor product of vector spaces. In this space we introduce a multiplication by

$$\begin{aligned}&((\sum_{i\in I} x_i\otimes y_i)\oplus z_1)((\sum_{j\in J} x_j\otimes y_j)\oplus z_2) \\ &= \sum_{i\in I}(x_i y_i\otimes \ell z_2 \oplus \tfrac{1}{2}\omega(z_2)x_i y_i) + \sum_{j\in J}(x_j y_j\otimes \ell z_1 \oplus \tfrac{1}{2}\omega(z_1)x_j y_j)\end{aligned} \tag{8.16}$$

where xy is the product of x and y in $\mathfrak{A}$.
We have to show that (8.16) is well defined (for the special case $\Gamma = \frac{1}{2}I$ and $\mathfrak{A}\circledast\mathfrak{A}$ instead of $\mathfrak{A}\otimes\mathfrak{A}$ cf. Gonshor [48]).
For this purpose we identify the elements $x\otimes y \oplus 0 \in \tilde{\mathfrak{Z}}$ with $x\otimes y \in \mathfrak{A}\otimes\mathfrak{A}$ and the elements $0\oplus z \in \tilde{\mathfrak{Z}}$ with $z \in \mathfrak{A}$. Then we can represent any element of $\tilde{\mathfrak{Z}}$ uniquely as the sum of an element of $\mathfrak{A}\otimes\mathfrak{A}$ and an element of $\mathfrak{A}$, and (8.16) is reduced to a sum of products of the form

$$(x\otimes y)\ z = xy\otimes \ell z \oplus \frac{1}{2}\omega(z)xy \quad . \tag{8.17}$$

The tensor product $\mathfrak{A}\otimes\mathfrak{A}$ can be interpreted as the factor space of the free module $\mathbb{R}(\mathfrak{A}\times\mathfrak{A})$ over the subspace

$$\mathfrak{h} = \langle (\lambda_1 x_1 + \lambda_2 x_2, y) - \lambda_1 (x_1, y) - \lambda_2 (x_2, y), (x, \lambda_1 y_1 + \lambda_2 y_2) -$$

$$- \lambda_1 (x, y_1) - \lambda_2 (x, y_2) \mid x, x_i, y, y_i \in \mathfrak{A}, \lambda_i \in \mathbb{R}, i=1,2 \rangle .$$

We introduce products of elements in $\mathbb{R}(\mathfrak{A} \times \mathfrak{A})$ with elements of $\mathfrak{A}$ by

$$(x,y)z = (xy, \ell z) + \frac{1}{2}\omega(z)xy$$

and the linear extension onto $\mathbb{R}(\mathfrak{A} \times \mathfrak{A})$. The space $\mathfrak{h}$ is invariant with respect to this multiplication since for instance

$$(\lambda_1 x_1 + \lambda_2 x_2, y) - \lambda_1 (x_1, y) - \lambda_2 (x_2, y) \; z$$

$$= ((\lambda_1 x_1 + \lambda_2 x_2) y, \ell z) + \frac{1}{2}\omega(z)(\lambda_1 x_1 + \lambda_2 x_2) y$$

$$- \lambda_1 (x_1 y, \ell z) - \frac{1}{2}\omega(z)\lambda_1 x_1 - \lambda_2 (x_2 y, \ell z) - \frac{1}{2}\omega(z)\lambda_2 x_2 y$$

$$= (\lambda_1 (x_1 y) + \lambda_2 (x_2 y), \ell z) - \lambda_1 (x_1 y, \ell z) - \lambda_2 (x_2 y, \ell z) \in \mathfrak{h} .$$

Similar relations hold for the other generators of $\mathfrak{h}$. Thus (8.17) is well defined and also (8.16) is well defined. The distribution law is easily checked and thus by (8.16) a multiplication in $\tilde{\mathfrak{Z}}$ is defined.

Definition 8.7. The algebra $\tilde{\mathfrak{Z}} = \mathfrak{A} \otimes \mathfrak{A} \oplus \mathfrak{A}$ with the multiplication given by (8.16) is the sex linked duplicate of the algebra $\mathfrak{A}$ (with respect to the linear mapping ℓ).

Remark. In the preceding definition we followed Gonshor [48] who had introduced the name "sex linked duplication" for $\mathfrak{A} \otimes \mathfrak{A} \oplus \mathfrak{A}$ with $\Gamma = \frac{1}{2}I$. We underline that the construction of the algebra $\mathfrak{A} \otimes \mathfrak{A} \oplus \mathfrak{A}$ works for any baric algebra $\mathfrak{A}$ with weight homomorphism ω and linear mapping $\ell : \mathfrak{A} \longrightarrow \mathfrak{A}$ with the property that $\omega(\ell x) = \frac{1}{2}\omega(x)$ for all $x \in \mathfrak{A}$. The algebra $\mathfrak{A}$ need not to be related with sex linkage, or even with genetics. In view of this and the fact that $\mathfrak{A} \otimes \mathfrak{A} \oplus \mathfrak{A}$ is built out of three copies of $\mathfrak{A}$ Holgate suggested to call $\mathfrak{A} \otimes \mathfrak{A} \oplus \mathfrak{A}$ the "asymmetric triplicate" of $\mathfrak{A}$.

The following assertion is easily verified.

Theorem 8.8. The zygotic algebra $\mathfrak{Z}$ for sex linked inheritance and the sex linked duplicate $\tilde{\mathfrak{Z}} = \mathfrak{A} \otimes \mathfrak{A} \oplus \mathfrak{A}$ of $\mathfrak{A}$ are isomorphic.

In this isomorphism the female zygotes $a_{ij} \in \mathfrak{Z}$ are represented by elements contained in $\mathfrak{A} \otimes \mathfrak{A} \oplus \langle 0 \rangle \cong \mathfrak{A} \otimes \mathfrak{A}$, i.e. by elements of the vector space corresponding to the (non commutative) duplicate $\mathfrak{A} \otimes_d \mathfrak{A}$ of $\mathfrak{A}$, cf. section 6C.

Let $\omega_d := \omega \otimes \omega$ be the tensor product of the mapping ω with itself. The ideal of $\mathfrak{A} \otimes \mathfrak{A} \oplus \mathfrak{A}$, isomorphic with $\mathfrak{B}$, has the representation

$$\tilde{\mathfrak{B}} := \{ x \oplus z \mid x \in \mathfrak{A} \otimes \mathfrak{A},\ z \in \mathfrak{A},\ \omega_d(x) = \omega(z) \} \quad . \qquad (8.18)$$

From theorem 8.6 follows with assistance of theorem 8.8 that $\tilde{\mathfrak{B}}$ is a baric algebra with weight homomorphism $\tilde{\varphi} : \mathfrak{B} \longrightarrow \mathbb{R}$ defined by

$$\tilde{\varphi}(x+z) = \omega_d(x) \qquad (= \omega(z)) \quad . \qquad (8.19)$$

From theorem 6.15 we obtain for $\tilde{\mathfrak{M}} := \ker \tilde{\varphi}$ the representation

$$\tilde{\mathfrak{M}} = (\mathfrak{A} \otimes \mathfrak{N} + \mathfrak{N} \otimes \mathfrak{A}) \oplus \mathfrak{N} \ . \qquad (8.20)$$

The following theorem has been shown by Gonshor [48] for Mendelian segregation rates in the male sex, i.e. $\Gamma = \frac{1}{2} I$, symmetric inheritance in the female sex, i.e. $\mathfrak{A}$ is commutative, and $\mathfrak{A} \circledast \mathfrak{A}$ instead of $\mathfrak{A} \otimes \mathfrak{A}$.

<u>Theorem 8.9.</u> Let $\mathfrak{A}$ be a Gonshor genetic algebra over $\mathbb{R}$ with a canonical basis $c_0, \dots, c_m$. If the linear mapping $\ell : \mathfrak{A} \longrightarrow \mathfrak{A}$ (8.15) is, with respect to this basis, represented by a lower triangular matrix, then the ideal $\tilde{\mathfrak{B}}$ of the sex-linked duplicate $\tilde{\mathfrak{Z}} = \mathfrak{A} \otimes \mathfrak{A} \oplus \mathfrak{A}$ of $\mathfrak{A}$ is a genetic algebra.

<u>Proof.</u> The algebra $\tilde{\mathfrak{B}}$ is baric with weight homomorphism $\tilde{\varphi}$ (8.19) and kernel $\tilde{\mathfrak{M}} = (\mathfrak{A} \otimes \mathfrak{N} + \mathfrak{N} \otimes \mathfrak{A}) \oplus \mathfrak{N}$. Since $\tilde{\mathfrak{Z}}$ and hence $\tilde{\mathfrak{B}}$ are commutative, we can exploit a criterion for genetic algebras from Gonshor [48] : $\tilde{\mathfrak{B}}$ is genetic, if $\tilde{\mathfrak{M}}^{\langle r+1 \rangle} = \langle 0 \rangle$ for some $r \in \mathbb{N}$, here $\mathfrak{M}^{\langle k \rangle}$ denotes the set of all linear combinations of left products of elements of $\tilde{\mathfrak{B}}$ with at least k factors from $\tilde{\mathfrak{M}}$, cf. theorem 3.25.
The algebra $\tilde{\mathfrak{M}}^{\langle k+1 \rangle}$ is the smallest ideal of $\tilde{\mathfrak{B}}$ which contains $\tilde{\mathfrak{M}}\,\tilde{\mathfrak{M}}^{\langle k \rangle}$.
Since $\mathfrak{A}$ is Gonshor genetic with canonical basis $c_0, \dots, c_m$ the linear spaces

$$\mathfrak{N}_k := \langle c_k, \dots, c_m \rangle \ , \quad k=1,\dots,m,\ \mathfrak{N}_{m+1} := \langle 0 \rangle$$

form a sequence of ideals of $\mathfrak{A}$, and

$$\mathfrak{N}\mathfrak{N}_k, \mathfrak{N}_k\mathfrak{N} \subseteq \mathfrak{N}_{k+1}, \quad k=1,\dots,m.$$

Now we suppose that $\mathscr{C}$ is represented by a lower triangular matrix with respect to the canonical basis. This is equivalent to assume the mapping $\mathscr{C}$ leaves the ideals $\mathfrak{N}_k$, k=1,...,m, of $\mathfrak{A}$ invariant, i.e. $\mathscr{C}\mathfrak{N}_k \subseteq \mathfrak{N}_k$, k=1,...,m. We are going to prove by induction on $k\in\mathbb{N}$ that

$$\widetilde{\mathfrak{M}}^{\langle k+1\rangle} \subseteq (\mathfrak{N}_k\otimes\mathfrak{N} + \mathfrak{N}\otimes\mathfrak{N}_k + \mathfrak{N}_{k+1}\otimes\mathfrak{A}) \oplus \mathfrak{N}_{k+1}, \quad k\in\mathbb{N}. \tag{8.21}$$

Using (8.16) we obtain for the second principal power of

$$\begin{aligned}
\widetilde{\mathfrak{M}}^2 &= ((\mathfrak{A}\otimes\mathfrak{N} + \mathfrak{N}\otimes\mathfrak{A}) \oplus \mathfrak{N})^2 \\
&= (\mathfrak{A}\mathfrak{N}\otimes\mathscr{C}\mathfrak{N} + \mathfrak{N}\mathfrak{A}\otimes\mathscr{C}\mathfrak{N}) \oplus \omega(\mathfrak{N})(\mathfrak{A}\mathfrak{N} + \mathfrak{N}\mathfrak{A}) \\
&= \mathfrak{A}\mathfrak{N}\otimes\mathscr{C}\mathfrak{N} + \mathfrak{N}\mathfrak{A}\otimes\mathscr{C}\mathfrak{N} \\
&\subseteq \mathfrak{N}\otimes\mathfrak{N} \\
&= \mathfrak{N}_1\otimes\mathfrak{N}_1 .
\end{aligned}$$

Furthermore

$$\begin{aligned}
\mathscr{B}\,\widetilde{\mathfrak{M}}^2 &\subseteq (\mathfrak{A}\otimes\mathfrak{A}\oplus\mathfrak{A})(\mathfrak{N}\otimes\mathfrak{N}) \\
&= \mathfrak{N}^2\otimes\mathscr{C}\mathfrak{A} \oplus \omega(\mathfrak{A})\mathfrak{N}^2 \\
&\subseteq \mathfrak{N}^2\otimes\mathfrak{A}\oplus\mathfrak{N}^2 \\
&\subseteq \mathfrak{N}_2\otimes\mathfrak{A}\oplus\mathfrak{N}_2 .
\end{aligned}$$

From the definition of $\widetilde{\mathfrak{M}}^{\langle 2\rangle}$ and the above relations we obtain

$$\widetilde{\mathfrak{M}}^{\langle 2\rangle} = (\mathfrak{N}_1\otimes\mathfrak{N}_1 + \mathfrak{N}_2\otimes\mathfrak{A}) \oplus \mathfrak{N}_2 ,$$

hence (8.21) is fulfilled for k=1. Assume (8.21) for one $k\in\mathbb{N}$. Then

$$\begin{aligned}
\widetilde{\mathfrak{M}}\,\widetilde{\mathfrak{M}}^{\langle k\rangle} &\subseteq [(\mathfrak{A}\otimes\mathfrak{N} + \mathfrak{N}\otimes\mathfrak{A})\oplus\mathfrak{N}][(\mathfrak{N}_k\otimes\mathfrak{N} + \mathfrak{N}\otimes\mathfrak{N}_k + \mathfrak{N}_{k+1}\otimes\mathfrak{A})\oplus\mathfrak{N}_{k+1}] \\
&= (\mathfrak{A}\mathfrak{N} + \mathfrak{N}\mathfrak{A})\otimes\mathscr{C}\mathfrak{N}_{k+1} + (\mathfrak{N}_k\mathfrak{N} + \mathfrak{N}\mathfrak{N}_k + \mathfrak{N}_{k+1}\mathfrak{A})\otimes\mathscr{C}\mathfrak{N} \\
&\subseteq \mathfrak{N}\otimes\mathfrak{N}_{k+1} + \mathfrak{N}_{k+1}\otimes\mathfrak{N} ,
\end{aligned}$$

and

$$\begin{aligned}
\widetilde{\mathscr{B}}(\widetilde{\mathfrak{M}}\widetilde{\mathfrak{M}}^{\langle k\rangle}) &\subseteq (\mathfrak{A}\otimes\mathfrak{A}\oplus\mathfrak{A})(\mathfrak{N}\otimes\mathfrak{N}_{k+1} + \mathfrak{N}_{k+1}\otimes\mathfrak{N}) \\
&\subseteq [(\mathfrak{N}\mathfrak{N}_{k+1} + \mathfrak{N}_{k+1}\mathfrak{N})\otimes\mathscr{C}\mathfrak{A}] \oplus \omega(\mathfrak{A})(\mathfrak{N}\mathfrak{N}_{k+1} + \mathfrak{N}_{k+1}\mathfrak{N}) \\
&\subseteq (\mathfrak{N}_{k+2}\otimes\mathfrak{A}) \oplus \mathfrak{N}_{k+1} ,
\end{aligned}$$

hence (8.21) is fulfilled for k+1 instead of k.
From (8.21) in view of $\mathfrak{N}_{m+1} = \langle 0 \rangle$ it follows that

$$\tilde{\mathfrak{M}}^{\langle m+2 \rangle} = \langle 0 \rangle \quad ,$$

hence $\tilde{\mathfrak{B}}$ is genetic. □

For the case of simple Mendelian segregation in both sexes Holgate [61] has constructed a canonical basis of the ideal $\tilde{\mathfrak{B}}$ of the sex linked duplicate of $\mathfrak{A}$ from a canonical basis of $\mathfrak{A}$. This procedure can be carried over to the general case. However in the following we can do without a canonical basis of $\tilde{\mathfrak{B}}$.

We add some remarks on the case of symmetric inheritance in the female sex. Then the segregation rates satisfy (8.9). We identify the genotypes a_{ij} and a_{ji}, $i \neq j$. Then the genetics is described by a (m+1)(m+4)/2-dimensional algebra over

$$\mathfrak{Z}_s := \left\{ \sum_{\substack{i,j=0\\ i\leq j}}^{m} \alpha_{ij} a_{ij} + \sum_{i=0}^{m} \alpha_i a_i \;\middle|\; \alpha_{ij}, \alpha_i \in \mathbb{R} \right\}$$

with multiplication given by

$$\begin{aligned} a_{ij} a_l &= \sum_{\substack{k,r=0\\ k\leq r}}^{m} (\gamma_{ijk}\gamma_{lr} + \gamma_{ijr}\gamma_{lk}) a_{kr} + \frac{1}{2} \sum_{k=0}^{m} \gamma_{ijk} a_k \\ a_{ij} a_{lp} &= a_l a_p = 0, \qquad i,j,l,p = 0,\dots,m. \end{aligned} \tag{8.22}$$

$\mathfrak{Z}_s$ is called the <u>zygotic algebra</u> for <u>symmetric sex linked inheritance</u>. It is easy to see that $\mathfrak{Z}_s$ is a dibaric algebra in the sense of Holgate [66]. The vector space

$$\mathfrak{b}_s := \left\{ \sum_{\substack{i,j=0\\ i\leq j}}^{m} \alpha_{ij} a_{ij} + \sum_{i=0}^{m} \alpha_i a_i \in \mathfrak{Z}_s \;\middle|\; \sum_{\substack{i,j=0\\ i\leq j}}^{m} \alpha_{ij} = \sum_{i=0}^{m} \alpha_i \right\}$$

is an ideal of $\mathfrak{Z}_s$ containing $\mathfrak{Z}_s^2$, and $\mathfrak{b}_s$ is a baric algebra with weight homomorphism

$$\varphi_s\Big(\sum_{\substack{i,j=0\\ i\leq j}}^{m} \alpha_{ij} a_{ij} + \sum_{i=0}^{m} \alpha_i a_i \Big) := \sum_{i=0}^{m} \alpha_i \;.$$

Let $\mathfrak{A} \circledast \mathfrak{A}$ be the vector space corresponding to the commutative duplicate of $\mathfrak{A}$, i.e. $\mathfrak{A} \otimes \mathfrak{A} /$ for $= \langle x \otimes y - y \otimes x \mid x, y \in \mathfrak{A} \rangle$.

Since $\mathcal{J}$ is invariant with respect to the multiplication in $\mathfrak{A} \otimes \mathfrak{A} \oplus \mathfrak{A}$ given by (8.16), the vector space

$$\mathfrak{Z}_s := \mathfrak{A} \otimes \mathfrak{A} \oplus \mathfrak{A}$$

with multiplication as in (8.16) becomes an algebra too; this algebra is the <u>commutative</u> <u>sex</u> <u>linked</u> <u>duplicate</u> of $\mathfrak{A}$.
The algebras $\mathfrak{Z}_s$ and $\tilde{\mathfrak{Z}}_s$ are isomorphic.
Furthermore the ideal $\tilde{\mathfrak{B}}_s$ of $\tilde{\mathfrak{Z}}_s$ isomorphic with $\mathfrak{B}_s$ is a genetic algebra if $\mathfrak{A}$ is a genetic algebra and the linear mapping $\mathfrak{C}$ is represented by a lower triangular matrix with respect to a canonical basis of $\mathfrak{A}$.

We return to the general case of possibly asymmetric inheritance in the female sex and we investigate sequences of plenary powers of elements $x \in \mathfrak{B}$, $\varphi(x) = 1$, cf. W.-B. [113] for some of the results.
Let

$$x^{[t]} = \sum_{i,j=0}^{m} \alpha_{ij}{}^{[t]} a_{ij} + \sum_{i=0}^{m} \alpha_i{}^{[t]} a_i, \qquad t \in \mathbb{N},$$

be the t-th plenary power of an element $x \in \mathfrak{B}$ with $\varphi(x) = 1$.
Then the coordinates of $x^{[t]}$ satisfy the recursions

$$\begin{aligned} \alpha_{kr}{}^{[t+1]} &= \Big(\sum_{i,j=0}^{m} \alpha_{ij}{}^{[t]} \gamma_{ijk}\Big)\Big(\sum_{l=0}^{m} \alpha_l{}^{[t]}\, {}_2\gamma_{lr}\Big), && k,r=0,\dots,m, \\ \alpha_k{}^{[t+1]} &= \sum_{i,j=0}^{m} \alpha_{ij}{}^{[t]} \gamma_{ijl}, && k=0,\dots,m. \end{aligned} \tag{8.23}$$

These $(m+1)^2 + m+1$ difference equations with initial data

$$\alpha_{kr}^{[1]}, \ \alpha_k^{[1]}, \quad k,r=0,\dots,m, \quad \sum_{k,r=0}^{m} \alpha_{kr}^{[1]} = \sum_{k=0}^{m} \alpha^{[1]} = 1, \tag{8.24}$$

are equivalent with the following m+1 difference equations

$$\alpha_k{}^{[t+2]} = \sum_{i,j=0}^{m} \alpha_i{}^{[t+1]} \gamma_{ijk} \sum_{l=0}^{m} \alpha_l{}^{[t]}\, {}_2\gamma_{lj}, \quad k=0,\dots,m \tag{8.25}$$

with the special initial data

$$\alpha_k^{[1]}, \alpha_k^{[2]} := \sum_{i,j=0}^{m} \alpha_{ij}{}^{[1]} \gamma_{ijk}, \quad k=0,\dots,m. \tag{8.26}$$

The solutions are related by the equations

$$\alpha_{kr}{}^{[t+1]} = \alpha_k{}^{[t+1]} \Big(\sum_{l=0}^{m} \alpha_l{}^{[t]}\, {}_2\gamma_{lr}\Big), \qquad k,r=0,\dots,m. \tag{8.27}$$

From (8.25) and (8.27) we obtain immediately conditions for an element $0 \neq e = \sum_{i,j=0}^{m} \eta_{ij} a_{ij} + \sum_{i=0}^{m} \eta_i a_i \in \mathfrak{B}$ to be an idempotent of $\mathfrak{B}$, namely

$$\eta_k = \sum_{i,j=0}^{m} \eta_i \gamma_{ijk} \sum_{l=0}^{m} \eta_l 2\gamma_{lh}, \qquad k=0,\dots,m, \tag{8.28}$$

and

$$\eta_{kr} = \eta_k \sum_{l=0}^{m} \eta_l 2\gamma_{lr}, \qquad k,r=0,\dots,m. \tag{8.29}$$

Thus the recursions and also the conditions for idempotents reduce to (m+1)-dimensional problems which both can be treated in an algebra which has the same underlying vector space as the algebra $\mathfrak{A}$. In this vector space we define a multiplication by

$$(a_i, a_j) \longmapsto a_i \circ a_j := a_i(2\ell a_j), \qquad i,j=0,\dots,m, \tag{8.30}$$

where xy is the product (8.14) in $\mathfrak{A}$ and ℓ is the linear mapping defined by (8.15). This new algebra over $\langle a_0,\dots,a_m \rangle_{\mathbb{R}}$ is denoted by $\mathfrak{A}^\circ$. The multiplication constants of $\mathfrak{A}^\circ$ with respct to the original basis $a_0,\dots,a_m$ are

$$\gamma^\circ_{ijk} = \sum_{l=0}^{m} \gamma_{ilk} 2\gamma_{jl}, \qquad i,j,k=0,\dots,m. \tag{8.31}$$

From (8.3) we see that the algebra $\mathfrak{A}^\circ$ is baric with weight homomorphism ω. By assumption the constants γ_{ijk} and γ_{ik}, $i,j,k=0,\dots,m$, are non negative and sum up to 1 and 1/2, respectively, and thus the algebra $\mathfrak{A}^\circ$, and of course $\mathfrak{A}$, have a genetic realization.
From theorem 6.29 and from (6.17) follows immediately

Theorem 8.10. Let $\mathfrak{A}$ be a Gonshor genetic algebra with a canonical basis $c_0,\dots,c_m$ and the left and right roots $\lambda_{000} = 1$, λ_{0ii}, λ_{i0i}, $i=1,\dots,m$. Let the linear mapping $\ell : \mathfrak{A} \to \mathfrak{A}$, defined in (8.15), be represented with respect to this basis by a lower triangular matrix with diagonal $\frac{1}{2}(1,\mu_1,\dots,\mu_m)$.
Then $\mathfrak{A}^\circ$ is a Gonshor genetic algebra with the left and right train roots

$$\lambda^\circ_{000} = 1, \quad \lambda^\circ_{0ii} = \mu_i \lambda_{0ii}, \quad \lambda^\circ_{i0i} = \lambda_{i0i}, \quad i=1,\dots,m. \tag{8.32}$$

Theorem 8.11. Assume the hypothesis of theorem 8.10. Then

(1) The algebra $\mathcal{B}$, cf. theorem 8.6, has always an idempotent.
If

$$\mu_i \lambda_{oii} + \lambda_{ioi} \neq 1 \qquad \text{for all } i=1,\dots,m, \tag{8.33}$$

then $\mathcal{B}$ has exactly one idempotent.

(2) If

$$|\mu_i||\lambda_{oii}|+|\lambda_{ioi}| < 1 \qquad \text{for all } i=1,\dots,m, \tag{8.34}$$

then the sequence of plenary powers converges for any initial element $x \in \mathcal{B}$ of weight 1 towards the idempotent of $\mathcal{B}$.

Proof. As we have seen there is a one-to-one correspondence between the idempotents of $\mathcal{B}$ and those of $\mathfrak{A}^o$ given by equation (8.28). Since the algebra $\mathfrak{A}^o$ has a genetic realization, it has always an idempotent in view of theorem 5.21. From theorem 5.18 it follows that $\mathfrak{A}^o$ has exactly one idempotent if the sums

$$\lambda^o_{oii} + \lambda^o_{ioi} = \mu_i \lambda_{oii} + \lambda_{ioi}, \quad i=1,\dots,m,$$

of corresponding left and right train roots of $\mathfrak{A}^o$ are different from 1.

II. Let (8.34) be fulfilled. Consider now the sequence of plenary powers for an element

$$x = x^{[1]} = \sum_{i,j=o}^{m} \alpha_{ij}^{[1]} a_{ij} + \sum_{i=o}^{m} \alpha_i^{[1]} a_i \in \mathcal{B}$$

of weight 1. The difference equations (8.23) for the coordinates of the plenary powers of x are equivalent with the difference equations (8.25) which define a sequence $\{z^{[t]}\}$ of elements in $\mathfrak{A}^o$ such that

$$z^{(t+2)} = z^{(t+1)} \circ z^{(t)}, \qquad t \in \mathbb{N}, \tag{8.35}$$

with

$$z^{(l)} := \sum_{i=o}^{m} \alpha_i^{[l]} a_i \qquad \text{for } l=1,2, \tag{8.36}$$

here the coefficients of $\alpha_i^{[l]}$, $i=o,\dots,m$, $l=1,2$, are given by (8.26). Since $x \in \mathcal{B}$ has weight 1, also the elements $z^{(1)}, z^{(2)} \in \mathfrak{A}^o$ have weight 1. From (8.35) it follows that all $z^{(t)}$, $t \in \mathbb{N}$, have weight 1. In view of (8.34) we know that $\mathfrak{A}$ has exactly one idempotent.
We can choose a canonical basis $c_o,\dots,c_m$ of $\mathfrak{A}$ such that c_o is the idempotent. From the hypothesis on the linear mapping ℓ it follows that

$c_o,\dots,c_m$ is also a canonical basis of $\mathfrak{A}^o$. Let λ^o_{ijk}, $i,j,k=0,\dots,m$, be the multiplication constants of $\mathfrak{A}^o$ with respect to this basis. Let the coordinates of $z^{(t)}$ be given by

$$z^{(t)} = \sum_{i=0}^{m} \zeta_i^{(t)} c_i \,, \qquad t \in \mathbb{N},$$

then

$$\zeta_o^{(t)} = 1, \qquad t \in \mathbb{N},$$

and from (8.35)

$$\zeta_k^{(t+2)} = \lambda^o_{kok}\zeta_k^{(t+1)} + \lambda^o_{okk}\zeta_k^{(t)} + \sum_{i,j=0}^{k-1} \zeta_i^{(t+1)} \lambda^o_{ijk}\zeta_j^{(t)} \,, \tag{8.37}$$
$$k=1,\dots,m, t \in \mathbb{N}.$$

We show by induction that the sequences of coordinates $\{\zeta_k^{(t)}\}$, $k=1,\dots,m$, converge to zero.
The first equation of (8.37) reduces to the linear homogeneous difference equation

$$\zeta_1^{(t+2)} = \lambda^o_{1o1}\zeta_1^{(t+1)} + \lambda^o_{o11}\zeta_1^{(t)} \,, \qquad t \in \mathbb{N}.$$

The roots ν_{11}, ν_{12} of the associated characteristic polynomial

$$p(\xi) = \xi^2 - \lambda^o_{1o1}\xi - \lambda^o_{o11}$$

are less than 1 in modulus if and only if

$$|\lambda^o_{1o1}| + |\lambda^o_{o11}| < 1 \quad .$$

In view of (8.32) and the assumption that (8.34) is valid, the latter inequality is fulfilled. Hence every solution $\{\zeta_1^{(t)}\}$ of the above difference equation converges towards zero.
Now assume it has been shown that all coordinate sequences $\{\zeta_1^{(t)}\}, \dots, \{\zeta_{k-1}^{(t)}\}$ converge to 0.
The k-th sequence of coordinates satisfy the linear inhomogeneous difference equation (8.37) for the index k. By the induction hypothesis the inhomogeneous term goes to zero for t going to infinity. In view of

$$|\lambda^o_{kok}| + |\lambda^o_{okk}| = |\lambda_{kok}| + |\mu_k||\lambda_{okk}| < 1$$

we have

$$|\nu_{ki}| < 1$$

for the roots ν_{ki}, $i=1,2$, of the characteristic polynomial associated with the homogeneous difference equation. Therefore $\{\zeta_k^{(t)}\}$ also converges to zero. □

For several special cases the conditions (8.33) and (8.34) simplify.

Theorem 8.12. Assume the hypothesis of theorem 8.10. If corresponding left and right train roots of $\mathfrak{A}$ coincide, i.e.

$$\lambda_{oii} = \lambda_{ioi} =: \lambda_i, \quad i=1,\dots,m, \tag{8.38}$$

(in particular if $\mathfrak{A}$ is commutative) then conditions (8.33) and (8.34) of theorem 8.11 are equivalent to

$$(\lambda_i, \mu_i) \neq (1/2, 1) \qquad i=1,\dots,m, \tag{8.39}$$

and

$$|\lambda_i| < 1/2 \text{ or } |\mu_i| < 1, \quad i=1,\dots,m, \tag{8.40}$$

respectively.

Proof. Since the Gonshor genetic algebra $\mathfrak{A}$ has a genetic realization, corresponding left and right train roots satisfy the relation

$$|\lambda_{oii} + \lambda_{ioi}| \leq 1, \qquad i=1,\dots,m,$$

cf. theorem 5.23. If (8.38) is valid we have

$$|\lambda_i| \leq 1/2, \qquad i=1,\dots,m.$$

Furthermore the linear mapping $2\mathcal{C}: \mathfrak{A} \longrightarrow \mathfrak{A}$ is represented by the column stochastic matrix $2\Gamma^T$ with respect to the natural basis $a_o,\dots,a_m$, hence

$$|\mu_i| \leq 1, \qquad i=1,\dots,m.$$

In view of these bounds for the complex numbers μ_i and λ_i condition (8.33), which reduces to

$$\mu_i \lambda_{oii} + \lambda_{ioi} = (1+\mu_i)\lambda_i \neq 1, \qquad i=1,\dots,m,$$

is satisfied if and only if (8.39) holds.
Furthermore (8.34) is equivalent to (8.40). □

If we have simple Mendelian segregation rates in the female sex, the algebra $\mathfrak{A}$ is commutative and every basis $c_o, c_1, \dots, c_m$ of $\mathfrak{A}$ with

$\omega(c_o) = 1$, $\omega(c_i) = 0$, $i=1,\dots,m$, is canonical.
Hence for all segregation rates in the male sex there exists a canonical basis of $\mathfrak{A}$ such that the linear mapping $\ell : \mathfrak{A} \longrightarrow \mathfrak{A}$ is represented by a lower triangular matrix. Thus the hypothesis of theorem 8.10 is always satisfied.
Furthermore in view of $\lambda_i = 1/2$, $i=1,\dots,m$, we obtain from theorems 8.11 and 8.12 the following

Theorem 8.13. Assume simple Mendelian segregation rates in the female sex. Let $\mu_o = 1, \mu_1, \dots, \mu_m$ be the eigenvalues of the stochastic matrix 2Γ.

(1) If

$$\mu_i \neq 1, \qquad i=1,\dots,m$$

then $\mathfrak{B}$ has exactly one idempotent.

(2) If

$$|\mu_i| < 1, \qquad i=1,\dots,m$$

then the sequence of plenary powers of every $x \in \mathfrak{B}$ of weight 1 converges towards the idempotent of $\mathfrak{B}$.

On the other side, let us assume simple Mendelian segregation rates in the male sex, i.e. $2\Gamma = I$, and $\mu_i = 1$, $i=0,\dots,m$. Hence theorem 8.11 yields

Theorem 8.14. Assume simple Mendelian segregation in the male sex and that the segregation rates in the female sex are such that the algebra $\mathfrak{A}$ is Gonshor genetic. Let $\lambda_{ooo} = 1, \lambda_{oii}, \lambda_{ioi}$, $i=1,\dots,m$, be the left and right train roots of $\mathfrak{A}$.

(1) If

$$\lambda_{oii} + \lambda_{ioi} \neq 1, \qquad i=1,\dots,m,$$

then $\mathfrak{B}$ has exactly one idempotent.

(2) If

$$|\lambda_{oii}| + |\lambda_{ioi}| < 1, \qquad i=1,\dots,m,$$

then the sequence of plenary powers of each element of weight 1 converges to the idempotent.

Finally let us assume arbitrary segregation rates in the male sex and additive segregation rates

$$\gamma_{ijk} = \alpha_{ik} + \beta_{jk}, \quad i,j,k=0,\dots,m, \tag{8.1}$$

in the female sex. Although we know from section 8A that the algebra is not Gonshor genetic in general, and consequently we theorem 8.11 we can give a sufficient condition for the convergence of the sequence of plenary powers of x of weight 1. The coordinates of the plenary powers

$$x^{[t]} = \sum_{i,j=0}^{m} \alpha_{ij}^{[t]} a_{ij} + \sum_{i=0}^{m} \alpha_i^{[t]} a_i , \qquad t \in \mathbb{N}$$

satisfy the quadratic difference equations (8.25). By assumption (8.1) these equations reduce to a system of linear equations

$$\alpha_k^{[t+2]} = \sum_{i=0}^{m} \alpha_i^{[t+1]} \alpha_{ik} + \sum_{l=0}^{m} \alpha_l^{[t]} \sum_{j=0}^{m} 2\gamma_{lj}\beta_{jk}, \quad k=0,\dots,m, t \in \mathbb{N},$$

which in matrix form can be written as

$$\alpha^{[t+2]} = \alpha^{[t+1]} A + \alpha^{[t]} 2\Gamma B, \qquad t \in \mathbb{N}.$$

These equations correspond to a first order difference equation with the matrix

$$M := \begin{pmatrix} 0 & 2\Gamma B \\ I & A \end{pmatrix} .$$

Hence we have

Theorem 8.15. Assume arbitrary segregation in the male and additive segregation rates in the female sex. Let $\nu_0 = 1, \nu_1, \dots, \nu_{2m+1}$ be the roots of the equation

$$\det(\lambda^2 I - \lambda A - 2\Gamma B) = 0. \qquad (8.41)$$

If $|\nu_i| < 1$, $i=1,\dots,2m+1$, then the sequence of plenary powers of elements of $\mathfrak{b}$ of unit weight converges towards the idempotent of $\mathfrak{b}$.

In particular this theorem covers the case of mutation. Then $a_0, \dots, a_m$ correspond to the different alleles at one locus on the X-chromosome. The matrices 2A and 2Γ describe mutation in the female and male sex, respectively, and B = A. If the mutation rates are identical in both sexes, i.e. $2A = 2\Gamma =: S$, with some row stochastic matrix S, then (8.41) can be factorized into

$$\det(\lambda I - \tfrac{1}{2}S) \det(\lambda I - S) = 0 .$$

Thus we have shown

Theorem 8.16. Let $a_0, a_1, \dots, a_m$ be the different alleles at a sex linked locus and let $0 \leq \sigma_{ik} \leq 1$ be the probability that an allele a_i mutates

into an allele a_k, $i,k=0,\dots,m$.
If the matrix $S = (\sigma_{ik})$ is regular then the sequence of frequencies converges towards the equilibrium.

We refer to the papers of Reiersøl [96] and Heuch [56] for investigations on several sex linked and partially sex linked loci with recombination.

C. Inbreeding systems

So far only populations with random mating between all individuals or all individuals of different sex have been described by algebras. In 1978 Holgate [70] designed an algebraic model for a population in which in every generation a fixed portion reproduces by self-fertilization while the remaining portion is subject to random mating. Recently Holgate [72] has extended this model to include further types of inbreeding, such as brother-sister-mating, mating between double first cousins,... and mating between 2^{k-2} - fold cousins of degree k-2.

For the traditional treatment of mathematical models for inbreeding populations we refer to Kempthorne [81] , Li [83] and especially to Karlin's book [79] on "equilibrium behavior of population genetic models with non random mating".
There, for several inbreeding systems (i.g. with two alleles at one locus or two loci) the difference equations for the relative frequencies of zygotes and couples of zygotes are given and their convergence behavior is discussed, whereby every genetic situation is treated separately. Now Holgate's algebraic models for inbreeding systems allow a simultaneous treatment of several genetic situations.

We begin with Holgate's model for mixed selfing and random mating populations:
Let the genotypes (in general zygotes) of a population be $a_0,\dots,a_m$.
Let $\mathcal{V} := \langle a_0,\dots,a_m\rangle_{\mathbb{R}}$ be the $\mathbb{R}$-vector space spanned by $a_0,\dots,a_m$. We assume that the genetics of such a population with random mating throughout would be described by an algebra $\mathfrak{A}$ over $\mathcal{V}$ which is baric with weight homomorphism ω given by $\omega(a_i) = 1$, $i=0,\dots,m$.
If the same population would allow only <u>self-fertilization</u>, then the genetic behavior would be described by a linear mapping $f: \mathcal{V} \longrightarrow \mathcal{V}$ with

$$f : x = \sum_{i=0}^{m} \alpha_i a_i \longmapsto \sum_{i=0}^{m} \alpha_i a_i^2 \quad . \tag{8.42}$$

Now if the portion Θ, $0 \leq \Theta \leq 1$ of the population is randomly mating and the remaining portion $1 - \Theta$ performs self-fertilization then the genetics of this population is described by an algebra $\mathfrak{A}_\Theta$ over with the multiplication

$$(x,y) \longmapsto x \circ y := \Theta xy + (1-\Theta)\,\omega(y)\mathcal{S}x \quad . \tag{8.43}$$

We shall discuss which properties of the algebra $\mathfrak{A}$ will be carried over to the algebra $\mathfrak{A}_\Theta$.

We can interpret $\mathfrak{A}_\Theta$ as a convex combination of the algebra $\mathfrak{A}$ and an algebra $\mathfrak{A}_1$ over $\mathcal{V}$ with multiplication given by

$$(x,y) \longmapsto \omega(y)\,\mathcal{S}x \ .$$

Since the linear mapping $\mathcal{S}$ is weight preserving, the algebra $\mathfrak{A}_\Theta$ is also baric with the same homomorphism ω.
Furthermore if $\mathfrak{A}$ has a genetic realization and $\mathcal{S}$ leaves the simplex

$$\gamma_m = \left\{ \sum_{i=0}^{m} \alpha_i a_i \;\middle|\; 0 \leq \alpha_i \leq 1, \quad i=0,\dots,m, \ \sum_{i=0}^{m} \alpha_i = 1 \right\}$$

invariant, then also $\mathfrak{A}_\Theta$ has a genetic realization.
In general the algebra $\mathfrak{A}_\Theta$ will be non commutative, even if $\mathfrak{A}$ is commutative. This is another example which proves the concept of non commutative algebras in genetics useful.

Let $\mathfrak{A}$ be a Gonshor genetic algebra with basis $c_0,\dots,c_m$ over $\mathbb{R}(\mathbb{C})$. Repeating the arguments of section 6D we have: If the linear mapping leaves the subspaces $\langle c_k,\dots,c_m\rangle$, $k=1,\dots,m$, of $\mathcal{V}$ ($\mathcal{V}_{\mathbb{C}}$) invariant, then the algebra $\mathfrak{A}_\Theta$ is also Gonshor genetic.

In this case Holgate [72] calls the self-fertilization mapping <u>compatible</u> with the canonical basis.
For the commutative case Holgate provides sufficient conditions for compatibility.

According to (8.43) the products of the elements of the canonical basis are computed as

$$\begin{aligned} c_i \circ c_0 &= \Theta c_i c_0 + (1-\Theta)\mathcal{S}c_i \ , && i=0,\dots,m, \\ c_i \circ c_j &= \Theta c_i c_j \ , && i=0,\dots,m,\ j=1,\dots,m. \end{aligned} \tag{8.44}$$

Thus we have to investigate the action of the linear mapping f on the basis elements. For this purpose we introduce the following notations. Let the multiplication table of $\mathfrak{A}$ be

$$c_i c_j = \sum_{k=0}^{m} \lambda_{ijk} c_k, \qquad i,j=0,\ldots,m.$$

The basis transformation $\{a_i\} \longrightarrow \{c_i\}$ is described by

$$a_i = \sum_{j=0}^{m} \sigma_{ji} c_j \quad \text{or} \quad c_i = \sum_{j=0}^{m} \tau_{ji} a_j, \qquad i=0,\ldots,m. \tag{8.45}$$

Since $\mathfrak{A}$ is baric with weight homomorphism ω, we have

$$\sigma_{oi} = 1, \qquad i=0,\ldots,m, \tag{8.46}$$

and

$$\sum_{j=0}^{m} \tau_{jo} = 1, \quad \sum_{j=0}^{m} \tau_{ji} = 0, \qquad i=1,\ldots,m. \tag{8.47}$$

Then the action of f on the canonical basis $c_o,\ldots,c_m$ is given by

$$\begin{aligned} f\,c_i &= f\Big(\sum_{j=0}^{m} \tau_{ji} a_j\Big) = \sum_{j=0}^{m} \tau_{ji} a_j^2 = \sum_{j=0}^{m} \tau_{ji} \Big(\sum_{l=0}^{m} \sigma_{lj} c_l\Big)\Big(\sum_{k=0}^{m} \sigma_{kj} c_k\Big) \\ &= \sum_{k,l=0}^{m} \rho(i,l,k)\, c_l c_k, \qquad i=0,\ldots,m \end{aligned} \tag{8.48}$$

where

$$\rho(i,l,k) := \sum_{j=0}^{m} \tau_{ji}\, \sigma_{lj}\, \sigma_{kj}, \qquad i,l,k=0,\ldots,m. \tag{8.49}$$

Since $c_o,\ldots,c_m$ is a canonical basis of $\mathfrak{A}$ the product $c_l c_k$ is contained in the subspace $\langle c_{\max(l,k)},\ldots,c_m\rangle$ of $\mathfrak{N}$ ($\mathfrak{N}_{\mathbb{C}}$). Thus the linear mapping f is compatible with the canonical basis if

$$\rho(i,l,k) = 0 \text{ for } i > \max(l,k), \qquad i,l,k=0,\ldots,m. \tag{8.50}$$

In this case we obtain

$$c_i = \sum_{s=i}^{m} \sum_{l,k=0}^{m} \rho(i,l,k)\, \lambda_{lks} c_s, \qquad i=0,\ldots,m.$$

In view of (8.46) and (8.47) we have

$$\rho(i,k,0) = \rho(i,0,k) = \sum_{j=0}^{m} \tau_{ji}\, \tau_{oj}\, \sigma_{kj} = \sigma_{ik}, \qquad i,k=0,\ldots,m.$$

Thus the coefficient of c_i in $\mathcal{f} c_i$ is given by

$$\varrho(i,o,i)\lambda_{oii} + \varrho(i,i,o)\lambda_{ioi} = \lambda_{oii} + \lambda_{ioi}, \quad i=0,\dots,m.$$

Hence the left and right train roots of $\mathfrak{A}_\Theta$ are

$$1, \Theta\lambda_{oii}, \Theta\lambda_{ioi} + \Theta\lambda_{ioi} + (1-\Theta)(\lambda_{oii} + \lambda_{ioi}) = (1-\Theta)\lambda_{oii} + \lambda_{ioi},$$

$$r=1,\dots,m.$$

Now we have collected sufficiently many relations to apply theorem 5.25.

Theorem 8.17. Let $\mathfrak{A}$ be a Gonshor genetic algebra over $\mathcal{V} = \langle a_o,\dots,a_m\rangle_{\mathbb{R}}$ with left and right train roots $\lambda_{ooo} = 1, \lambda_{oii}, \lambda_{ioi}, i=1,\dots,m$.
Let $\mathfrak{A}_\Theta$ be the algebra over $\mathcal{V}$ given by (8.43).
If $\mathfrak{A}$ has a canonical basis $c_o,\dots,c_m$ such that the corresponding basis transformation (8.45) satisfies condition (8.50) then $\mathfrak{A}_\Theta$ is a Gonshor genetic algebra with left and right train roots $1, \Theta\lambda_{oii}$, $(1-\Theta)\lambda_{oii} + \lambda_{ioi}$, $i=1,\dots,m$.
If in addition $|\lambda_{oii} + \lambda_{ioi}| < 1$, $i=1,\dots,m$, then the sequence of plenary powers converges for every $x \in \mathfrak{A}_\Theta$ of weight 1 towards the idempotent of $\mathfrak{A}_\Theta$.

For several concrete genetic situations described by commutative genetic algebras Holgate has checked whether the self-fertilization mapping is compatible with a canonical basis. He found that this is indeed the case for the zygotic algebra of diploid individuals with an autosomal locus with several alleles and simple Mendlian segregation rates and in the zygotic algebra for 2r-ploid individuals with one locus with two alleles and a mixture of chromosome- and chromatid segregation. If several loci of diploid individuals are considered then the self-fertilization mapping is compatible with the canonical basis if the coupling between the loci is strict or if the loci are totally uncoupled. Already in the case of two coupled loci with two alleles at each locus and a recombination probability ϱ, $0 < \varrho < \frac{1}{2}$, the self-fertilization mapping is no longer compatible. But it turns out, that $\mathcal{f}$ is weakly compatible in the terminology of Holgate, cf. [72].

The essential ideas of Holgate's paper on inbreeding systems are the process of duplication and, if possible, the linearizing procedure of a quadratic operator in a higher dimensional vector space.

Let $a_o,\dots,a_m$ be the gametes produced by a population and assume that

the inheritance of this population under random mating can be described by a gametic algebra $\mathcal{G}$.
Let

$$\mathcal{G}^{\otimes 1} := \mathcal{G} \otimes_d \mathcal{G}$$

denote the (non commutative) duplicate of $\mathcal{G}$, and let

$$\mathcal{G}^{\otimes i} := \mathcal{G}^{\otimes (i-1)} \otimes_d \mathcal{G}^{\otimes (i-1)}, \qquad i > 1,$$

denote the i-fold duplicate of $\mathcal{G}$. Then $\mathcal{G}^{\otimes 1}$ is the zygotic and $\mathcal{G}^{\otimes 2}$ the copular algebra of the population under random mating, cf. section 6C.
The self-fertilization mapping $\mathcal{f}_1 := \mathcal{f}$, cf. (8.42), is a linear mapping of the zygotic algebra $\mathcal{G}^{\otimes 1}$ into itself. The brother-sister-mating can be described by a linear mapping $\mathcal{f}_2$ of the copular algebra $\mathcal{G}^{\otimes 2}$ into itself (with respect to couples of zygotes this linear mapping is represented by a so called generation matrix, cf. e.g. [79]).
Furthermore the mating between 2^{i-2} - fold duplicate $\mathcal{G}^{\otimes i}$ into itself, i=3.
Let $\Theta_o,\dots,\Theta_k \in \mathbb{R}$ with

$$0 \leq \Theta_i \leq 1, \qquad i=0,\dots,k, \quad \sum_{i=0}^{k} \Theta_i = 1,$$

and assume that a fraction Θ_o of the population practise random mating, a fraction Θ_1 reproduces by self-fertilization, a fraction Θ_2 reproduces by brother sister mating, and a fraction Θ_i reproduces by mating between 2^{i-2} - fold cousins of degree i-2, i=3,...,k.
Then the mating system is described by the <u>inbreeding algebra</u> $\mathcal{G}_\Theta$, $\Theta := (\Theta_o,\dots,\Theta_k)$, over the vector space underlying $\mathcal{G}^{\otimes k}$ with the multiplication

$$(x,y) \longmapsto x \circ y := \Theta_o xy + \sum_{i=1}^{k} \Theta_i\, \omega^{\otimes k}(y)\, \mathcal{f}_i^{\otimes k-i}(x), \tag{8.51}$$

here xy is the product of x and y in $\mathcal{G}^{\otimes k}$, $\omega^{\otimes k}$ denotes the k-fold tensor product of ω with itself, and $\mathcal{f}_i^{\otimes k-i}$ is (k-i)-fold tensor product of $\mathcal{f}_i$ with itself, which is a linear mapping of $\mathcal{G}^{\otimes k}$ into itself.
Hence the algebra $\mathcal{G}_\Theta$ is a convex-combination of the algebra $\mathcal{G}^{\otimes k}$ and k further algebras whose multiplications are determined by the inbreeding mappings $\mathcal{f}_i$, i=1,...,k.

Now assume that the gametic algebra $\mathcal{G}$ is a Gonshor genetic algebra.

With theorem 6.19 it follows that the k-fold duplicate $\mathcal{G}^{\otimes k}$ is Gonshor genetic, too.
Generalizing the concept of compatibility of the self-fertilization mapping $\mathcal{f}_1 = \mathcal{f}(8.42)$, one would call the inbreeding mappings $\mathcal{f}_1,\dots,\mathcal{f}_k$ compatible with $\mathcal{G}^{\otimes k}$ (or with $\mathcal{G}$) if there exists a canonical basis of $\mathcal{G}^{\otimes k}$ such that the linear mappings $\mathcal{f}_i^{\otimes\, k-i}$, i=1,...,k, are represented by lower triangular matrices.
It is easy to see that in this case the inbreeding algebra $\mathcal{G}_\Theta$ is Gonshor genetic.
But we know already from the inbreeding algebra for selfing with two coupled loci that not all relevant inbreeding algebras have this property. Therefore it was necessary to find a new concept for the investigation of sequences of plenary powers.

Holgate [72] calls the inbreeding mappings $\mathcal{f}_1,\dots,\mathcal{f}_k$ weakly compatible with $\mathcal{G}^{\otimes k}$ (or $\mathcal{G}$), if the submanifold of $\mathcal{G}^{\otimes k}$, consisting of all $x \in \mathcal{G}^{\otimes k}$ with $\omega^{\otimes k}(x) = 1$ can be embedded in a higher dimensional vector space $\mathcal{L}(\mathcal{G}_\Theta)$ such that to the operator $\mathcal{E}: x \longmapsto x^2$ of $\mathcal{G}_\Theta$ corresponds a linear operator $\hat{\mathcal{E}}$ of $\mathcal{L}(\mathcal{G}_\Theta)$ into itself.

If in an actual genetic situation with gametic algebra $\mathcal{G}$ which is Gonshor genetic, the inbreeding mappings are weakly compatible with $\mathcal{G}$, then the convergence behavior of plenary powers of elements of weight 1 is determined by the eigenvalues of $\hat{\mathcal{E}}$.

Holgate [72] shows that in the case of two linked loci the self-fertilization mapping $\mathcal{f}$ is weakly compatible with the gametic algebra and he computed the characteristic roots of $\hat{\mathcal{E}}$. In addition, the self-fertilization mapping is weakly compatible with the gametic algebra for an arbitrary, finite number of linked loci, cf. Holgate [71] ,[72].

9. CONCLUDING REMARKS AND BERNSTEIN ALGEBRAS

A. Open problems

In the preceding two sections we have, with the exception of migration, considered all essential genetic situations which, have so far been described by algebras.
Since we can construct new algebras, describing complex situations, from "simple" algebras modelling partial aspects, the field of applications for genetic algebras can still be greatly extended beyond the frame-work of this monograph and beyond the scope of the present literature.
It would be helpful to have more detailed theorems on the existence of 1- or more-parametric families of idempotents and on the convergence of plenary powers of elements of weight 1. Furthermore one has to study the possibility of periodic solutions in cases where several train roots equal 1/2 in modulus.

First steps towards an investigation of spatially distributed populations have been made.
It would be desirable to have a theory for a large number of loci, possibly with a continuous distribution of loci.

A problem, unsolved so far, is to generalize the concept of algebras in genetics to include selection.

Finally one could try to model biological systems with several species by algebras.

B. Bernstein algebras

We feel compelled to include a section on Bernstein algebras in this monograph. These algebras have been defined by Holgate [68] after Ljubič [84] had investigated, in great detail, the problem posed by S. Bernstein [13] ,[14] ,[15] , of the conditions required to ensure that a population attains equilibrium after one generation.

One considers an infinitely large population with discrete generations. Let $a_o,\ldots,a_m$ be the different genotypes. At each moment the state of the population is described by the frequency of genotypes, i.e. by an element $(\alpha_o,\ldots,\alpha_m)^T$ of the m-dimensional simplex

$$\tilde{\gamma}_m = \left\{ (\alpha_o,\ldots,\alpha_m)^T \in \mathbb{R}^{m+1} \,\middle|\, 0 \le \alpha_i \le 1,\ \ i=0,\ldots,m,\ \sum_{i=1}^{m} \alpha_i = 1 \right\}.$$

Let the state of the population in the next generation be given by $\tilde{\Psi}(\alpha_o,\ldots,\alpha_m)^T$, where

$$\tilde{\Psi} : \tilde{\mathcal{S}}_m \longrightarrow \tilde{\mathcal{S}}_m$$

is a quadratic operator, i.e. an operator in which each component is bilinear in the coordinates of the argument.
The operator $\tilde{\Psi}$ is called an <u>evolutionary operator</u>, cf. Ljubič [84]. Such an evolutionary operator satisfied <u>Bernstein's principle</u> if

$$\tilde{\Psi}^2 = \tilde{\Psi} \quad . \tag{9.1}$$

Since $0 \notin \tilde{\mathcal{S}}_m$, an evolutionary operator cannot be the zero mapping. Hence the evolutionary operators satisfying Bernstein's principle are exactly the idempotent quadratic operators of $\tilde{\mathcal{S}}_m$.

The biological interpretation is that a population satisfying Bernstein's principle is in equilibrium after one generation. Thus the Bernstein principle is a generalization of part of the Hardy-Weinberg law (cf. Edwards [25]).

Bernstein's aim was to classify the idempotent evolutionary operators. For m=2 (i.e. populations with 3 genotypes) he gave a complete solution and for $m > 2$ he found two classes of evolutionary operators satisfying his principle, cf. [13] , [14] ,[15]. Ljubič [84] ,[86] considers so called two-level populations satisfying Bernstein's principle and provides, sometimes with additional hypotheses, a canonical form of the evolutionary operators.

To each population with genotypes $a_o,\ldots,a_m$ where the inheritance is described by an evolutionary operator $\tilde{\Psi}$, one can associate, in a canonical way, an algebra, cf. Ljubič [84] and Holgate [68] :
Let $\mathcal{V}$ be the free $\mathbb{R}$-module spanned by $a_o,\ldots,a_m$,

$$\mathcal{V} := \langle a_o,\ldots,a_m \rangle_{\mathbb{R}} \quad .$$

Trivially the evolutionary operator $\tilde{\Psi}$ induces a mapping Ψ of the simplex $\mathcal{S}_m$ into itself,

$$\mathcal{S}_m = \left\{ \sum_{i=0}^{m} \alpha_i a_i \;\middle|\; 0 \le \alpha_i \le 1, \quad i=0,\ldots,m, \ \sum_{i=0}^{m} \alpha_i = 1 \right\} \quad .$$

This mapping can be extended in a canonical way to the whole vector space $\mathcal{V}$. This extension is again denoted by Ψ .

By

$$(x,y) \longmapsto xy := \frac{1}{2}\left[\Psi(x+y) - \Psi(x) - \Psi(y)\right] \tag{9.2}$$

we introduce a multiplication in $\mathcal{V}$. Then $\mathcal{V}$ becomes a commutative algebra $\mathfrak{A}$ over $\mathbb{R}$. Let $\omega : \mathcal{V} \longrightarrow \mathbb{R}$ be the mapping defined by

$$\omega(a_i) := 1, \qquad i=0,\dots,m,$$

and its linear extension onto $\mathcal{V}$. Since Ψ is a quadratic operator we have

$$\omega(\Psi(x)) = \omega^2(x) \quad , \qquad x \in \mathcal{V} \quad . \tag{9.3}$$

Now we apply the linear mapping ω to (9.2) and we obtain, with the help of (9.3),

$$\begin{aligned}\omega(xy) &= \frac{1}{2}\left[\omega(\Psi(x+y)) - \omega(\Psi(x)) - \omega(\Psi(y))\right] \\ &= \frac{1}{2}\left[\omega^2(x+y) - \omega^2(x) - \omega^2(y)\right] \\ &= \frac{1}{2}\left[\omega^2(x) + 2\,\omega(x)\,\omega(y) + \omega^2(y) - \omega^2(x) - \omega^2(y)\right] \\ &= \omega(x)\,\omega(y).\end{aligned}$$

Thus $\mathfrak{A}$ is a baric algebra with weight homomorphism ω .

If the evolutionary operator $\tilde{\Psi}$ satisfies Bernstein's principle (9.1) then the operator Ψ obeys the following law

$$\Psi^2(x) = \omega^2(x)\,\Psi(x), \qquad x \in \mathcal{V} . \tag{9.4}$$

Then $\mathfrak{A}$ is a Bernstein algebra in the following sense

<u>Definition 9.1.</u> Let $\mathfrak{A}$ be a finite-dimensional, commutative baric algebra over a field $\mathbb{K}$, char $\mathbb{K} \neq 2$, with weight homomorphism ω . Then $\mathfrak{A}$ is called a <u>Bernstein algebra</u> if the plenary powers of any element $x \in \mathfrak{A}$ satisfy the train equation

$$x^{[3]} - \omega^2(x)\; x^{[2]} = 0 \quad . \tag{9.5}$$

<u>Remark.</u> Holgate [68] requires that the underlying field of a Bernstein algebra is the field of complex numbers, while Ljubič [87] restricts his definition to algebras over the reals. Since all algebraic properties remain true for an arbitrary field $\mathbb{K}$ of characteristic different from 2, we have generalized the definition in this direction.

Furthermore in Holgate's terminology equation (9.5) must be satisfied only for all elements of non zero weight. But subsequent to his definition Holgate proved that in every Bernstein algebra (in his sense) the third plenary power $y^{[3]}$ vanishes for every element y of zero weight, and hence equation (9.5) is satisfied for all elements of the algebra. Therefore our definition is not restrictive in this direction.

First, let us give two examples for Bernstein algebras arising in genetics.

Example 9.2. (1) The gametic algebra $\mathfrak{G}$ for simple Mendelian inheritance (cf. example 1.1 and theorem 7.2) satisfies the train equation

$$x^2 - \omega(x)x = 0$$

i.e.

$$x^2 = \omega(x)x.$$

Squaring the latter equation we obtain

$$(x^2)^2 = \omega^2(x)x^2$$

which is equivalent to formula (9.5), hence $\mathfrak{G}$ is a Bernstein algebra. (2) The zygotic algebra $\mathfrak{Z}$ for simple Mendelian inheritance (cf. theorem 7.3) is a Bernstein algebra since the plenary powers satisfy the train equation given by (7.4).

The proof of the following lemma is similar to that of lemma 3.4 which was proposed by Holgate [67].

Lemma 9.3. A Bernstein algebra $\mathfrak{A}$ over $\mathbb{K}$ possesses exactly one weight homomorphism.

Proof. By definition 9.2 there exists a non trivial homomorphim $\omega : \mathfrak{A} \longrightarrow \mathbb{K}$. Define $\mathfrak{N} := \ker\omega$.
Let $\varphi : \mathfrak{A} \longrightarrow \mathbb{K}$ be a further non trivial homomorphism. Since φ is compatible with the multiplication in $\mathfrak{A}$, formula (9.5) implies that $\varphi((y^2)^2) = \varphi^4(y) = 0$ for all $y \in \mathfrak{N}$, hence $\varphi(y) = 0$ for all $y \in \mathfrak{N}$.
Now let $x \in \mathfrak{A} \setminus \mathfrak{N}$. Then $\omega(x) \neq 0$.
Furthermore $\omega(\frac{x^2}{\omega(x)} - x) = 0$, hence $\frac{x^2}{\omega(x)} - x \in \mathfrak{N}$.

Therefore $\varphi(\frac{x^2}{\omega(x)} - x) = 0$, i.e. $\varphi(x)(\frac{\varphi(x)}{\omega(x)} - 1) = 0$.
It follows that either $\varphi(x) = 0$ or $\varphi(x) = \omega(x)$. The first case can be

excluded since φ was assumed to be non trivial, hence $\varphi = \omega$. □

The following result is due to Ljubič [87].

Lemma 9.4. Let $\mathfrak{A}$ be a Bernstein algebra with weight homomorphism ω. Then the set of idempotents of $\mathfrak{A}$ is given by

$$\mathfrak{J} = \{ x^2 \mid x \in \mathfrak{A} \text{ with } \omega(x) = 1 \} \quad .$$

Proof. I. Let $e \in \mathfrak{A}$ be an idempotent, i.e. $e \neq 0$ and $e^2 = e$. We substitute e into (9.5) and obtain

$$e - \omega(e)e = 0 \quad .$$

Since $e \neq 0$ it follows that $\omega(e) = 1$, hence $e = e^2 \in \mathfrak{J}$.
II. Take $x \in \mathfrak{A}$ with $\omega(x) = 1$. Then $\omega(x^2) = \omega^2(x) = 1$, therefore $x^2 \neq 0$. From (9.5) it follows that $(x^2)^2 = x^2$. Hence x^2 is an idempotent of $\mathfrak{A}$. □

Since every baric algebra possesses elements of unit weight, lemma 9.4 implies the following statement given by Holgate [68].

Corollary 9.5. Every Bernstein algebra possesses at least one idempotent.

In the following theorem 9.6(1)-(4) we have collected several results on Bernstein algebras, first stated and proved by Ljubič in terms of operators (cf. [84], theorem 4.1). Later on, after Holgate [68] had defined a Bernstein algebra, Ljubič has formulated these results in terms of algebras (cf. [84],[88]). We will give the algebraic formulation of his results and we will prove them in terms of algebras, too, whereby we use a method, the method of linearizing polynomial identities, first introduced by Holgate [68] in this context. (For an extensive discussion of this method we refer to section 3C.)
Furthermore several of these result have been stated and proved in terms of algebras by Holgate [68].

Theorem 9.6. Let $\mathfrak{A}$ be a Bernstein algebra over $\mathbb{K}$ with weight homomorphism ω, $\mathfrak{N} := \ker\omega$. Then the following statements are true:
(1) With respect to an idempotent $e \in \mathfrak{A}$ (whose existence is guaranteed by corollary 9.5) $\mathfrak{A}$ splits into the direct sum

$$\mathfrak{A} = \mathfrak{E} \oplus \mathfrak{U} \oplus \mathfrak{Z} \tag{9.6}$$

where

$$\mathfrak{E} := \langle e \rangle_{\mathbb{K}},\ \mathfrak{U} := \{ ey \mid y \in \mathfrak{N} \},\ \mathfrak{Z} := \{ x \in \mathfrak{A} \mid ex = 0 \} .$$

(2)
$$\omega(z) = 0 \quad \text{for all } z \in \mathfrak{Z}\ , \tag{9.7}$$
$$eu = \tfrac{1}{2}u \quad \text{for all } u \in \mathfrak{U}\ . \tag{9.8}$$

(3)
$$(y^2)^2 = 0 \quad \text{for all } y \in \mathfrak{N}\ , \tag{9.9}$$
$$u^3 = 0 \quad \text{for all } u \in \mathfrak{U}\ . \tag{9.10}$$
$$u(uz) = 0 \quad \text{for all } u \in \mathfrak{U} \text{ and } z \in \mathfrak{Z}\ , \tag{9.11}$$
$$uz^2 = 0 \quad \text{for all } u \in \mathfrak{U} \text{ and } z \in \mathfrak{Z}\ , \tag{9.12}$$
$$u^2(uz) = 0 \quad \text{for all } u \in \mathfrak{U} \text{ and } z \in \mathfrak{Z}\ , \tag{9.13}$$
$$(uz)z^2 = 0 \quad \text{for all } u \in \mathfrak{U} \text{ and } z \in \mathfrak{Z}\ , \tag{9.14}$$
$$2(uz)^2 + u^2z^2 = 0 \quad \text{for all } u \in \mathfrak{U} \text{ and } z \in \mathfrak{Z}\ . \tag{9.15}$$

(4) The decomposition of every $x \in \mathfrak{A}$ and its square according to (9.6) is given by

$$x = \omega(x)e + u + z \tag{9.16}$$
$$x^2 = \omega^2(x)e + (\omega(x)u + 2uz + z^2) + u^2 \tag{9.17}$$

(5)
$$\mathfrak{U}\mathfrak{Z} \subseteq \mathfrak{U}\ , \tag{9.18}$$
$$\mathfrak{U}^2 \subseteq \mathfrak{Z}\ , \tag{9.19}$$
$$\mathfrak{Z}^2 \subseteq \mathfrak{U}\ , \tag{9.20}$$
$$\mathfrak{U}\mathfrak{Z}^2 = \langle 0 \rangle . \tag{9.21}$$

<u>Proof.</u> I. Let $e \in \mathfrak{A}$ be an idempotent, then $\omega(e) = 1$ (cf. lemma 9.4), hence $e \in \mathfrak{N}$ and we have the decomposition

$$\mathfrak{A} = \mathfrak{E} \oplus \mathfrak{N}\ . \tag{9.22}$$

Since $\mathfrak{N}$ is an ideal of $\mathfrak{A}$, the linear transformation $\mathcal{L}_e : \mathfrak{A} \longrightarrow \mathfrak{A}$, $x \longmapsto ex$, induces a linear transformation $\mathcal{L}_{e|\mathfrak{N}}$ of $\mathfrak{N}$ into itself. Hence $\mathfrak{N}$ splits into the direct sum of

$$\operatorname{Im} \mathcal{L}_{e|\mathfrak{N}} = \mathfrak{U} \text{ and } \ker \mathcal{L}_{e|\mathfrak{N}}\ .$$

Since e is an idempotent, the left transformation $\mathcal{L}_e$ acts as identity on $\mathfrak{E}$. Hence $\mathfrak{Z} = \ker \mathcal{L}_e$ has no component in $\mathfrak{E}$ and, in view of (9.22), $\mathfrak{Z} \subseteq \mathfrak{N}$, which is equivalent to (9.7). Furthermore

$$\mathfrak{Z} = \ker \mathcal{L}_e = \ker \mathcal{L}_{e|\mathfrak{N}}$$

and we obtain the decomposition (9.6).

II. Equation (9.7) has been shown in part I of the present proof. We choose an element $y \in \mathfrak{N}$ and consider all elements of the form $e + \eta y$, $\eta \in \mathbb{K}$.
These elements satisfy equation (9.5), i.e.

$$(e + \eta y)^{[3]} - (e + \eta y)^{[2]} = 0 \quad . \tag{9.23}$$

We compute the second and third plenary power of $e + \eta y$ and substitute them into (9.23). Then we collect terms with equal powers of η and obtain

$$\eta^4 (y^2)^2 + 4\eta^3 (ey)y^2 + \eta^2 \left[2ey^2 + 4(ey)^2 - y^2\right] + 2\eta\left[2e(ey) - ey\right] = 0.$$

Since this equation is satisfied for all $\eta \in \mathbb{K}$, it follows that equation (9.9) as well as the following identities are satisfied:

$$(ey)y^2 = 0, \qquad y \in \mathfrak{N}, \tag{9.24}$$

$$2ey^2 + 4(ey)^2 - y^2 = 0, \qquad y \in \mathfrak{N}, \tag{9.25}$$

$$2e(ey) - ey = 0, \qquad y \in \mathfrak{N}. \tag{9.26}$$

The last equation says that the action of $\mathcal{L}_e$ on $\mathfrak{U}$ is given by (9.8). (Ljubič [87] considered $\mathcal{L}_{2e} = 2\mathcal{L}_e$ instead of $\mathcal{L}_e$, which induces an idempotent mapping on $\mathfrak{N}$.)

III. Take $u \in \mathfrak{U}$ and $z \in \mathfrak{Z}$ and consider all elements of the form $\eta u + \zeta z$, $\eta, \zeta \in \mathbb{K}$. Since all these elements are contained in $\mathfrak{N}$, they satisfy equations (9.24), (9.9) and (9.25). We compute $(\eta u + \zeta z)^2$ and substitute this expression into these equations while we are using (9.8) and the definition of $\mathfrak{Z}$. Then we obtain

$$\tfrac{1}{2}\eta^3 u(u^2) + \eta^2\zeta u(uz) + \tfrac{1}{2}\eta\zeta^2 uz^2 = 0$$

$$\eta^4 (u^2)^2 + 4\eta^3\zeta u^2(uz) + 2\eta^2\zeta^2\left[2(uz)^2 + u^2z^2\right] + 4\eta\zeta^3(uz)z^2 + \zeta^4(z^2)^2 = 0$$

$$2\eta^2 eu^2 + 2\eta\zeta\left[2e(uz) - uz\right] + \zeta^2\left[2ez^2 - z^2\right] = 0.$$

Since η and ζ range over $\mathbb{K}$ the factors of $\eta^3, \eta^2\zeta, \ldots, \zeta^2$ must vanish. From the first and second equation we obtain (9.10) - (9.15) as well as (9.9) with u and z, respectively, instead of y. From the third equation we obtain the following identities

$$eu^2 = 0 \quad , \qquad u \in \mathfrak{U}, \tag{9.27}$$

$$2e(uz) - uz = 0 \quad , \qquad u \in \mathfrak{U},\ z \in \mathfrak{Z} \quad , \tag{9.28}$$

$$2ez^2 - z^2 = 0 \quad , \qquad z \in \mathfrak{Z} \quad . \tag{9.29}$$

Equation (9.22) says that $u^2 \in \mathfrak{Z}$ for all $u \in \mathfrak{U}$, and equations (9.28) and (9.29) imply that uz, $z^2 \in \mathfrak{U}$ for all $u \in \mathfrak{U}$ and $z \in \mathfrak{Z}$.
IV. From the decomposition (9.6) of the Bernstein algebra we know that every element $x \in \mathfrak{A}$ can be represented as (9.16) with uniquely determined $u \in \mathfrak{U}$ and $z \in \mathfrak{Z}$.
With (9.8) and the definition of $\mathfrak{Z}$ we obtain

$$x^2 = \omega^2(x)e + \omega(x)u + 2uz + z^2 + u^2 \quad .$$

From the interpretation of (9.27) - (9.29) just given it follows that (9.17) yields the decomposition of x^2 according to (9.6).
V. In view of the definition of $\mathfrak{U}$ equation (9.28) is equivalent to relation (9.18). Take u_1, $u_2 \in \mathfrak{U}$ and consider $u_1 + u_2 \in \mathfrak{U}$.
This element satisfies (9.27), i.e.

$$eu_1^2 + 2e(u_1u_2) + eu_2^2 = 0 \quad .$$

In view of (9.27) with u_1 and u_2, respectively, instead of u, we obtain

$$e(u_1u_2) = 0 \qquad \text{for all } u_1,u_2 \in \mathfrak{U}\,, \tag{9.30}$$

which is equivalent to (9.19).
We substitute $z = z_1 + z_2$, with z_1, $z_2 \in \mathfrak{Z}$, in (9.29) and (9.12), and apply (9.29) and (9.12) to z_1 and z_2, respectively, then we obtain

$$2e(z_1z_2) - z_1z_2 = 0 \qquad \text{for all } z_1,z_2 \in \mathfrak{Z}\,, \tag{9.31}$$

and

$$u(z_1z_2) = 0 \qquad \text{for all } u \in \mathfrak{U} \text{ and } z_1,z_2 \in \mathfrak{Z}. \tag{9.32}$$

Hence relations (9.20) and (9.21) are satisfied. □

Remark. In addition to (9.15) Ljubič [87] claims that every summand on the left hand side of (9.15) vanishes separately, but we have not succeeded in proving this assertion.

The converse of theorem 9.6(1) - (3) is also true.
This result, in terms of operators, was given by Ljubič [84].
It can be proved by equating $x^3 - \omega^2(x)x^2$ to zero.

Theorem 9.7. Every baric algebra $\mathfrak{A}$ which possesses an idempotent e with respect to which it is decomposed according to (9.6) and whose elements satisfy (9.7) - (9.15) is a Bernstein algebra.

The following result has been pointed out by Holgate [68] and Ljubič [87] ,[88].

Lemma 9.8. Let $\mathfrak{A}$ be a Bernstein algebra with weight homomorphism ω, $\mathfrak{N} := \ker\omega$, and $e \in \mathfrak{A}$ an idempotent.
Then the set $\mathfrak{J}$ of idempotents of $\mathfrak{A}$ is given by

$$\mathfrak{J} = \{e + u + u^2 \mid u \in \mathfrak{U}\} \qquad (9.33)$$

where

$$\mathfrak{U} = \{ey \mid y \in \mathfrak{N}\} \quad .$$

Proof. From lemma 9.4 we know that every idempotent $e_1 \in \mathfrak{A}$ is of unit weight. Hence its decomposition according to (9.16) is given by

$$e_1 = e + u + z$$

with uniquely determined $u \in \mathfrak{U}$ and $z \in \mathfrak{Z}$, furthermore (9.17) implies

$$e_1^2 = e + (u + 2uz + z^2) + u^2 \; .$$

In view of decomposition (9.6) the idempotent condition $e_1^2 = e_1$ is equivalent to

$$u^2 = z \qquad (9.34)$$

and

$$u + 2uz + z^2 = u \quad . \qquad (9.35)$$

For $z = u^2$ equations (9.10) and (9.9), with $y = u$, say that

$$uz = 0 \qquad \text{and} \qquad z^2 = 0 \quad .$$

Hence equation (9.35) is satisfied for all $z = u^2$. Therefore the set of idempotents of $\mathfrak{A}$ is given by (9.33).□

The decomposition (9.6) of a Bernstein algebra $\mathfrak{A}$ in general depends on the choice of the idempotent $e \in \mathfrak{A}$. The following lemma clarifies the relationship between two such decompositions.

Lemma 9.9. Let $\mathfrak{A}$ be a Bernstein algebra with weight homomorphism ω, $\mathfrak{N} := \ker\omega$.
Let $e_1, e_2 \in \mathfrak{A}$ be idempotents. Define $\mathfrak{U}_i := \mathcal{L}_{e_i}\mathfrak{N}$, $i=1,2$.
Then there are uniquely determined $u_i \in \mathfrak{U}_i$, $i=1,2$, such that

$$e_1 = e_2 + u_2 + u_2^2 \; , \qquad e_2 = e_1 + u_1 + u_1^2 \; . \qquad (9.36)$$

Furthermore

$$e_1e_2 = e_1 + \frac{1}{2}u_1 = e_2 + \frac{1}{2}u_2 \quad , \tag{9.37}$$

and

$$u_1^2 = u_2^2 = - u_1u_2 \quad . \tag{9.38}$$

<u>Proof.</u> The representation (9.36) of e_1 and e_2 follows from lemma 9.8 with e_2 and e_1, respectively, instead of e. With theorem 9.6 we obtain on the one hand

$$e_1e_2 = e_1(e_1 + u_1 + u_1^2) = e_1 + \frac{1}{2}u_1$$

and on the other hand

$$e_1e_2 = (e_2 + u_2 + u_2^2)e_2 = e_2 + \frac{1}{2}u_2 \quad .$$

Hence (9.37) is valid. Furthermore (9.37) implies

$$e_1 = e_2 + \frac{1}{2}u_2 - \frac{1}{2}u_1, \quad e_2 = e_1 + \frac{1}{2}u_1 - \frac{1}{2}u_2 \quad .$$

Now we compare this representation of e_1 and e_2 with that given by (9.36). Then

$$u_2 + u_2^2 = \frac{1}{2}u_2 - \frac{1}{2}u_1, \quad u_1 + u_1^2 = \frac{1}{2}u_1 - \frac{1}{2}u_2$$

which implies that

$$u_2^2 = - \frac{1}{2}(u_1 + u_2) = u_1^2 \ .$$

In view of (9.9) for u_1 instead of y, the square of the above equation vanishes, therefore

$$(u_1 + u_2)^2 = 0 \quad ,$$

and

$$2u_1u_2 = - (u_1^2 + u_2^2) = - 2u_1^2 \ .$$

Hence (9.39) is satisfied. □

Although the decomposition $\mathfrak{A} = \mathfrak{E} \oplus \mathfrak{U} \oplus \mathfrak{Z}$ of a Bernstein algebra depends on the choice of the idempotent $e \in \mathfrak{A}$, the dimension of $\mathfrak{U} = \mathcal{L}_e\mathfrak{N}$ does not depend on e, i.e. this number is an invariant of the algebra $\mathfrak{A}$. This fact, which was pointed out by Ljubič [72], can be derived from lemma 9.8.

Let r := dim $\mathfrak{U}$. Since $\mathfrak{U}^2 \subseteq \mathfrak{Z}$ and $\mathfrak{U} \cap \mathfrak{Z} = \langle 0 \rangle$, every idempotent of $\mathfrak{A}$, represented by $e + u + u^2$ with an $u \in \mathfrak{U}$, cf. (9.33), is uniquely determined by $u \in \mathfrak{U}$. It follows that the set of idempotents of $\mathfrak{A}$ is an r-parametric family over $\mathbb{K}$. Hence r must be independent of e.

Now we can follow Ljubič [84] and associate with each Bernstein algebra $\mathfrak{A} = \mathfrak{E} \oplus \mathfrak{U} \oplus \mathfrak{Z}$ a pair of integers, the type (r+1,d) of the algebra, where

$$r = \dim \mathfrak{U} \qquad \text{and} \qquad d = \dim \mathfrak{Z} \ .$$

Clearly the sum of the two integers of the type of $\mathfrak{A}$ equals the dimension of $\mathfrak{A}$.

In this monograph we can not present the whole classification of Bernstein algebras via their type, which was given by Ljubič (cf. [84], [87], [88]). We only mention the following result of Ljubič [87],[88], compare also Holgate [68] for the case r+d+1 = 3.

Theorem 9.10. Let $\mathfrak{A}$ be an (m+1)-dimensional algebra over $\mathbb{K}$ with weight homomorphism ω.

(1) If $\mathfrak{A}$ is of type (1,m), then $\mathfrak{A}$ has exactly one idempotent e and the multiplication in $\mathfrak{A}$ is given by
$$x_1x_2 = \omega(x_1)\,\omega(x_2)e \qquad \text{for all } x_1,x_2 \in \mathfrak{A}\ . \tag{9.39}$$

(2) If $\mathfrak{A}$ is of type (m+1,0) then every element of $\mathfrak{A}$ of unit weight is an idempotent and the multiplication in $\mathfrak{A}$ is given by
$$x_1x_2 = \tfrac{1}{2}(\,\omega(x_1) + \omega(x_2)x_1) \qquad \text{for all } x_1,x_2 \in \mathfrak{A}\ . \tag{9.40}$$

Proof. With the help of an idempotent $e \in \mathfrak{A}$ the Bernstein algebra splits into $\mathfrak{A} = \mathfrak{E} \oplus \mathfrak{U} \oplus \mathfrak{Z}$.

I. Let the type of $\mathfrak{A}$ be (1,m), i.e. dim $\mathfrak{U} = 0$, hence $\mathfrak{U} = \langle 0 \rangle$. From lemma 9.8 we obtain that $\mathfrak{A}$ has exactly one idempotent, namely the element e chosen above. Let $x_1,x_2 \in \mathfrak{A}$. Then there are uniquely determined $z_1,z_2 \in \mathfrak{Z}$ such that

$$x_i = \omega(x_i)e + z_i, \qquad i=1,2.$$

Then

$$\begin{aligned} x_1x_2 &= \omega(x_1)\,\omega(x_2)e + \omega(x_1)ez_2 + \omega(x_2)ez_1 + z_1z_2 \\ &= \omega(x_1)\,\omega(x_2)e \end{aligned}$$

in view of z_1,z_2 $\ker \mathcal{L}_e = \mathfrak{Z}$ and $z_1z_2 \in \mathfrak{U} = \langle 0 \rangle$.

II. Let $\mathfrak{A}$ be of type (m+1,0), i.e. dim $\mathfrak{U}$ = m. Hence $\mathfrak{Z} = \langle 0 \rangle$ and

$\mathfrak{A} = \mathcal{E} \oplus \mathfrak{U}$.

Since $\mathfrak{U}^2 \subseteq \mathfrak{Z}$ by (9.19), lemma 9.8 implies that in the present case the set of idempotents of $\mathfrak{A}$ is given by $\{e + u \mid u \in \mathfrak{U}\}$.

In view of the decomposition of $\mathfrak{A}$ this says that every element of $\mathfrak{A}$ of unit weight is an idempotent.

Let $x_1, x_2 \in \mathfrak{A}$. Then there are uniquely determined $u_1, u_2 \in \mathfrak{U}$ such that

$$x_i = \omega(x_i)e + u_i, \quad i=1,2.$$

Because $u_1 u_2 \in \mathfrak{Z} = \langle 0 \rangle$ we have

$$\begin{aligned} x_1 x_2 &= \omega(x_1)\omega(x_2)e + \tfrac{1}{2}\omega(x_1)u_2 + \tfrac{1}{2}\omega(x_2)u_1 \\ &= \tfrac{1}{2}\omega(x_1)\left[\omega(x_2)e + u_2\right] + \tfrac{1}{2}\omega(x_2)\left[\omega(x_1)e + u_1\right] \\ &= \tfrac{1}{2}\left[\omega(x_1)x_2 + \omega(x_2)x_1\right] \quad . \square \end{aligned}$$

Theorem 9.10 implies that for every integer m+1 there exists up to isomorphy, exactly one algebra of type (1,m) and one of type (m+1,0). Ljubič [88] calls the former algebra the <u>constant</u> one and the latter the <u>unit</u> one. For m+1=3 Holgate [68] proposed calling the former one the <u>equipotent</u> algebra, while he mentioned that the latter one is the elementary algebra, which we have called the gametic algebra for simple Mendelian inheritance, cf. section 7A.

Indeed if we compare (9.40) and (7.1) we see that for every $m \in \mathbb{N}$ the Bernstein algebra of type (m+1,0) is the gametic algebra for simple Mendelian inheritance over a field $\mathbb{K}$, char $\mathbb{K} \neq 2$, with m+1 alleles. Furthermore we have the following characterization.

<u>Lemma 9.11.</u> The Bernstein algebra $\mathfrak{A}$ over $\mathbb{K}$ (char $\mathbb{K} \neq 2$) of type (m+1,0) coincides with the (m+1)-dimensional train algebra of rank 2.

<u>Proof.</u> I. Let $\mathfrak{A}$ be the Bernstein algebra of type (m+1,0). Then the multiplication in $\mathfrak{A}$ is given by (9.40), hence

$$x^2 = \omega(x)x \qquad \text{for all } x \in \mathfrak{A} \quad .$$

II. The converse implication follows from a result of Etherington [28] which we have mentioned in section 7A and which we will prove here. Let $\mathfrak{A}$ be a train algebra of rank 2. Then the train equation reads

$$x^2 - \omega(x)x = 0, \qquad x \in \mathfrak{A} \quad . \tag{9.41}$$

We substitute $x + y$, $x,y \in \mathfrak{A}$, into (9.41), and we obtain

$$2xy - \omega(x)y - \omega(y)x = 0.$$

Hence, in view of char $\mathbb{K} \neq 2$, the multiplication in $\mathfrak{A}$ is given by (9.40), i.e. $\mathfrak{A}$ is a Bernstein algebra of type m+1.□

In addition to the preceding lemma we have the following

Theorem 9.12. Let $\mathfrak{A}$ be a train algebra over $\mathbb{K}$, char $\mathbb{K} \neq 2$, of rank 3 with principle train roots $\lambda_0 = 1, \lambda_1 = 0$. Then

(1) $\mathfrak{A}$ is a Bernstein algebra

(2) For every decomposition $\mathfrak{A} = \mathfrak{E} \oplus \mathfrak{U} \oplus \mathfrak{Z}$ with respect to an idempotent (cf. (9.6)), we have in addition to (9.7)-(9.21)

$$\mathfrak{Z}^2 = \langle 0 \rangle \tag{9.42}$$

$$(uz)z = 0 \quad \text{for all } u \in \mathfrak{U},\ z \in \mathfrak{Z}. \tag{9.43}$$

Proof. I. Under the above assumptions the rank equation of $\mathfrak{A}$ reads

$$x^3 - \omega(x)x^2 = 0. \tag{9.44}$$

Take $x,y \in \mathfrak{A}$ and consider all elements of the form $\xi x + \eta y$, $\xi,\eta \in \mathbb{K}$. These elements satisfy (9.44), i.e.

$$(\xi x + \eta y)^3 - (\xi\omega(x) + \eta\omega(y))(\xi x + \eta y)^2 = 0.$$

Hence

$$\xi^3[x^3 - \omega(x)x^2] + \xi^2\eta[2x(xy) + x^2y - 2\omega(x)xy - \omega(y)x^2]$$
$$+\xi\eta^2[2(xy)y + xy^2 - 2\omega(y)xy - \omega(x)y^2] + \eta^3[y^3 - \omega(y)y^2] = 0 .$$

Since ξ,η range over $\mathbb{K}$, all terms in brackets must vanish, in particular

$$2x(xy) + x^2y - 2\omega(x)xy - \omega(y)x^2 = 0, \quad x,y \in \mathfrak{A}. \tag{9.45}$$

For $y = x^2$ equation (9.45) reads

$$2x^4 + x^{[3]} - 2\omega(x)x^3 - \omega^2(x)x^2 = 0 . \tag{9.46}$$

From (9.44) we know that

$$x^3 = \omega(x)x^2 \text{ and } x^4 = \omega^2(x)x^2.$$

We substitute these expressions into (9.46) and obtain

$$x^{[3]} - \omega^2(x)x^{[2]} = 0 \quad ,$$

i.e. $\mathfrak{A}$ is a Bernstein algebra.

II. Consider the decomposition $\mathfrak{A} = \mathfrak{E} \oplus \mathfrak{U} \oplus \mathfrak{Z}$ and linearize (9.44) with $x = \omega(x)e + u + z$, $u \in \mathfrak{U}$, $z \in \mathfrak{Z}$. Then, in view of theorem 9.6, formula (9.44) reduces to

$$-\frac{1}{2}\omega(x)z^2 + u^2z + 2(uz)z + z^3 = 0 \ .$$

Since $\omega(x)$ ranges over $\mathbb{K}$, we have

$$z^2 = 0$$

and

$$2(uz)z + u^2z + z^3 = 0 \quad .$$

With $z = z_1 + z_2$, the former equation yields

$$z_1z_2 = 0, \qquad z_1, z_2 \in \mathfrak{Z} \ ,$$

i.e. (9.42) is valid. Hence, in view of $u^2z \in \mathfrak{Z}^2$ and $z^3 = z^2z$ the latter equation reduces to (4.43).□

A well investigated class of Bernstein algebras are the so called normal ones, which have been introduced by Ljubič, cf. [87] and [88]. He called a Bernstein algebra $\mathfrak{A}$ over $\mathbb{K}$ of type (r+1,d) with weight homomorphism ω, normal, if the vector space of all linear forms $\varphi : \mathfrak{A} \longrightarrow \mathbb{K}$ satisfying

$$(xy) = \frac{1}{2}(\omega(x)\varphi(y) + \omega(y)\varphi(x))$$

is of dimension r+1. Ljubič gave several equivalent characterizations of these algebras, one of which we shall use as definition here (without proving its equivalence to the original definition). Subsequently we will state several properties to these algebras which were mentioned by Ljubič [87] .

Definition 9.13. Let $\mathfrak{A}$ be a Bernstein algebra with weight homomorphism ω. $\mathfrak{A}$ is called normal, if the following identity is valid

$$x^2y = \omega(x)xy \qquad \text{for all } x,y \in \mathfrak{A} \ . \tag{9.47}$$

Note, that in terms of left transformations of $\mathfrak{A}$ the identity (9.47) reads

$$\mathcal{L}_{x^2} = \omega(x)\,\mathcal{L}_x \qquad \text{for all } x \in \mathfrak{A} \ . \tag{9.48}$$

Theorem 9.14. Let $\mathfrak{A}$ be a Bernstein algebra with weight homomorphism ω. Decompose $\mathfrak{A} = \mathfrak{E} \oplus \mathfrak{U} \oplus \mathfrak{Z}$ with respect to an idempotent $e \in \mathfrak{A}$, cf. (9.6). Then the following properties are equivalent.

(1) $\mathfrak{A}$ is normal,

(2) $\mathfrak{U}\mathfrak{Z} = \langle 0 \rangle, \quad \mathfrak{Z}^2 = \langle 0 \rangle,$ (9.49)

(3) The square of every $x = \omega(x)e + u + z \in \mathfrak{A}$ is given by

$$x^2 = \omega^2(x)e + \omega(x)u + u^2. \qquad (9.50)$$

Proof. Let $x,y \in \mathfrak{A}$ be arbitrary. Then there are uniquely determined $u_i \in \mathfrak{U}$, $z_i \in \mathfrak{Z}$, $i=1,2$ (depending on x and y) such that $x = \omega(x)e + u_1 + z_1$ and $y = \omega(y)e + u_2 + z_2$. With theorem 9.6 follows that

$$\begin{aligned}\omega(x)xy &= \omega^2(x)\omega(y)e \\ &+ [\omega(x)(\tfrac{1}{2}\omega(x)u_2 + \tfrac{1}{2}\omega(y)u_1 + u_1z_2) + \omega(x)(u_2z_1 + z_1z_2)] \\ &+ \omega(x)u_1u_2\end{aligned}$$

and

$$\begin{aligned}x^2y &= \omega^2(x)\omega(y)e \\ &+ [\omega(x)(\tfrac{1}{2}\omega(x)u_2 + \tfrac{1}{2}\omega(y)u_1 + u_1z_2) + \omega(y)(u_1z_1 + \tfrac{1}{2}z_1^2) \\ &\qquad + (u_1^2u_2 + u_1^2z_2 + 2(u_1z_1)z_2 + z_1^2z_2] \\ &+ [\omega(x)u_1u_2 + 2(u_1z_1)u_2].\end{aligned}$$

In every case the $\mathfrak{E}$ - components of $\omega(x)xy$ and x^2y are equal.

I. Assume $\mathfrak{A}$ is normal. Then, in addition, the $\mathfrak{U}$ - and $\mathfrak{Z}$- components of $\omega(x)xy$ and x^2y must coincide for all $x,y \in \mathfrak{A}$, hence in particular the equation

$$\begin{aligned}&\omega(x)[u_2z_1 + z_1z_2] \\ &= \omega(y)[u_1z_1 + \tfrac{1}{2}z_1^2] + [u_1^2u_2 + (u_1z_1)z_2 + z_1^2z_2 + u_1^2z_2]\end{aligned}$$

must be satisfied for all $\omega(x), \omega(y) \in \mathbb{K}$ and all $u_i \in \mathfrak{U}$, $z_i \in \mathfrak{Z}$, $i=1,2$. Since $\omega(x)$ and $\omega(y)$ range over $\mathbb{K}$ independently, all three terms in square brackets must vanish, in particular

$$u_2z_1 + z_1z_2 = 0 \qquad \text{for all } u_2 \in \mathfrak{U},\ z_1,z_2 \in \mathfrak{Z}.$$

For $z_2 = 0$, this equation implies

$$u_2z_1 = 0 \qquad \text{for all } u_2 \in \mathfrak{U},\ z_1 \in \mathfrak{Z},$$

i.e.

$$\mathfrak{U}\mathfrak{Z} = \langle 0 \rangle ,$$

and with $u_2 = 0$ it follows that

$$z_1 z_2 = 0 \qquad \text{for all } z_1 z_2 \in \mathfrak{Z} ,$$

i.e.

$$\mathfrak{Z}^2 = \langle 0 \rangle .$$

Hence (2) is satisfied.

II. Assume (2). Then the $\mathfrak{U}$ - and $\mathfrak{Z}$ - components of $\omega(x)xy$ and x^2y, in the decomposition given above, both reduce to $\frac{1}{2}\omega(x)(\omega(y)u_1 + \omega(x)u_2)$ and $\omega(x)u_1u_2$, respectively, hence $\mathfrak{A}$ is normal.
III. The implication (2) $\Longrightarrow$ (3) is obvious in view of (9.17).
IV. Assume (3). On comparing this representation of x^2 with that given in (9.7) it follows that

$$\omega(x)uz + z^2 = 0 \qquad \text{for all } \omega(x) \in \mathbb{K},\ u \in \mathfrak{U},\ \text{and } z \in \mathfrak{Z} .$$

Hence

$$uz = 0 \qquad \text{for all } u \in \mathfrak{U},\ z \in \mathfrak{Z} ,$$

and

$$z^2 = 0 \qquad \text{for all } z \in \mathfrak{Z} .$$

Substituting $z = z_1 + z_2$, $z_1, z_2 \in \mathfrak{Z}$ into the latter equation we obtain

$$z_1 z_2 = 0 \qquad \text{for all } z_1, z_2 \in \mathfrak{Z} .$$

Hence (2) is valid. □

Remark. Since the definition 9.13 of a normal Bernstein algebra is independent of idempotents of $\mathfrak{A}$, statements (2) and (3) of theorem 9.14 are valid for every idempotent $e \in \mathfrak{A}$.

Example 9.15. The Bernstein algebras of type (1,m) and (m+1,0) are normal ones.
This can easily be seen, since in the former case $\mathfrak{U} = \langle 0 \rangle$ while in the latter case $\mathfrak{Z} = \langle 0 \rangle$. Hence in both cases (9.18) and (9.20) imply (9.49), which is one of the given characterizations of a normal Bernstein algebra, cf. theorem 9.14.

Theorem 9.16. Every normal Bernstein algebra $\mathfrak{A}$ is a special train algebra Hence, in particular, it is a train algebra, and its rank is either 2 or 3, and its train roots are $\lambda_0 = 1$ or $\lambda_0 = 1$, $\lambda_1 = 0$, respectively.

Proof. Let ω be the weight homomorphism of $\mathfrak{A}$, and $\mathfrak{N} := \ker\omega$.
Let $e \in \mathfrak{A}$ be an idempotent, then $\mathfrak{A}$ splits into $\mathfrak{A} = \mathfrak{E} \oplus \mathfrak{U} \oplus \mathfrak{Z}$ and $\mathfrak{N} = \mathfrak{U} \oplus \mathfrak{Z}$, from theorem 9.14 it follows that

$$\mathfrak{N}^2 = (\mathfrak{U} \oplus \mathfrak{Z})^2 = \mathfrak{U}^2 + \mathfrak{U}\mathfrak{Z} + \mathfrak{Z}^2 = \mathfrak{U}^2$$

and

$$\mathfrak{N}^3 = \mathfrak{U}^2(\mathfrak{U} \oplus \mathfrak{Z}) = \mathfrak{U}^2\mathfrak{U} + \mathfrak{U}^2\mathfrak{Z} = \langle 0 \rangle$$

in view of $\mathfrak{U}^2 \subseteq \mathfrak{Z}$ (9.19).
Hence $\mathfrak{N}$ is nilpotent of an index not greater than 3. Furthermore $\mathfrak{N}^2$ is an ideal of $\mathfrak{A}$ since

$$\mathfrak{A}\mathfrak{N}^2 = (\mathfrak{E} \oplus \mathfrak{U} \oplus \mathfrak{Z})\mathfrak{U}^2 = \mathfrak{E}\mathfrak{U}^2 + \mathfrak{U}^2 + \mathfrak{U}^2\mathfrak{Z} = \langle 0 \rangle$$

in view of $\mathfrak{U}^2 \subseteq \mathfrak{Z}$, the definition of $\mathfrak{Z}$ (cf. theorem 9.6) and theorem 9.14.
Thus $\mathfrak{A}$ satisfies all properties required for a special train algebra, cf. definition 3.27. In particular $\mathfrak{A}$ is a train algebra. Let $x = \omega(x)e + u + z \in \mathfrak{A}$, then theorem 9.14 says

$$x^2 = \omega^2(x)e + \omega(x)u + u^2 \quad ,$$

furthermore

$$x^3 = \omega^3(x)e + \omega^2(x)u + \omega(x)u^2$$

hence the equation

$$x^3 - \omega(x)x^2 = 0 \tag{9.44}$$

is satisfied for all $x \in \mathfrak{A}$. Then the rank equation of $\mathfrak{A}$ is either (9.44) or

$$x^2 - \omega(x)x = 0 \quad .$$

Hence the rank of $\mathfrak{A}$ is either 3 or 2, and the principal train roots are $\lambda_o = 1$, $\lambda_1 = 0$ or $\lambda_o = 1$, respectively. □

Since every special train algebra is a genetic algebra (cf. theorem 3.28) the above theorem implies that every normal Bernstein algebra is genetic, which was pointed out by Ljubič [87] . In addition we can specify a canonical basis and calculate the train roots.

Theorem 9.17. Let $\mathfrak{A}$ be a normal Bernstein algebra of type $(r+1,d)$, $r+d = m$.

(1) Then $\mathfrak{A}$ is a genetic algebra with train roots 1, 1/2 (r-times) and 0 (d-times).
(2) Decompose $\mathfrak{A} = \mathfrak{E} \oplus \mathfrak{U} \oplus \mathfrak{Z}$ with respect to an idempotent $e \in \mathfrak{A}$. Then $e, c_1, \dots, c_r, c_{r+1}, \dots, c_m$ with linearly independent elements

$c_1,\dots,c_r \in \mathfrak{U}$ and $c_{r+1},\dots,c_m \in \mathfrak{Z}$ form a canonical basis of $\mathfrak{A}$.

Proof. Consider the above mentioned decomposition of $\mathfrak{A}$ and choose $c_1,\dots,c_m$ as indicated. Then $e^2 = e$ and, in view of (9.8) and the definition of $\mathfrak{Z}$,

$$ec_i = \tfrac{1}{2}c_i, \quad i=1,\dots,r,$$

$$ec_j = 0 \quad , \quad j=r+1,\dots,m.$$

Furthermore in view of $\mathfrak{U}^2 \subseteq \mathfrak{Z}$, $\mathfrak{U}\mathfrak{Z} = \langle 0 \rangle$ and $\mathfrak{Z}^2 = \langle 0 \rangle$ we have

$$c_i c_j = \sum_{k=r+1}^{m} \lambda_{ijk} c_k \qquad \text{for } i,j=1,\dots,r,$$

$$c_i c_j = 0 \qquad \text{for } i=1,\dots,r,\ j=r+1,\dots,m \text{ or } i,j=r+1,\dots,m.$$

Then $\mathfrak{A}$ is a genetic algebra with $e,c_1,\dots,c_m$ as a canonical basis (cf. definition 3.12) and the train roots are 1, 1/2, 0 with the stated multiplicaties.

After we have seen that the concept of normal Bernstein algebras is very restrictive in a certain sense, cf. theorem 9.15, we come back to the general case and deduce further invariants of Bernstein algebras.
First we will follow Holgate [68], who has chosen another decomposition of a Bernstein algebra $\mathfrak{A}$.
Again, let $e \in \mathfrak{A}$ be an idempotent and decompose $\mathfrak{A}$ into $\mathfrak{A} = \mathfrak{E} \oplus \mathfrak{U} \oplus \mathfrak{Z}$ according to (9.6). In view of $\mathfrak{U}^2 \subseteq \mathfrak{Z}$, cf. (9.19), there exists a complementary subspace $\mathfrak{W}$ such that $\mathfrak{Z} = \mathfrak{U}^2 \oplus \mathfrak{W}$. Hence $\mathfrak{A}$ splits into

$$\mathfrak{A} = \mathfrak{E} \oplus \mathfrak{U} \oplus \mathfrak{U}^2 \oplus \mathfrak{W} .$$

(Note that in Holgate's decomposition $\mathfrak{W}$ need not to be in $\mathfrak{Z} = \ker \mathcal{L}_e$.)
In some sense this decomposition of $\mathfrak{A}$ is more appropriate, since it contains $\mathfrak{A}^2$ as a direct summand, cf. the following theorem which is essentially due to Holgate [68].

Theorem 9.18. Let $\mathfrak{A}$ be a Bernstein algebra with weight homomorphism ω , $\mathfrak{N} := \ker\omega$. Let $e \in \mathfrak{A}$ be an idempotent. Then

$$\mathfrak{C} := \mathfrak{E} \oplus \mathfrak{U} \oplus \mathfrak{U}^2, \tag{9.52}$$

with $\mathfrak{U} := \mathcal{L}_e \mathfrak{N}$, is an ideal of $\mathfrak{A}$, furthermore

$$\mathfrak{C}^2 = \mathfrak{C} \tag{9.53}$$

and

$$\mathfrak{A}^2 = \mathfrak{C} \quad . \tag{9.54}$$

Proof. Decompose $\mathfrak{A} = \mathfrak{L} \oplus \mathfrak{U} \oplus \mathfrak{Z}$. Then in view of $\mathfrak{L}^2 = \mathfrak{L}$, $\mathfrak{L}\mathfrak{U} = \mathfrak{U}$, $\mathfrak{L}\mathfrak{Z} = \langle 0 \rangle$, and (9.16)-(9.20) we have

$$\mathfrak{A}^2 = (\mathfrak{L}\oplus\mathfrak{U}\oplus\mathfrak{Z})^2 = \mathfrak{L}^2 + \mathfrak{L}\mathfrak{U} + \mathfrak{U}^2 + \mathfrak{L}\mathfrak{Z} + \mathfrak{U}\mathfrak{Z} + \mathfrak{Z}^2$$

$$= \mathfrak{L} + \mathfrak{U} + \mathfrak{U}^2 .$$

From formular (9.51) we know that this sum is a direct one, hence (9.54) is true. Furthermore $\mathfrak{A}\mathfrak{C} \subseteq \mathfrak{A}^2 = \mathfrak{C}$, thus $\mathfrak{C}$ is an ideal of $\mathfrak{A}$. In view of $\mathfrak{U}^2 \subseteq \mathfrak{Z}$ and $\mathfrak{U}\mathfrak{Z}$, $\mathfrak{Z}^2 \subseteq \mathfrak{U}$ it follows immediately that $\mathfrak{C}^2 = \mathfrak{C}$. □

Although the subspace $\mathfrak{U}$ of a Bernstein algebra $\mathfrak{A}$, in general, depends on the choice of an idempotent $e \in \mathfrak{A}$, the core $\mathfrak{C}$ does not depend on e, in view of $\mathfrak{A}^2 = \mathfrak{C}$. Since $\dim \mathfrak{U} = r$ is an invariant number of $\mathfrak{A}$ and $\mathfrak{A}^2 = \mathfrak{C}$, it follows that $\dim \mathfrak{U}^2$, too, is an invariant number of $\mathfrak{A}$. This number gives, besides the type of a Bernstein algebra, a further possibility for a classification of Bernstein algebras.

We will mention here that Holgate [68] has introduced the concept of orthogonal Bernstein algebras, whereby this property may be defined by the additional demand $\mathfrak{U}^3 = \langle 0 \rangle$. It turns out that the core of an orthogonal Bernstein algebra is a special train algebra whose rank, in general, is 4, while its principle train roots are $\lambda_0 = 1$, $\lambda_1 = 1/2$, $\lambda_2 = 0$.

In the following we take up an idea of Etherington, which he introduced in his study of commutative train algebras of rank 2 and 3 (cf. [28]), namely the idea of constructing ideals of a baric algebra such that the factor algebra is a train algebra of "low" rank.

Lemma 9.19. Let $\mathfrak{A}$ be a Bernstein algebra over $\mathbb{K}$ with weight homomorphism ω , $\mathfrak{N} := \ker \omega$. Define

$$\mathfrak{P} := \langle x^2 - \omega(x)x \mid x \in \mathfrak{A} \rangle_{\mathbb{K}} \quad . \tag{9.55}$$

Then

(1) $\mathfrak{P}$ has the representation $\mathfrak{P} := (\mathfrak{U}\mathfrak{Z} + \mathfrak{Z}^2) \oplus \mathfrak{Z}$ (9.56)
for every decomposition $\mathfrak{A} = \mathfrak{L} \oplus \mathfrak{U} \oplus \mathfrak{Z}$.

(2) $\mathfrak{P}$ is an ideal of $\mathfrak{A}$ with $\mathfrak{A} > \mathfrak{N} \geq \mathfrak{P} \geq \mathfrak{N}^2$ and $\mathfrak{A}/\mathfrak{P}$ is a train algebra of rank 2.

Proof. In view of (9.16) and (9.17) we have

$$x^2 - \omega(x)x = [2uz + z^2] + [u^2 - \omega(x)z] \qquad (9.57)$$

for all $x = \omega(x)e + u + u \in \mathfrak{A}$, in which the first bracket term lies in $\mathfrak{U}\mathfrak{Z} + \mathfrak{Z}^2 \subseteq \mathfrak{U}$, while the second bracket term is in $\mathfrak{U}^2 + \mathfrak{Z} = \mathfrak{Z}$, cf. (9.18)-(9.20).

Furthermore the linear subspace $\mathfrak{P}$ of $\mathfrak{A}$ spanned by all elements (9.57) has a decomposition $\mathfrak{P} = \mathfrak{U}_1 \oplus \mathfrak{Z}$, where $\mathfrak{U}_1$ is spanned by $2uz + z^2$, $u \in \mathfrak{U}$, $z \in \mathfrak{Z}$.

In view of

$$2uz = [2(u+u_1)z + z^2] - [2u_1z - z^2] \in \mathfrak{U}_1 \quad \text{for all } u,u_1 \in \mathfrak{U}, z \in \mathfrak{Z},$$

and

$$2z_1z_2 = (z_1 + z_2)^2 - z_1^2 - z_2^2 \in \mathfrak{U}_1 \qquad \text{for all } z_1,z_2 \in \mathfrak{Z},$$

we have

$$\mathfrak{U}_1 = \mathfrak{U}\mathfrak{Z} + \mathfrak{Z}^2,$$

hence $\mathfrak{P}$ is given by (9.56).

II. With theorem 9.6 we obtain

$$\begin{aligned}\mathfrak{A}\mathfrak{P} &= (\mathfrak{E} \oplus \mathfrak{U} \oplus \mathfrak{Z})((\mathfrak{U}\mathfrak{Z} + \mathfrak{Z}^2) \oplus \mathfrak{Z}) \\ &= \mathfrak{E}(\mathfrak{U}\mathfrak{Z} + \mathfrak{Z}^2) + \mathfrak{E}\mathfrak{Z} + \mathfrak{U}(\mathfrak{U}\mathfrak{Z} + \mathfrak{Z}^2) + \mathfrak{U}\mathfrak{Z} + (\mathfrak{U}\mathfrak{Z} + \mathfrak{Z}^2)\mathfrak{Z} + \mathfrak{Z}^2 \\ &\subseteq (\mathfrak{U}\mathfrak{Z} + \mathfrak{Z}^2) \oplus \mathfrak{U}^2 \subseteq (\mathfrak{U}\mathfrak{Z} + \mathfrak{Z}^2) \oplus \mathfrak{Z}.\end{aligned}$$

Hence $\mathfrak{P}$ is an ideal of $\mathfrak{A}$. By definition $\mathfrak{P} \subseteq \mathfrak{N}$. Furthermore, in view of

$$\mathfrak{N}^2 = (\mathfrak{U} \oplus \mathfrak{Z})^2 = (\mathfrak{U}\mathfrak{Z} + \mathfrak{Z}^2) \oplus \mathfrak{U}^2 ,$$

$\mathfrak{P}$ contains $\mathfrak{N}^2$.

Since $\mathfrak{P}$ is an ideal of $\mathfrak{A}$, the factor space $\mathfrak{A}/\mathfrak{P}$ is an algebra, and by the definition of $\mathfrak{P}$, the rank equation of this algebra is

$$\bar{x}^2 + \bar{\omega}(\bar{x})\bar{x} = \bar{0}, \qquad \bar{x} \in \mathfrak{A}/\mathfrak{P},$$

with $\bar{\omega}: \mathfrak{A}/\mathfrak{P} \longrightarrow \mathbb{K}$ defined by

$$\bar{\omega}(\bar{x}) := \omega(x+\mathfrak{P}) = \omega(x) . \square$$

Let $\mathfrak{A}$ be an (m+1)-dimensional Bernstein algebra. Decompose $\mathfrak{A} = \mathfrak{E} \oplus \mathfrak{U} \oplus \mathfrak{Z}$ with respect to an idempotent $e \in \mathfrak{A}$. Since the definition (9.55) of the subspace $\mathfrak{P}$ of $\mathfrak{A}$ only depends on the uniquely determined weight homomorphism ω of $\mathfrak{A}$, $\dim \mathfrak{P}$ is a characteristic number of $\mathfrak{A}$. Since furthermore $\dim \mathfrak{Z}$ is determined by m-r, and

$r = \dim \mathfrak{U}$ is an invariant of $\mathfrak{A}$, the integer $\dim(\mathfrak{U}\mathfrak{Z} + \mathfrak{Z}^2)$, in addition, is independent of the choice of an idempotent of $\mathfrak{A}$. Therefore this number, too, can be used for a classification of Bernstein algebras.

Hence we can associate with every (m+1)-dimensional Bernstein algebra $\mathfrak{A}$ its type (r+1,d) and the integers $\dim(\mathfrak{U}\mathfrak{Z} + \mathfrak{Z}^2)$ and $\dim \mathfrak{U}^2$, i.e. three integers in view of r+d=m.

But not all triples of integers can occur, since they must fulfil the inequalities

$$\dim(\mathfrak{U}\mathfrak{Z} + \mathfrak{Z}^2) \leq r, \quad \dim \mathfrak{U}^2 \leq d = m - r,$$

and there are further restrictions implied by (9.9) - (9.15).

In the following we will classify all 3-dimensional Bernstein algebra via these three integers.

It turns out that apart from the trivial cases of type (1,2) and (3,0) there are two classes of Bernstein algebras. One of them forms a 1-parameter family over $\mathbb{K}$. All algebras of this class are normal and all, with the exception of one, are isomorphic to the zygotic algebra for simple Mendelian inheritance with two alleles. The algebras of the second class are neither normal nor train algebras. They form a 2-parametric family over $\mathbb{K}$.

For a similar classification we refer to Holgate's paper [68], section 4. Note that for the last mentioned case the multiplication table suggested by Holgate contains 3 parameters. The reason for this is that in his decomposition (9.51) the subspace $\mathfrak{M}$ need not be in $\mathfrak{Z}$.

Table 9.20. Classification of all 3-dimensional Bernstein algebras

Let $\mathfrak{A}$ be a 3-dimensional Bernstein algebra. Decompose $\mathfrak{A} = \mathfrak{E} \oplus \mathfrak{U} \oplus \mathfrak{Z}$ with resect to an idempotent $e \in \mathfrak{A}$. Define $r := \dim \mathfrak{U}$, $d := \dim \mathfrak{Z}$ and choose a basis e, c_1, c_2 of $\mathfrak{A}$ with $c_i \in \mathfrak{U}$, $i=1,r$, and $c_j \in \mathfrak{Z}$, $j=r+1,2$.
Then we can distinguish four different classes of algebras:

	type $(r+1,d)$	$\dim(\mathfrak{U}\mathfrak{Z} + \mathfrak{Z}^2)$	$\dim \mathfrak{U}^2$	e^2	ec_1	ec_2	c_1^2	c_1c_2	c_2^2	properties
(1)	(1,2)	0	0	e	0	0	0	0	0	normal the constant (or equipotent) algebra
(2)	(2,1)	0	0	e	$\frac{1}{2}c_1$	0	0	0	0	normal
			1				αc_2			normal and isomorphic with the zygotic algebra for simple Mendelian inheritance with 2 alleles
(3)	(2,1)	1	0	e	$\frac{1}{2}c_1$	0	0	βc_1	γc_1	not normal no train algebra $(\mathfrak{U} \oplus \mathfrak{Z})^l = \langle c_1 \rangle$, $l \geq 2$
(4)	(3,0)	0	0	e	$\frac{1}{2}c_1$	$\frac{1}{2}c_2$	0	0	0	normal the unit algebra isomorphic with the gametic algebra for simple Mendelian inheritance with 2 alleles

with $\alpha \neq 0$ and $(\beta,\gamma) \neq (0,0)$.

Proof of 9.20:

I. The Bernstein algebras of type (1,2) and (3,0) have been discussed in theorem 9.10 and example 9.15. It turns out that in both cases $\mathfrak{U}\mathfrak{Z} + \mathfrak{Z}^2 = \langle 0 \rangle$ and $\mathfrak{U}^2 = \langle 0 \rangle$. Furthermore the multiplication table follows directly from (9.39) and (9.40), respectively.

II. Now let $\mathfrak{A}$ be of type (2,1).

Then $\mathfrak{U} = \langle c_1 \rangle$, $\mathfrak{Z} = \langle c_2 \rangle$ and $e^2 = e$, $ec_1 = \frac{1}{2}c_1$, $ec_2 = 0$.

(i) If $\dim(\mathfrak{U}\mathfrak{Z} + \mathfrak{Z}^2) = 0$, then $\mathfrak{A}$ is normal, cf. theorem 9.14. Furthermore in view of $c_1c_2 \in \mathfrak{U}\mathfrak{Z}$ and $c_2^2 \in \mathfrak{Z}^2$ it follows that

$$c_1c_2 = c_2^2 = 0 \quad .$$

If $\dim \mathfrak{U}^2 = 0$, then in view of $c_1^2 \in \mathfrak{U}^2$ we have $c_1^2 = 0$.

If $\dim \mathfrak{U}^2 = 1$, then there exists an $\alpha \in \mathbb{K}$, $\alpha \neq 0$, such that $c_1^2 = \alpha c_2$.

Hence, in case (2) there exists a 1-parameter family of Bernstein algebras.

Let $\alpha \neq 0$. Then we can assume without loss of generality that $\alpha = 1$, because otherwise we could choose $\tilde{c}_2 := \frac{1}{\alpha}c_2$ instead of c_2. One easily checks that with respect to the new basis

$$a_0 := e, \quad a_1 := e - \frac{1}{2}c_1, \quad a_2 := e - c_1 + c_2$$

the multiplication table of $\mathfrak{A}$ reads

	a_0	a_1	a_2
a_0	a_0	$\frac{1}{2}a_0 + \frac{1}{2}a_1$	a_1
a_1		$\frac{1}{4}a_0 + \frac{1}{2}a_1 + \frac{1}{4}a_2$	$\frac{1}{2}a_1 + \frac{1}{2}a_2$
a_2			a_2

Hence this multiplication table of $\mathfrak{A}$ coincides with that of the zygotic algebra $\mathfrak{Z} = \langle AA, Aa, aa \rangle_{\mathbb{K}}$ for simple Mendelian inheritance with two alleles A and a, if we put $a_0 = AA$, $a_1 = Aa$, $a_2 = aa$, cf. section 7A.

(ii) Assume $\dim(\mathfrak{U}\mathfrak{Z} + \mathfrak{U}^2) = 1$, then $\mathfrak{A}$ is not normal, cf. theorem 9.14. Furthermore there exist parameters $\beta, \gamma \in \mathbb{K}$, at least one of them different from zero, such that

$$c_1c_2 = \beta c_1 \quad \text{and} \quad c_2^2 = \gamma c_1 \quad .$$

Since $u^3 = 0$ for all $u \in \mathfrak{U}$ (9.10), we must have $c_1^2 = 0$, hence $\dim \mathfrak{U}^2 = 0$. Take $x = \omega(x)e + \xi_1c_1 + \xi_2c_2 \in \mathfrak{A}$. Then the left transformation ℓ_x

of $\mathfrak{A}$ is represented by the matrix

$$L_x = \begin{pmatrix} \omega(x) & 0 & 0 & 0 \\ \frac{1}{2}\xi_1 & \frac{1}{2}\omega(x) & + \beta\xi_2 & 0 \\ 0 & & 0 & \beta\xi_1 + \gamma\xi_2 \end{pmatrix}$$

with respect to the basis e, c_1, c_2. From section 3 we know that the identity

$$x(x - \omega(x))(x - (\tfrac{1}{2}\omega(x) + \beta\xi_2))(x - (\beta\xi_1 + \gamma\xi_2)) = 0$$

is satisfied for all $x \in \mathfrak{A}$. Furthermore, in view of $(\beta,\gamma) \neq (0,0)$ the rank polynomial of $\mathfrak{A}$, which is a divisor of the left hand side of the above equation, contains at least one factor which depends on a coordinate ξ_1 or ξ_2 of x. Hence $\mathfrak{A}$ is not a train algebra.
For the principal powers of $\mathfrak{N} = \mathfrak{U} \oplus \mathfrak{Z}$ we obtain at once

$$\mathfrak{N}^l = \langle c_1 \rangle \qquad \text{for all } l \geq 2. \ \square$$

Holgate, in his aforementioned paper, has taken up Bernstein's classification of idempotent evolutionary operators acting on the simplex $\tilde{\mathfrak{F}}_2$ of $\mathbb{R}^3$ (cf. [15]), and he has deduced a "canonical" form of the multiplication tables of the algebras corresponding to these evolutionary operators. We will mention how these algebras fit into the classification given above: Holgate's algebras $\mathfrak{b}_0$ and $\mathfrak{b}_3$ correspond to the Bernstein algebras of type (1,2) and (3,0), respectively. The algebras $\mathfrak{b}_{11}$ and $\mathfrak{b}_2$ are isomorphic to algebras of class (2) (via $e = b_0$, $c_1 = b_1$, $c_2 = b_2$ in the first case, and $e = b_0$, $c_1 = b_1 + b_2$, $c_2 = b_2$ in the second case).
Furthermore the algebra $\mathfrak{b}_{12}$ belongs to class (3) (via $e = b_0$, $c_1 = b_2$, $c_2 = b_1 + (\alpha - 1)b_2$).

Finally we mention a generalization of Bernstein algebras proposed by Abraham [3]:
An algebra $\mathfrak{A}$ is called a <u>k-th order Bernstein algebra</u> if $\mathfrak{A}$ satisfies definition 9.1 with

$$x^{[k+2]} - \omega^{2^k}(x)\,x^{[k+1]} = 0$$

instead of (9.5).

A population which is described by a k-th order Bernstein algebra is in equilibrium after k+1 generations of random mating.

REFERENCES

G [1] Abraham, V. M.: Linearising quadratic transformations in genetic algebras. Thesis, Univ. of London 1976.

G [2] Abraham, V. M.: A note on train algebras. Proc. Edinburgh Math. Soc. (2) 20 (1976) 53-58.

G [3] Abraham, V. M.: Linearising quadratic transformations in genetic algebras I. Proc. London Math. Soc. (to appear).

G [4] Abraham, V. M.: Linearising quadratic transformations in genetic algebras II. Proc. London Math. Soc. (to appear).

[5] Albert, A. A.: Structure of algebras. American Mathematical Society, New York 1939.

[6] Albert, A. A.: The radical of a non-associative algebra. Bull. Amer. Math. Soc. 48 (1942) 891-897.

[7] Albert, A. A.: Nonassociative algebras I: Fundamental concepts and isotopy. Ann. of Math. 43 (1942) 685-708.

[8] Albert, A. A.: On Jordan algebras of linear transformations. Trans. Amer. Math. Soc. 59 (1946) 524-555.

[9] Albert, A. A.: A Structure theory for Jordan algebras. Ann. of Math. 48 (1947) 546-567.

[10] Andreoli, G.: Algebre non associative e sistemi differenziati di Riccati in un problema di Genetica. Ann. Mat. Pura Appl. 49 (1960) 97-116.

[11] Ballonoff, P. (Editor): Genetic and social structure: Mathematical structuralism in population genetics and social theory. Dowden, Hutchinson & Ross. Inc. Stroundsburg, Pennsylvania 1974.

[12] Bennett, J. H.: On the theory of random mating. Ann. Eugen. (Ann. Hum. Gen.) 18 (1954) 311-317.

[13] Bernstein, S.: Demonstration mathématique de la loi d'hérédité de Mendel. Comptes rendus de l'Acad. des Sci. Paris 177 (1923) 528-531.

[14] Bernstein, S.: Principe de stationarité et généralisation de la loi de Mendel. Comptes rendus de l'Acad. des Sci. Paris 177 (1923) 581-584.

[15] Bernstein, S.: Solution of a mathematical problem connected with the theory of heredity. U. Zap. Naun.-Issed. Kafedr Ukrain. Otd. Mat. 1 (1924) 83-115. (Russisch). Englische Übersetzung: Ann. Math. Statist. 13 (1942) 53-61.

G [16] Bertrand, M.: Algèbres non associative et algèbres génétiques. Mémorial des Sciences Mathématiques 162, Gauthiers Villars, Paris 1966.

[17] Bertrand, M.: Propriétés de certaines algèbres non associatives. Comptes Rendus Acad. Sci. Paris 266 A (1968) 855-858.

G [18] Bertrand, M.: Résolution, a l'aide de résults d'algèbres non associative, de certaines problèmes d'algèbre génétique Publ. Inst. Statist. Univ. Paris 20 (3-4)(1971) 5-84.

[19] Braun, H., Koecher, M.: Jordan-Algebren. Springer-Verlag, Berlin, Heidelberg, New York 1966.

[20] Crow, J. F., Kimura, M.: An introduction to population genetics theory. Harper & Row, New York, Evanston, London 1970.

[21] Dickson, L. E.: Linear algebras. Trans. Amer. Math. Soc. 13 (1912) 59-73.

[22] Dickson, L. E.: Linear algebras. Cambridge Tracts in Mathematics and Mathematical Physics No 16, 1914.

[23] Dmitriew, N. A., Dynkin, E. B.: Über die charakteristischen Zahlen stochastischer Matrizen. Doklady A.N.S.S.S.R. 49 (1945) 159-162. (Russisch).

[24] Dmitriew, N. A., Dynkin, E. B.: Die charakteristischen Wurzeln stochastischer Matrizen. Izvestiya A.N.S.S.S.R. Ser. mat. 10 (1946) 167-184. (Russisch).

[25] Edwards, A. W. F.: Foundation of mathematical genetics. Cambridge Univ. Press, Cambridge, London, New York, Melbourne 1977.

[26] Etherington, I. M. H.: On non-associative combinations. Proc. Roy. Soc. Edinburgh 59 (1939) 153-162.

G [27] Etherington, I. M. H.: Genetic algebras. Proc. Roy. Soc. Edinburgh 59 (1939) 242-258.

G [28] Etherington, I. M. H.: Commutative train algebras of ranks 2 and 3. J. London Math. Soc. 15 (1940) 136-149. Corrigendum ibid. 20 (1945) 238.

G [29] Etherington, I. M. H.: Non-associative algebra and the symbolism of genetics. Proc. Roy. Soc. Edinburgh B 61 (1941)24-42.

G [30] Etherington, I. M. H.: Duplication of linear algebras. Proc. Edinburgh Math. Soc. (2) 6 (1941) 222-230.

G [31] Etherington, I. M. H.: Special train algebras. Quart. J. Math. Oxford, Ser. (2) 12 (1941) 1-8.

G [32] Etherington, I. M. H.: Non-commutative train algebras of ranks 2 and 3. Proc. London Math. Soc. (2) 52 (1951) 241-252.

G [33] Fielden, Ch. M.: Inbreeding - the three locus case. Symposium on Mathematical genetics. Adv. Appl. Prob. 8 (1976) 613-650.

[34] Fisher, R. A.: The theory of linkage in polysomic inheritance. Phil. Trans. Roy. Soc. London B 233 (1947) 55-87.

G [35] Fortini, P., Barakat, R.: Genetic algebras for tetraploidy with several loci. J. Math. Biol. (to appear).

[36] Frobenius, G.: Über vertauschbare Matrizen. Sitzungsberichte der Preussischen Akad. der Wissenschaften zu Berlin, 1. Halbjahresband (1896) 601-614.

[37] Gantmacher, F. R.: Matrizenrechnung I. VEB Deutscher Verlag der Wissenschaften, Berlin 1970.

[38] Gantmacher, F. R.: Matrizenrechnung II. VEB Deutscher Verlag der Wissenschaften, Berlin 1971.

[39] Geiringer, H.: On the probability theory of linkage in Mendelian heredity. Ann. Math. Stat. 15 (1944) 25-57.

[40] Geiringer, H.: Further remarks on linkage theory in Mendelian heredity. Ann. Math. Stat. 16 (1945) 390-393.

[41] Geiringer, H.: Contribution to the heredity theory of multivalents. J. Math. Phys. 26 (1947) 246-278.

[42] Geiringer, H.: Chromatid segregation of tetraploids and hexaploids. Genetics 34 (1949) 665-684.

[43] Gerstenhaber, M.: On nilalgebras and linear varieties of nilpotent matrices II. Duke Math. J. 27 (1960) 21-31.

[44] Glivenko, V.: Algèbre mendelienne. Comptes rendus (Doklady) de l'Acad. des Sci. de l'URSS 4 (13)(1936) 385-386.

G [45] Gonshor, H.: Special train algebras arising in genetics. Proc. Edinburgh Math. Soc. (2) 12 (1960) 41-53.

G [46] Gonshor, H.: Special train algebras arising in genetics II. Proc. Edinburgh Math. Soc. (2) 14 (1965) 333-338.

G [47] Gonshor, H.: Contributions to genetic algebras. Proc. Edinburgh Math. Soc. (2) 17 (1971) 289-298.

G [48] Gonshor, H.: Contributions to genetic algebras II. Proc. Edinburgh Math. Soc. (2) 18 (1973) 273-279.

[49] Greub, W. H.: Multilinear Algebra. Springer Verlag, Berlin, Heidelberg, New York 1967.

[50] Haldane, J. B. S.: Theoretical genetics of autopolyploids. J. Genetics 22 (1930) 359-372.

[51] Herstein, I. N.: Noncommutative rings. The Mathematical Association of America, John Wiley, New York 1968.

G [52] Heuch, I.: Sequences in genetic algebras for overlapping generations. Proc. Edinburgh Math. Soc. (2) 18 (1972) 19-29.

G [53] Heuch, I.: k loci linked to a sex factor in haploid individuals. Biometr. Z. 13 (1972) 57-68.

G [54] Heuch, I.: The linear algebra for linked loci with mutation. Math. Biosciences 16 (1973) 263-271.

G [55] Heuch, I.: Genetic algebras and time continuous models. Theor. Pop. Bio. 4 (1973) 133-144.

G [56] Heuch, I.: Partial and complete sex linkage in infinite populations. J. Math. Biol. 1 (1975) 331-343.

G [57] Heuch, I.: Genetic algebras for systems with linked loci. Math. Biosciences 34 (1977) 35-47. Erratum, ibid. 37 (1977)279.

G [58] Heuch, I.: An explicit formula for frequency changes in genetic algebras. J. Math. Biol. 5 (1977) 43-53.

G [59] Heuch, I.: Genetic algebras considered as elements in a vector space. SIAM J. Appl. Math. 35 (1978) 695-703

G [60] Heuch, I.: The genetic algebra for polyploidy with an arbitrary amount of double reduction. J. Math. Biol., 6 (1978) 343-352.

G [61] Holgate, P.: Genetic algebras associated with polyploidy. Proc. Edinburgh Math. Soc. (2) 15 (1966) 1-9. Corrigendum, ibid. 17 (1970) 120.

G [62] Holgate, P.: Sequences of powers in genetic algebras. J. London Math. Soc. 42 (1967) 489-496.

G [63] Holgate, P.: Jordan algebras arising in population genetics. Proc. Edinburgh Math. Soc. (2) 15 (1967) 291-294. Corrigendum, ibid. 17 (1970) 120.

G [64] Holgate, P.: The genetic algebra of k linked loci. Proc. London Math. Soc. (3) 18 (1968) 315-327.

G [65] Holgate, P.: Interaction between migration and breeding studied by means of genetic algebras. J. Appl. Prob. 5 (1968) 1-8.

G [66] Holgate, P.: Genetic algebras associated with sex linkage. Proc. Edinburgh Math. Soc. (2) 17 (1970) 113-120.

G [67] Holgate, P.: Characterizations of genetic algebras. J. London Math. Soc. (2) 6 (1972) 169-174.

G [68] Holgate, P.: Genetic algebras satisfying Bernstein's stationarity principle. J. London Math. Soc. (2) 9 (1975) 613-623.

G [69] Holgate, P.: Direct products of genetic algebras and Markov chains. J. Math. Biol. 3 (1976) 289-295.

G [70] Holgate, P.: Selfing in genetic algebras. J. Math. Biol. 6 (1978) 197-206.

G [71] Holgate, P.: Canonical multiplication in the genetic algebra for linked loci. Linear Alg. Appl. 26 (1979) 281-287.

G [72] Holgate, P.: Genetic algebras for some inbreeding systems. Proc. Roy. Soc. Edinburgh A (to appear).

[73] Holgate, P.: Les grades de certaines algèbres génétiques. Submitted to Comptes Rendus Acad. Sci. Paris.

G [74] Holgate, P.: Population Algebras. Preprint, Univ. of London 1979.

[75] Jacobson, N.: The theory of rings. American Mathematical Society, New York 1943.

[76] Jacobson, N.: Lie algebras. Interscience Publishers, New York, London, Sydney 1962.

[77] Jacobson, N.: Structure of rings. American Mathematical Society, Colloquium Publications, Providence, 2nd ed. 1964.

[78] Jennings, H. S.: The numerical results of diverse systems of breeding, with respect to two pairs of characters, linked or independent, with special relation to the effects of linkage. Genetics 2 (1917) 97-154.

[79] Karlin, S.: Equilibrium behavior of population genetic models with non-random mating. Gordon and Breach, Science Publishers, New York, London, Paris 1969.

[80] Karpelevič, F. I.: Über die charakteristischen Wurzeln von Matrizen mit nicht-negativen Elementen. Izvestiya A.N.S.S.S.R. Ser. mat. 15 (1951) 361-383. (Russisch)

[81] Kempthorne, O.: An introduction to genetic statistics. John Wiley & Sons, New York 1957.

[82] Kostizin, V. A.: Sur les coefficients mendéliens d'hérédité. Comptes rendus de l'Acad. des Sci. 206 (1938) 883-885. Auch in: Scudo, F. M. Ziegler, J. R.: The golden age of theoretical ecology : 1923-1940. Lecture Notes in Biomathematics No 22, Springer-Verlag, Berlin, Heidelberg, New York 1978.

[83] Li, C. C.: Population genetics. The University of Chicago Press, Chicago, London 1955.

[84] Ljubič, Ju. I.: Basic concepts and theorems of evolution genetics of free populations. Uspehi Mat. Nauk 26, no 5 (161) (1971) 51-116. (Russisch). Englische Übersetzung: Russian Math. Surveys 26, no 5 (1971) 51-123.

[85] Ljubič, Ju. I.: On the mathematical theory of heredity. Doklady A.N.S.S.S.R. 209, no 5 (1973) 1028-1030. (Russisch). Englische Übersetzung: Soviet Math. Dokl. 14, no 2 (1973) 579-581.

[86] Ljubič, Ju. I.: Two-level Bernstein Populations. Mat. Sbornik 95, no 4 (137) (1974) 606-628. (Russisch) Englische Übersetzung: Math. USSR Sbornik 24 no 4 (1974) 593-615.

[87] Ljubič, Ju. I.: Bernsteinsche Algebren. Uspehi Mat. Nauk. 32, no 6 (198) (1977) 261-262. (Russisch).

G [88] Ljubič, Ju. I.: Algebraic methods in evolutionary genetics. Biom. J. 20 (1978) 511-529.

[89] Marcus, M., Minc, H.: A survey of matrix theory and matrix inequalities. Allyn and Bacon, Boston 1964.

[90] McCoy, N. H.: On the characteristic roots of matric polynomials. Bull. Amer. Math. Soc. 42 (1926) 592-600.

[91] de Oliveira, G. N.: A method for constructing stochastic matrices. Estudos Gerais Universitarios de Moçambique Lourenço Marques, Vol. II, Serie I (1966) 1-7.

[92] de Oliveira, G. N.: Sobre matrizes estocástias e duplamente estocástias. Coimbra 1968. Auch in: Rev. Fac. Ci. Univ. Coimbra 41 (1968) 15-221.

[93] Perfect, H.: On positive stochastic matrices with real characteristic roots. Proc. Cambridge Phil. Soc. 48 (1952) 271-276.

[94] Perfect, H.: Methods of constructing certain stochastic matrices. Duke Math. J. 20 (1953) 395-404.

[95] Perfect, H.: Methods of constructing certain stochastic matrices II. Duke Math. J. 22 (1955) 305-311.

G [96] Raffin, R.: Axiomatisation des algèbres génétiques. Acad. R. de Belgique Bull. Cl. Sci. (5) 37 (1951) 359-366.

[97] Rees, D.: Linear Systems of algebras. Ann. of Math. 51 (1950) 123-160.

G [98] Reiersøl, O.: Genetic algebras studied recursively and by means of differential operators. Math. Scand. 10 (1962) 25-44.

[99] Riordan, J.: Combinatorial identitics. John Wiley, New York 1968.

[100] Robbins, R. B.: Some applications of mathematics to breeding problems III. Genetics 3 (1918) 375-389.

G [101] Schafer, R. D.: Structure of genetic algebras. Amer. J. Math. 71 (1949) 121-135.

[102] Schafer, R. D.: An introduction to non-associative algebras. Acad. Press. New York, London 1966.

[103] Serebrovskij, A.: On the properties of the Mendelian equations. Doklady A.N.S.S.S.R. 2 (1934) 33-36. (Russisch). Englische Übersetzung: ibid. 2 (1934) 36-39.

[104] Seyffert, W.: Theoretische Untersuchungen über die Zusammensetzung tetrasomer Populationen. II. Selbstbefruchtung. Z. Vererbungslehre 90 (1959) 356-374.

[105] Seyffert, W.: Theoretische Untersuchungen über die Zusammensetzung tetrasomer Populationen. I. Panmixie. Biometr. Z. 2 (1960) 1-44.

[106] Shikata, M.: Group representation of genetic recombinations. Bull. Math. Biophys. 26 (1949) 91-100.

[107] Strobeck, C.: The algebra of recombination. Adv. Appl. Probability 8 (1976) 27-29.

[108] Suleǐmanova, H. R.: Stochastische Matrizen mit reellen charakteristischen Zahlen. Doklady A.N.S.S.S.R. 66 (1949) 343-345. (Russisch).

[109] Suleǐmanova, H. R.: Über die charakteristischen Zahlen von stochastischen Matrizen. Ucenie Zap. Mosk. Ped. Inst. 71 (1953) 167-177. (Russisch).

[110] Suttles, D.: A counterexample to a conjecture of Albert. Notices Amer. Math. Soc. A 19 (1972) 566.

[111] Varga, R. S.: Matrix iterative analysis. Prentice-Hall, New Jersey 1962.

G [112] Wörz-Busekros, A.: The zygotic algebra for sex-linkage. J. Math. Biol. 1 (1974) 37-46.

G [113] Wörz-Busekros, A.: The zygotic algebra for sex-linkage II. J. Math. Biol. 2 (1975) 359-371.

G [114] Wörz-Busekros, A.: Polyploidy with an arbitrary mixture of chromosome- and chromatid segregation. J. Math. Biol., 6 (1978) 353-365.

INDEX

GPSR Compliance

The European Union's (EU) General Product Safety Regulation (GPSR) is a set of rules that requires consumer products to be safe and our obligations to ensure this.

If you have any concerns about our products, you can contact us on ProductSafety@springernature.com

In case Publisher is established outside the EU, the EU authorized representative is:

Springer Nature Customer Service Center GmbH
Europaplatz 3
69115 Heidelberg, Germany

Batch number: 10704763

Printed by Printforce, the Netherlands